Subsurface Upgrading of Heavy Crude Oils and Bitumen

Subsurface Upgrading of Heavy Crude Oils and Bitumen

by

Cesar Ovalles

CRC Press
Taylor & Francis Group
Boca Raton London New York

CRC Press is an imprint of the
Taylor & Francis Group, an **informa** business

CRC Press
Taylor & Francis Group
6000 Broken Sound Parkway NW, Suite 300
Boca Raton, FL 33487-2742

First issued in paperback 2021

ISBN 13: 978-1-03-223890-6 (pbk)
ISBN 13: 978-1-138-74444-8 (hbk)

Library of Congress Cataloging-in-Publication Data

Names: Ovalles, Cesar, author.
Title: Subsurface upgrading of heavy crude oils and bitumen / Cesar Ovalles.
Description: Boca Raton : CRC Press, Taylor & Francis Group, 2019. | Includes
bibliographical references.
Identifiers: LCCN 2019016624 | ISBN 9781138744448 (hardback : alk. paper)
Subjects: LCSH: Petroleum--Refining. | Petroleum--Analysis. | Bitumen.
Classification: LCC TP690 .O875 2019 | DDC 665.5/3--dc23
LC record available at https://lccn.loc.gov/2019016624

Visit the Taylor & Francis Web site at
http://www.taylorandfrancis.com

and the CRC Press Web site at
http://www.crcpress.com

To my wife, Luisa (Lulú), for her patience and love from beginning to the end of this project that we called life.

To my children, Cesar Arturo and Manuela, hoping that I have been able to inspire them to achieve higher goals in their lives and careers.

Contents

Foreword

Subsurface Upgrading of Heavy Crude Oils and Bitumen is a timely book that seeks to address one of the thorniest problems that faces us today. Nature and geological forces have seen fit to provide us with a tremendous hydrocarbon resource in the form of heavy oil and bitumen, but at the same time the means to develop much of this resource in an economic and environmentally sound fashion is tantalizingly just out of reach. In this book, Cesar Ovalles brings together a deep knowledge of the chemistry and physical properties of heavy oil with a discussion of the various means, both proposed and tested, to move heavy oil resources to market using *in situ* upgrading technologies. For the first time in one volume, Cesar links the lessons learned from upgrading in an oil refinery setting to the practical realities of heavy oil production and asks the question, "Why can't we do upgrading in the oil reservoir?" If this were possible, much of the cost of an upgrader on the surface may be eliminated, while environmentally undesirable materials would be left in the reservoir. In addition, transportation costs could be reduced, and the environmental impact from the energy consumed to move heavy oil to market could be lessened. The audience for this book, students, academics, and industrial practitioners, will find the material a thought-provoking introduction to sub-surface upgrading.

Michael E. Moir, Ph.D.

Preface

Nowadays, all heavy crude oils and bitumen (HO/B) upgrading is carried out on the surface at refinery centers throughout the world. Conventional oil refineries are built for processing lighter crudes, so specially designed and relatively expensive "bottom-of-the-barrel" processes are used which, in turn, increase capital and operating expenditures. However, over the last 20 years, several technologies and scientific developments have appeared over the horizon that make the subsurface upgrading (SSU) of HO/B not only commercially available but proven at certain field conditions.

The primary objective of this book is to present an "in-depth" account and critical review of the progress of industry and academia in subsurface upgrading of heavy crude oils and bitumens. Examples were taken from many different parts of the planet from Canada to Venezuela and from China and Russia to the Middle East. The global outlook for exploring, extracting, and processing HO/B is one of long-term sustainable growth. We need critical data and insights to maximize the opportunity of making these essential raw materials into an integral part of the future of energy.

This book is intended to be an introduction for students, engineers, scientists, and practitioners in the area of enhanced oil recovery via SSU and also a resource and reference to be used during their research and technology development work. Throughout the following chapters, several topics are covered in order to describe the state of the art of subsurface upgrading of HO/B. Readers who are interested in more profound knowledge and theoretical fundamentals are welcome to consult the original references. Several authors' unpublished results are presented as well. For any inquiries about this material, feel free to contact the author.

This work is aimed to be a stand-alone monograph, so there are three chapters dedicated to the composition of HO/B and fundamentals of petroleum production and refining. But before we enter the main topics, the Introduction describes what heavy crude oils and bitumens are and the world proved reserves. Also, the definition of upgrading, the advantages and disadvantages of SSU, and the routes evaluated in the literature are presented. This introductory chapter finishes by discussing the impact on surface facilities and the environmental concerns of the subsurface upgrading processes.

Chapter 2 describes the reservoirs and the general characteristics in which heavy crude oils and bitumens are found in nature. Next, this chapter presents the physical and chemical aspects of HO/B composition such as elemental analysis, SARA, simulated distillation, the Boduszynski's Continuum Model of petroleum, and the relationships between viscosity and asphaltenes.

Chapter 3 discusses the basic concepts associated with heavy oil and bitumen production and transportation. This chapter presents an introduction to the enhanced oil recovery technologies, well issues, and oil, water, and gas separation processes. It is dedicated to the downstream professionals and refining engineers who are not familiar with the petroleum upstream technologies. The idea is that by reading these concepts and fundamentals, the understanding of the SSU processes can be made more accessible and straightforward. If more information is needed, please check the references included at the end of the chapter.

Chapter 4 describes the chemistry and industrial processes of refining and transformation of heavy crude oils and bitumens into higher value products, as they are currently employed in downstream facilities. This chapter covers the two main routes for HO/B upgrading, i.e., carbon rejection and hydrogen addition, as well as the stability of upgraded products and residue conversion. It is aimed at the upstream and midstream practitioners who are relatively unfamiliar with upgrading chemical pathways and to whom misinformation may generate both extremes: Lack of confidence on what is a feasible or unjustifiable expectation on the potential of some *in-situ* upgrading options.

Chapters 5 to 9 discuss the five routes for underground crude oil upgrading that have been reported in the open and patent literature. Specifically, Chapter 5 covers the physical separation of the HO/B into lighter and heavy fractions to increase rates and percentages of recovery and at the same time, improve the properties of the produced oil. Two physical separation routes have been reported in the literature, i.e., steam distillation and solvent deasphalting.

Chapter 6 presents the thermal conversion routes for the SSU of HO/B. In these concepts and processes, reservoir conditions are increased up to or above cracking temperature so oil-containing chemical bonds (carbon–carbon, sulfur–carbon, etc.) are broken with the concomitant permanent reduction of the viscosity and increasing in distillable materials in the produced oil.

Chapter 7 discusses the thermal hydrogen addition routes. In this pathway, hydrogen gas or hydrogen-donating compounds are used in the presence of an energy source for subsurface upgrading of HO/B. Generally, most of the hydrocarbon conversion is thermal, and the presence of the hydrogen source improves the quality and stability of the upgraded products.

In Chapter 8, the use of catalysts and hydrogen-donating compounds is discussed for the SSU of heavy crude oils and bitumens in the presence of an energy source (steam or other). As with the thermal-only processes, most of the hydrocarbon-bond cracking is due to the high temperature and residence times. By injecting catalysts downhole, the hydrogenation reaction is greatly enhanced, and lower viscosity, and higher API, hydrodesulfurization, and residue conversion can be obtained in comparison with thermal-only processes.

Chapter 9 presents the *in-situ* combustion thermal recovery methods for the production and subsurface upgrading of HO/B. In these processes, air or oxygen is injected downhole to create a combustion front that propagates from the injector to the producing well. At the same time, some of the heavy oil is burned, a residue is left behind, and the generated heat permanently reduces the viscosity of the remaining HO/B to increase the oil rate and total recovery.

Chapter 10 is dedicated to the subsurface upgrading routes that are currently being developed and not included in Chapters 5 through 9. Even though the production phenomena and the upgrading chemistry are basically the same, these new SSU routes bring something novel not found in any of the ones described in the previous chapters. For example, the use of electromagnetic and sonic energies is presented to enhance the thermal and catalytic cracking, hydrogen transfer, hydrogenation, or solvent deasphalting of HO/B. This chapter finishes with a summary of the key success factors, risks, and challenges that subsurface upgrading processes will face in the next decades.

While world demand for petroleum continues to rise, there have recently been increasing initiatives from environmental groups concerned about the long-term impact of extracting heavy crude oils and bitumens. These concerns arise not just from the direct effects to the environment but also from the fact that HO/B production has a decreased energy returned on energy invested. This situation means that they produce more greenhouse gases and other pollutants than do the same quantities of lighter crudes. Thus, new and improved technologies must be developed that reduce not only the cost of producing heavy crude oils and bitumens but reduce their environmental impact. This scenario is where subsurface upgrading could become a "big player" and part of the solution to the world energy appetite.

Finally, it can be foreseen that petroleum is going to continue playing a significant role in our lives in the near to medium-term future. While technologies are being invented to reduce our dependence on fossil fuels, it will be several decades before they become commonplace and affordable. Even if the world switched to an energy source independent of petroleum, crude oil and bitumens are integral parts of modern life beyond simply feedstocks for gasoline and other fuels. Objects as diverse as plastics, pharmaceuticals, and cosmetics use various aspects of petroleum-derived feeds as starting material and chemical reactions. In fact, our tremendous reliance on petroleum for manufacturing is one more reason not to merely burn it. Thus, subsurface upgrading processes could contribute to the sustainable production of HO/B beyond its current use as energy source and transportation fuel.

Acknowledgments

The author wishes to thank Chevron Energy Technology Company for the permission to publish this book. My gratitude to the Measurement and Chemistry and Heavy Oil Focus Areas for providing the funding. Special thanks to Michael E. Moir for his unconditional support and encouragement throughout the writing of this book.

My eternal gratitude to my parents, Josefina and Omar for being my role models and to my wife Luisa and children Manuela and Cesar A. for their patience and encouragement.

The author gratefully acknowledges the insightful technical discussions and debates with my Chevron colleagues and friends Estrella Rogel, Francisco Lopez Linares, Ian Benson, Ronald Behrens, Gunther Dieckmann, Don Kuehne, Brian Littlefield, John Segerstrom (r.i.p.), Art Inouye, Mridul Kumar, Jack Stevenson, Bruce E. Reynolds, Ajit Pradhan, Babak Fayyaz-Najafi, Tayseer Abdel-Halim, Hussein Alboudwarej, and Ed Chilton. To all, the author's most profound gratitude.

Many thanks to Harris Morazan, Lori Thomas, Janie Vein, and Kyle Hench for their technical assistance in many lab experiments and for making realities the most "crazy" ideas.

Finally, the author greatly acknowledges his non-Chevron collaborators Lante Carbognani, Pedro Vaca, Berna Hascakir, Raicelina Ramos, Parviz Rahimi, his undergraduate and graduate students, and to all the individuals that he has worked and interacted throughout his career in the petroleum industry in Venezuela and USA. It has been an incredible journey indeed!

About the Author

Cesar Ovalles graduated with a Licentiate degree in Chemistry from Simón Bolívar University and a Ph.D. in Chemistry from Texas A&M University. Right after graduation, he worked for 16 years at Petróleos de Venezuela Sociedad Anónima-Instituto de Tecnología Venezolano del Petróleo (PDVSA–INTEVEP). In 2006, he joined Chevron and is currently Technical Team Leader of the Heavy Oil Characterization group. He is involved in the areas of petroleum chemistry, heavy and extra-heavy crude oil upgrading (surface and subsurface), and asphaltene characterization. After his 28 years of industrial experience, Dr. Ovalles has published 62 papers in peer-reviewed scientific journals, 4 books, has been awarded 18 patents, and presented 116 papers at scientific and technical conferences. Additionally, he has published 14 articles in Venezuelan journals and 92 Technical Reports for an outstanding number of 306 total scientific productions. Cesar has also served as Associate Editor of Revista de la Sociedad Venezolana de Catalisis from 1996 to 2000 and Vision Tecnologica (Technical Journal of PDVSA–INTEVEP) from 2000 to 2002. In 2014, he won the Outstanding Technical Achievement Award from Hispanic Engineer National Achievement Awards Conference and the National STAR-Hispanic in Technology-Corporate Award from the Society of Hispanic Professional Engineers. Cesar is married to his college sweetheart, Luisa (Lulú), and is the father of two grown children. His son, Cesar Arturo, is a computer science graduate currently working for Safeway Supermarkets and his daughter, Manuela, is a microbiologist working for Mission Bio, a start-up company in South San Francisco.

1 Introduction

There is a growing consensus that the world's energy appetite will continue to expand over the next 25 years. Depending on the source, it is expected that the crude oil production will increase from 20% [Organization of the Petroleum Exporting Countries 2015] to 30% [International Energy Agency 2016] in comparison with current levels. There are few realistic substitutes for oil-derived products for fuels for trucks and planes and feedstocks for the chemicals industry. These three sectors account for all of the growth in global oil consumption. Thus, this forecast solidifies the central position of petroleum in the global energy mix for many decades to come. In particular, the world reserves of heavy crude oils and bitumens (HO/B) represent between 50% [British Petroleum 2016, Guzman 2015] and 70% [Faergestad 2016] of all hydrocarbons available on the planet. In the period 2015–2040, HO/B will continue to be one of the most important sources of energy especially with the addition of new projects and the anticipated return of stable market conditions [Organization of the Petroleum Exporting Countries 2015, International Energy Agency 2016].

As the world supply of light and easily extractable crude oils continues to decrease and demand continues to increase, the price that consumers are willing to pay for a barrel of crude increases as well [Petroleum.co.uk 2016]. As a result, heavier crude oil that was once uneconomical to extract due to high upfront costs has become profitable to produce and has started to emerge as a real factor in the oil and gas industry. Due to technological developments over the last decade, these advancements have given companies the ability to exploit heavy oil reservoirs economically and efficiently on a global scale. Thus, unconventional heavy oils are still projected to be the most significant contributor to world's oil supply (36%), higher than light, tight oil and condensate (26%), liquefied natural gas (25%), and biofuels (13%) [Petroleum.co.uk 2016]. The global outlook for exploring, extracting, and processing heavy crude oil is one of long-term sustainable growth.

HO/B production and upgrading technologies have three things in common. They have high capital and operating expenses (CAPEX and OPEX), decreased energy returned on energy invested, and a relatively larger impact on the environment. Depending on the geographic location, the cost of surfaces facilities for HO/B production has escalated in the last decade [Saniere *et al.* 2004]. Nowadays the cost of a 100 MBD unit is estimated to be on the order of tens of billions of dollars.

In the same way, the OPEX associated with the production, transporting, and refining of HO/B is relatively high due steam generation, light hydrocarbons used for diluent, and upgrading costs [Gray 2015]. Also, the need for trained and skillful operators throughout the value chain is of paramount importance to maintain reliability and maximize economic benefits.

While world demand for petroleum continues to climb, there has recently been increased pressure applied by environmental groups concerned about the long-term impact of producing and processing HO/B. Environmental concerns arise not just from the direct impact on the environment but also from the higher production of greenhouse gases (GHG) and other contaminants than the conventional production of lighter crudes [Huc 2011]. In other words, the extraction and use of HO/B are expected to have a significant impact on the problem of GHG emissions throughout the world [Petroleum.co.uk 2016]. To make important changes in the market penetration for HO/B and make its exploitation competitive with other hydrocarbon feedstocks (i.e., light tight oils), there is a need to reduce capital and operating costs significantly by using cutting-edge and emerging technologies while mitigating the environmental impacts as well as expected higher technology risks of the overall process.

Another important environmental aspect of HO/B production is the relatively higher water use and handling in comparison with the production of medium and light crude oils. The need to use steam injection processes for enhanced heavy oil recovery requires a constant source of fresh water since salt-containing aqueous fluids cannot be used because of severe, irreversible damage to the boilers [Donaldson 1979, Morrow *et al.* 2014]. In all cases, the demand for fresh water must be compatible with regional demand from municipalities and agricultural interests. On the other end of the process, large amounts of produced water must be separated from the oil, treated, and disposed in accordance with existing pollution-control regulations [Donaldson 1979]. In turn, this situation increases the cost and environmental concerns of HO/B projects.

Furthermore, heavy oil production facilities have a relatively large areal footprint. Because oil flow rates are generally low, many more wells are required to develop a HO/B field than the equivalent conventional oil field [Smalley 2000]. In new and environmentally sensitive areas such as the Arctic tundra or South America rainforest or wetlands the use of horizontal well technology has reduced the number of wells, allowing several wells to be drilled from a single pad. In this way, the effect of HO/B production on the environment is minimized [Smalley 2000].

As described in the next chapters, heavy crude oils and bitumens have high viscosity, a low amount of distillable material (gasoline or diesel), and high levels of undesirable constituents such as sulfur, nitrogen, and heavy metals such as vanadium and nickel. These chemical and physical properties make the production, transportation, and refining of HO/B a very challenging and costly task when using conventional technologies [Ovalles and Rechsteiner 2015, Speight 2007]. Thus, HO/B properties must be improved before the oil can be transformed into fuels, lubricant oils, plastics, petrochemicals, etc.

Nowadays, all the upgrading is carried out on the surface at refinery centers throughout the world. However, conventional oil refineries are built for processing lighter crudes, so specially designed and relatively expensive "bottom-of-the-barrel" processes are used which, in turn, increase CAPEX and OPEX. However, over the last 20 years, several technologies and scientific developments have appeared over the horizon that make the subsurface upgrading (SSU) of heavy crude oils and bitumens not only commercially available but proven at field conditions [Huc 2011]. This chapter presents an introduction to this topic. Next, definitions of these essential raw materials are discussed.

1.1   WHAT ARE HEAVY CRUDE OILS AND BITUMEN?

Before we start discussing subsurface upgrading, we must define what heavy crude oils and bitumens (HO/B) are. These essential raw materials are complex mixtures of naturally occurring liquids whose physical and chemical characteristics vary widely. Also, a continuously variable spectrum of properties can be found, not only geographically between hydrocarbon accumulations, but also laterally and vertically within a given petroleum-containing reservoir. In general, HO/B have high viscosities and low hydrogen contents and contain high amounts of heptane-insolubles (asphaltenes), hetero-atoms (i.e., nitrogen, sulfur, and oxygen), and metals, mainly nickel and vanadium. A detailed description of HO/B is presented in Chapter 2.

There have been several definitions of heavy crude oils and bitumens reported in the literature. They are based on the fundamental physical and chemical properties of HO/B that determine not only the production value but also how they must be transported and refined. These key properties are density, viscosity, and chemical analysis (sulfur, total acid number, metal content, etc.). From these properties, density and its equivalent, specific gravity (ratio of the density of sample to the density of water at the same temperature) are perhaps the most important properties of crude oils and their fractions [Boduszynski 2015]. The American Petroleum Institute (API) introduced the API gravity parameter [ASTM 2002] to expand the narrow range of specific gravity (Sp. Gr.) values and make it easier to use (Equation 1.1).

$$\text{API Gravity} = \frac{141.5}{\text{Sp. Gr. 60F / 60F}} - 131.5 \qquad (1.1)$$

Another important characteristic of HO/B is the viscosity at reservoir or gas-free conditions. However, measuring viscosity in gas-containing solutions is usually difficult to perform, so gas-free or "dead oil" values are recommended for ease and simplicity of measurement [Danyluk 1982]. As shown in Figure 1.1, the crude oil viscosity (measured at 50°C) increases with decreasing API gravity (increasing density). Similar relationships have been reported by other authors for "live" (gas-containing) and gas-free conditions [Smalley 2000, Speight 2007, Gray 2015].

Based on API gravity and viscosity, an international working group at the 2nd International Conference on Heavy Crude and Tar Sands [UNITAR 1982] adopted definitions for heavy crude oils and "tar sand" oils. These definitions are used throughout this book as they have received worldwide acceptance since 1982. Thus, heavy crude oils have an API between 10°API and 20°API and a gas-free viscosity at original reservoir temperature between 100 and 10,000 mPa s (cP). Additionally, "tar sand" oils, or as most commonly known, oilsands, have an API gravity lower than 10°API and a gas-free viscosity at original reservoir temperature greater than 10,000 cP. For the purpose of this book, the terms "tar sand" oils, oilsands, and bitumens are used interchangeably. A summary of the definitions of heavy crude oils and bitumens used in this book can be seen in Figure 1.2. These are based on API gravity, density, and viscosity as the primary analytical characterization tools.

However, several considerations must be made when working with the definition of heavy crude oils and bitumens described above. Firstly, the 10,000 cP cut-off in viscosity is mostly an arbitrary demarcation between heavy crude oils and bitumens [Danyluk 1982, U.S. Geological Survey 2006]. It originally arose from the California's heavy oil industry and had intrinsic limitations due to the

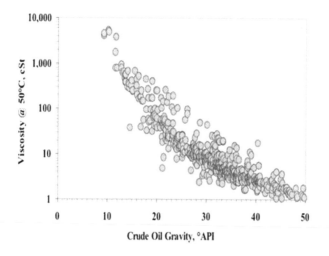

FIGURE 1.1 Crude oil API gravity versus viscosity of 50°C. Reprinted from Boduszynski [2015].

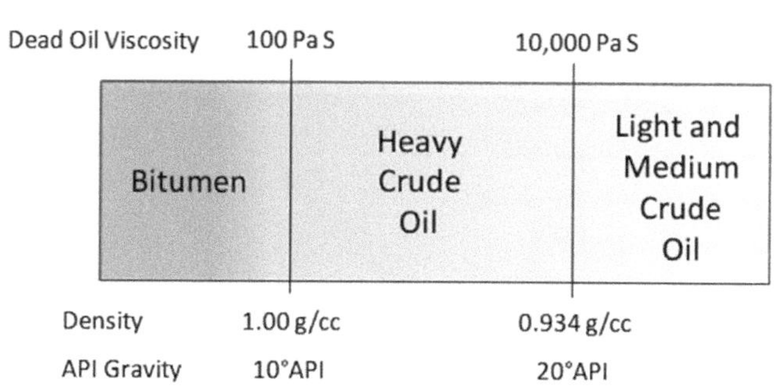

FIGURE 1.2 Definition of natural bitumens and heavy oil based on viscosity, density, and API gravity.

heterogeneity of naturally occurring reservoirs. Secondly, the implementation of a viscosity-based definition requires standardized sampling, measurement, and reporting protocols. This topic was also addressed by UNITAR in their original document [Gibson 1982].

In the last decades, the term "tar sand" has gained wide acceptance and is applied to a variety of rock types that contain heavy-petroleum/highly viscous materials [Danyluk 1982, Gray 2015]. However, it should be mentioned that "tar" is generally used in the petroleum industry to describe a heavy residual product coming from refining processes. The term "sand" is used to describe unconsolidated mineral matter and is composed mainly of silica. Thus, the combined term "tar sand" should be used with caution as these natural materials are found to contain not only silica containing sand but also sandstones and carbonate rocks.

Finally, it is important to mention that HO/B contain percentage levels of heteroatoms such as oxygen-, nitrogen-, and sulfur-bearing compounds [Gray 2015, Speight 2007, Huc 2011]. Thus, they are defined as "sour" crudes and should be handled accordingly. Additionally, heavy crude oils and bitumens contain trace level amounts (from 1 to ~1,000 ppm) of heavy metals (vanadium, nickel, iron, molybdenum, etc.) and present significant acidity as measured by their total acid number. All of these characteristics make these raw materials unique and require specially designed upgrading processes. These topics are fully described in Chapter 2.

1.2 WORLD RESERVES OF HEAVY CRUDE OILS AND BITUMEN

The world reserves of heavy crude oils and bitumens (HO/B) represent between 50% [British Petroleum 2016, Guzman 2015] and 70% [Faergestad 2016] of all hydrocarbons available on the planet. The regional distributions of heavy crude oils and bitumens are shown in Figures 1.3 and 1.4, respectively [Meyer *et al.* 2007]. As seen, HO/B can be found on all continents except Antarctica. The Middle East and South America have the most extensive volumes of heavy oil with more than 2,000 billion barrels (BBBL) of original oil in place (OOIP), followed by North America with 650 BBBL.

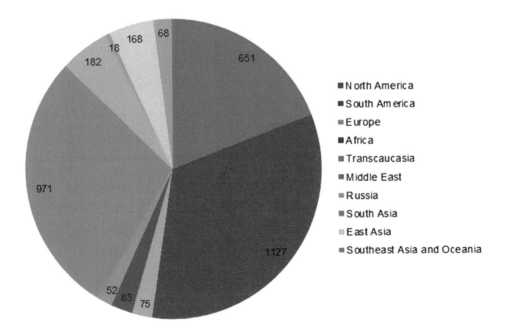

FIGURE 1.3 Regional distribution of heavy crude oils around the world. Numbers are *original oil in place in billion barrels.* Data from Meyer *et al.* [2007].

In the same way, North (~2,400 BBBL) and South America (2,260 BBBL) have the largest OOIP of bitumens (Figure 1.4). Large resource deposits are also found in eastern Siberia (~400 BBBL) but insufficient data are available to make more than nominal volume estimates [Meyer *et al.* 2007].

In the Americas, Canada has proven reserves of ~175 BBBL of which more than 20% is surface mineable bitumens, and the remaining 80% is considered "*In-situ* bitumen reserves" [Gray 2015]. That is, the latter is too deep to mine and can only be produced using thermal processes such as steam-assisted gravity drainage among other technologies [Gray 2015]. The heavy crude oil and bitumen deposits in Canada are located (Figure 1.5) in the Athabasca, Wabasca, Peace River, and Cold Lake regions of Alberta and the Lloydminster region spanning the provinces of Alberta and Saskatchewan.

Venezuela has proven reserves of ~1,000 BBBL [Schenk *et al.* 2009]. The USGS estimated that the Orinoco Oil Belt has a mean volume of 513 BBBL of technically recoverable heavy oil [Schenk *et al.* 2009]. This reservoir is located in the southeastern part of Venezuela at ~300 km southeast of Caracas (see Figure 1.6). The reservoir is north of the Orinoco River and runs from west to east ~500 km in length and ~75–100 km wide. Also, there is a significant amount of heavy crude oil reserves in the Boscan Field located west of the Maracaibo Lake (see right-hand side insert in Figure 1.6).

Tatneft, the Tatarstan state oil company, controls deposits that are between 50 and 250 meters in depth and which contain a conservative estimate of 50 million tons of tar sands resources, with potentially up to seven billion tons of recoverable tar sands oil [Tafneft 2017].

New heavy oil resources have been identified in Colombia (~110 BBBL), Ecuador, Peru, and other countries along the Andes. Similarly, Mexico, Brazil, China, Russia, Kazakhstan, and countries in the Middle East, such as Kuwait and Iran, have reported increases in heavy oil resources and production [Meyer *et al.* 2007].

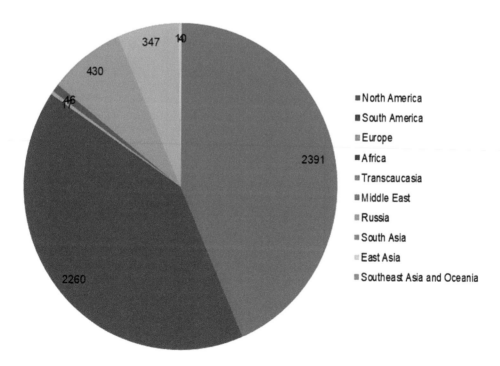

FIGURE 1.4 Regional distribution of bitumens around the world. Numbers are *original oil in place in billion barrels*. Data from Meyer *et al.* [2007].

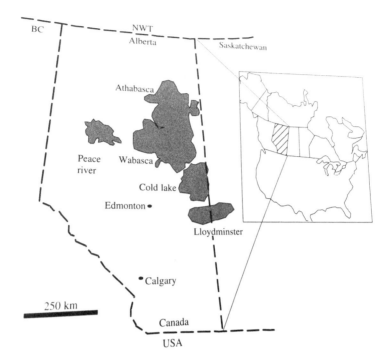

FIGURE 1.5   Geographic locations of heavy crude oil and bitumen deposits in Canada. Courtesy of Natural Resources Canada, Geological Survey of Canada.

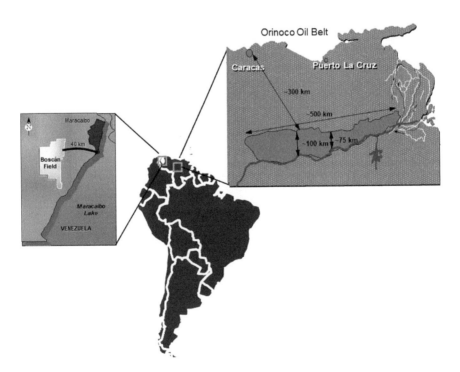

FIGURE 1.6   Geographic locations of heavy crude oil deposits in Venezuela.

1.3 WHAT IS UPGRADING AND WHY UPGRADE HEAVY OILS?

In general, the upgrading of HO/B and their fractions can be effectively carried out at the surface, and several examples have been commercially demonstrated. Low, medium, and high residuum conversion processes (such as Visbreaking, Delayed Coking, LC-Fining, hydrotreatment, etc.) have been developed and commercialized over the past 60 years [Speight 2007, Gray 2015, Huc 2011, Ramirez-Corredores 2017]. Therefore, one of the fundamental questions that is addressed throughout this book is, why can't upgrading be done subsurface? However, before we try to answer this question, let's look at the definition and economics of upgrading.

1.3.1 Definition and Levels of Upgrading

From the broad point of view, upgrading is any fractionation or chemical treatment of heavy crude oils and bitumens that produces *permanent changes* in the physical and chemical properties of these materials. Conventional oil refineries are designed to process lighter crudes (API > 20°API or densities lower than 950 kg/m³). Thus, HO/B upgrading is the term used to describe the processes that convert them to distillable liquids with similar quality to those typically handled by conventional refining units [Gray 2015].

There are three different categories of upgrading processes defined by the percentage of residue conversion that can be achieved. This later parameter measures the upgrading of a heavy crude oil or bitumen based on the conversion of the fraction with a boiling point greater than 538°C⁺ (1,000°F⁺) and is calculated with the following Equation 1.2:

% conv. 1000°F+

$$= \frac{\text{wt.\% Distillables } 1000°F + \text{inproduct} - \text{wt.\% Distillables } 1000°F + \text{in Feed}}{\text{wt.\% Distillables } 1000°F + \text{inproduct}} \times 100 \quad (1.2)$$

According to Figure 1.7, upgrading processes are classified as achieving low, medium, and high levels of conversion (Figure 1.7). In low-level upgrading processes, also known as "partial upgrading," the objective is to the reduce viscosity to allow shipment by pipelines without adding a solvent, external heat sources, or using other means of transportation (e.g., trucking). These processes are carried out in the production facilities of the upstream part of the petroleum business where conversion of the 538°C⁺ residue can reach values up to 30% (see Figure 1.7). These processes should not make the crude oils unstable to solid precipitation and should comply with pipelines specifications. Examples of these processes are Visbreaking and partial solvent deasphalting, among others [Luhning 2002].

Medium- and high-level upgrading processes are carried out at downstream facilities and are aimed at increasing the amount of distillable materials and at the same time, reducing the content of contaminants such as sulfur, nitrogen, and metals. Medium- and high-level upgrading processes are differentiated by the amount of 538°C⁺ (1,000°F⁺) residue conversion they can achieve with simultaneous production of high-value distilled products. Medium-conversion processes yield between 30% and 70% residue conversion (see Figure 1.7). Examples of these processes are Delayed Coking, LC-Finning, hydrovisbreaking, etc. [Noguchi 1991]. More detailed explanations of these processes are given in Chapter 4.

High-level upgrading processes yield between 70% and 105% in distillable materials with the production of high volumes of naphtha, middle distillates, and Vacuum Gas Oils. They can achieve almost complete removal of heteroatoms and metals, and due to the relatively larger amount of hydrogen added, the residue conversion can be higher than 100%. More details on the chemistry occurring during these processes are described in Chapter 4.

1.3.2 What Is Subsurface Upgrading?

Now that the general term of upgrading is defined, what specifically is subsurface upgrading? For the purposes of this book, subsurface upgrading is any underground, *in-situ*, or downhole recovery

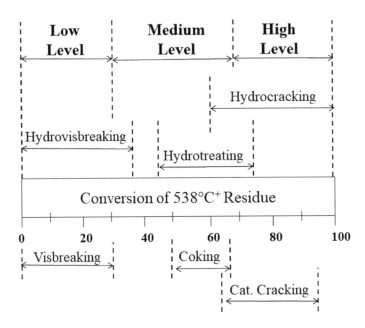

FIGURE 1.7 Levels of upgrading of HO/B measured as percentage of conversion of the 538°C⁺ residue (1,000°F⁺).

processes of the heavy crude oils and bitumens that lead to permanent changes in these materials. From the terminology described in the previous section, subsurface upgrading is a low conversion or partial upgrading process (see Figure 1.7). The goal is to increase the downhole mobility of the HO/B by reducing their viscosity, to increase oil rate and cumulative production, and/or to improve oil recovery and transportation via pipelines and increase the value of the upgraded crude oils. The advantages and disadvantages of subsurface upgrading are discussed in detail in Section 1.5

1.3.3 THE ECONOMICS OF UPGRADING

The decision to process heavy crude oil means having less-valuable products and relatively higher investment and more operational challenges than that found for conventional light oils. Refineries are willing to refine HO/B and invest in new infrastructure if the raw materials are sufficiently discounted to offset additional costs. As environmental regulations become even more strict, the discount rises because of the extra effort required to eliminate sulfur and metals and to sell residuals. Thus, the economics for subsurface and surface upgrading heavy crude oils and bitumens depend on the price differential between light (i.e., high API crude such as Brent or West Texas Intermediate) and heavy crude oils (low API such as Maya or Hamaca crudes), the existence of alternate markets for the residue fractions, the capital and operating costs of the upgrading unit [Gray 2015], and the environmental regulations and carbon taxes enacted in the producing and refining centers. Those costs are very variable and depend on many economic and geopolitical factors.

However, it can be estimated that for each API degree increase by upgrading, there is a net increase of $0.30–0.50 per barrel in the value of the upgraded crude oil [Smalley 2000]. The higher price is due mainly to the higher amount of distillable materials and lower residue content. In the same fashion, a reduction of one weight percent of the sulfur content led to a value increase of $0.40–0.90 per barrel of the upgraded HO/B. This higher price is due to the cost reduction on the disposal of sulfur-containing materials in downstream operations.

Another important property is the acidity of the HO/B as measured by the total acid number in mg of KOH/g of crude oil, also known as TAN number [ASTM 2006]. An increase in the value of

$0.50–1.50 per barrel for each TAN-number reduction of the upgraded HO/B can be estimated due to decreases in the transportation and corrosion costs.

The economics of the HO/B upgrading also depends on the viscosity of the products. The minimum requirement for transportation via pipelines is between 250 and 400 cSt at 37.8°C [Muñoz 2016], so for each barrel of reduction in diluent consumption, there is a net saving of $50–60 per barrel.

Finally, in the production of HO/B, 85% of GHG emissions are due to the combustion of natural gas for steam and power [Huc 2011]. The current regulatory environment has the potential to make heavy oil and bitumens portfolios less competitive than those based on conventional and lighters crudes. Also, in near-future regulatory scenarios, HO/B producers may need to reduce the price of their crude to compete with lower carbon intensity crudes, thus, increasing the risk of becoming uneconomic. These economic factors make subsurface upgrading a means of maintaining heavy crude oils and bitumens as an important part of the world energy supply in the future.

1.4 WHAT IS IN- AND OUT-OF-SCOPE?

In Table 1.1, the list of in- and out-of-scope topics for the subsurface upgrading of HO/B is presented. As mentioned in the previous sections, this book focuses on heavy oils as defined as API gravity lower than 20° and with the presence of at least one subsurface upgrading component in the process.

In-scope are the following areas: Underground, *in-situ*, downhole, and subsurface recovery processes of heavy crude oils and bitumens that produce upgrading of the chemical and physical properties of HO/B. The literature review discusses physical or laboratory experiments, numerical simulations, and field tests in the area. Also, unpublished data from the author's research are included as well.

Also, all routes reported in the literature for the subsurface upgrading of heavy oils (physical separation, thermal conversion, hydrogen addition, *In-Situ* Combustion, and hybrid concepts) are considered as well as different sources of energies, hydrogen, and downhole compatible catalysts (Table 1.1). Positive and negative environmental effects are evaluated and taken into consideration during all phases of the literature review. The downhole addition of oxidants (e.g., H_2O_2, $KMnO_4$, O_3, etc.) is discussed as well.

TABLE 1.1

In- and Out-of-Bound Topics for the Subsurface Upgrading of Heavy Crude Oils and Bitumens

In-Bounds	Out-of-Bounds
• Heavy oil reservoirs (< 20°API) • A subsurface upgrading of HO/B component • Any of the subsurface upgrading routes reported in the literature, i.e., physical separation, thermal, hydrogen addition, *In-Situ* Combustion, and hybrid concepts • Different sources of energy (steam, combustion, electrical, radiofrequency) • Different sources of hydrogen (water, H_2, syngas, CH_4, solvent) • Use of supported or dispersed catalysts downhole • A surface component related with integration with subsurface • Environmental advantages and disadvantages (leave contaminant behind, CO_2 sequestration, etc.) • Adding oxidants downhole (e.g., H_2O_2, $KMnO_4$, O_3, etc.)	• Medium, light, or condensate reservoirs (> 20°API) • Surface-only upgrading processes • In-well upgrading methods that use the casing as chemical reactor • *In-situ* processes that do not upgrade product (EOR, IOR, etc.) • *In-situ* bio-upgrading using bacteria, microbes, or enzymes • Any explosion-related downhole energy source using conventional or nuclear explosives

The following areas are considered out-of-scope (Table 1.1): Patent and open literature in cold and thermal enhanced oil recovery processes (EOR or IOR) that do not perform subsurface upgrading of heavy oils as measured in terms of increases in API gravity and distillable materials or permanent viscosity reduction. Surface-only processes ("in-field" or refinery), in-well processes, and *in-situ* bio-upgrading using bacteria, microbes, or enzymes.

Also, out-of-scope is the use of any explosion-related downhole energy source such as conventional or nuclear explosives. However, it is borderline if the upgrading is carried out in the pipeline from the producing facility but only when combined with *in-situ* and not as stand-alone processes.

1.5 ADVANTAGES OF SUBSURFACE UPGRADING

In general, underground upgrading processes have always been of interest to the petroleum industry mainly because of the inherent advantages compared with above-ground counterparts. In principle, the main objective of subsurface upgrading (SSU) is to reduce the viscosity of the heavy crude oils and bitumens to increase the rate of oil production (BBL/day) [Kapadia *et al.* 2015]. In this regard, SSU has similar objectives to any other enhanced oil recovery process but the difference is that permanent chemical and physical changes of the upgraded crude oil have occurred. As mentioned in Section 1.3.2, the viscosity reduction of the upgraded crude oil may come accompanied by the generation of additional distillable materials, reductions in sulfur, asphaltene content, TAN, metal content, etc.

In most cases, subsurface upgrading of heavy crude oils and bitumens has led to increases in the cumulative oil produced with the concomitant increase in the percentages of oil recovery of the reservoirs. This situation may lead to a possible transformation of unproven to proven oil reserves as defined by SPE [Society of Petroleum Engineers 1997], which in turn, translates into an increase of company assets.

SSU processes can reduce the environmental impact of HO/B production processes by lowering the greenhouse gases emissions due to reduction of the steam-to-oil ratio (i.e., less fuels burned to generate power and steam). Similarly, the production of an upgraded crude oil with higher API gravity and lower H/C molar ratio leads to a reduction on the carbon index so lower carbon taxes can be imposed. Furthermore, SSU technologies can leave downhole a series of contaminants that are challenging and costly to handle on the surface. Examples of such pollutants are asphaltenes, polyaromatic hydrocarbons, heavy ends, sulfur, vanadium, and nickel, etc.

By using SSU technologies, there is the possibility of reducing or eliminating CAPEX of the upgrading unit. As mentioned, HO/B production and upgrading technologies have high capital expenses with estimates in the order of tens of billions of US$ for a 100 MBD unit [Saniere *et al.* 2004].

In the same way, the use of SSU may lead to significant reductions in the operating costs associated with the production, transportation, and refining of HO/B due to decreases in consumption of costly light and medium petroleum oils used as diluents (~20–30%Vol). Additionally, the lower viscosity of the upgraded crude oil translates into lower lifting and transportation costs from underground to the refining centers with the additional economic benefits.

Finally, another advantage of subsurface upgrading processes is the use of the porous media (mineral formation) as a natural "catalytic reactor" to further improve the chemical and physical properties of upgraded crude oil [Muraza and Galadima 2015, Guo *et al.* 2016]. As we will see in the next chapters, there are advantages in the presence of the mineral formations in terms of availability of catalytic sites and behaving as an adsorbent of added catalytic materials.

1.6 CHALLENGES TO OVERCOME

Several issues have prevented the development, field testing, and commercialization of HO/B subsurface upgrading processes. One of the most significant problems is that downhole processes are difficult to control and monitor under reservoir conditions. For surface upgrading, either at the well site or refinery, distillation columns, reactors, heat exchangers, and other process units are well instrumented and monitored on an hourly or daily basis. In this way, engineers and operators are able to upgrade heavy oil reliably and safely from months to years with the concomitant economic benefits.

On the other hand, downhole processes are equipped with much less monitoring, and well sites are in remote and desolated areas. Thus, SSU processes run with less control and follow-up than the above-the-ground counterparts. Additionally, each well(s) and reservoir(s) may require individual treatment due to their unique geology and depositional environments. For example, the presence of downhole heterogeneities may affect the efficiency of the upgrading. Solvents, catalysts, and other chemicals can be lost due to "thief zones" located within the reservoir. Thus, previous reservoir characterization is of paramount importance for the success of the SSU process. All of these characteristics make SSU operation intrinsically more complex and sometimes less reliable than surface-only upgrading technologies. However, the development of new and state-of-the-art technologies for downhole monitoring (fiber optics, electromagnetic sensors, temperature resistant electronics, etc.) has improved the chances of developing successful SSU processes in the future.

In HO/B reservoirs, permeabilities and porosities of the mineral formation are relatively higher than those found in conventional medium- and light-oil analogs [Kapadia *et al.* 2015]. However, downhole mixing conditions are still very poor, and the diffusion of solvents, catalysts, and other chemicals within the porous media is very challenging [Pourabdollah and Mokhtari 2013]. Therefore, to achieve the same results, SSU processes must proceed at slower rates than those at surface conditions. The poor mixing and diffusion phenomena can be compensated with longer underground residence times (from days to months) in comparison with above-ground counterparts (from seconds to a few hours).

Additionally, the temperatures of heavy oil reservoirs are typically lower (5–90°C) than those used in surface processes, which in turn, reduces the effectiveness of injected solvents, catalysts, and/or additives. In general, the temperatures of physical separation (>150°C), hydrogen transfer or hydrogenation (> 380°C), and combustion (> 450°C) are much higher than those used in SSU. In some cases, there is the need for external energy sources (steam, electromagnetic, air injection, etc.) to bring the operating conditions up to those necessary for *in-situ* upgrading. These topics are covered in Chapter 10.

Finally, subsurface upgrading processes have the additional complication of bringing new equipment and operating practices to surface facilities. These equipment and practices have been used in refining centers for many years, but upstream operators are not familiar with them. This item must be taken into consideration when designing field tests of SSU processes.

Despite all the previous challenges and shortcoming, the size of the prize for SSU is quite vast, and several HO/B assets contain significant accumulations of heavy oil that could benefit from subsurface upgrading technologies. A few of these assets also have transportation challenges requiring upgrading before they can be successfully exploited.

1.7   UPGRADING ROUTES EVALUATED IN THE LITERATURE

The subsurface upgrading of heavy crude oils and bitumens (SSU-HO/B) is a fascinating and challenging subject mainly because it can affect the economics of *all* operations across the petroleum value chain (Figure 1.8). From a general point of view, SSU-HO/B borrows upgrading concepts from refining (Downstream) and applies them to oil production operations located Upstream and in Transportation (Figure 1.8). In the Upstream portion and more specifically in Exploration (Figure 1.8), the development and commercialization of a HO/B *in-situ* upgrading technology may lead to a transformation of unproven into proven reserves with significant economic benefits. As mentioned earlier, the production of upgraded crude oils with lower viscosity leads to increases in oil rates and makes their transportation via pipelines more economical. In Downstream (Figure 1.8),

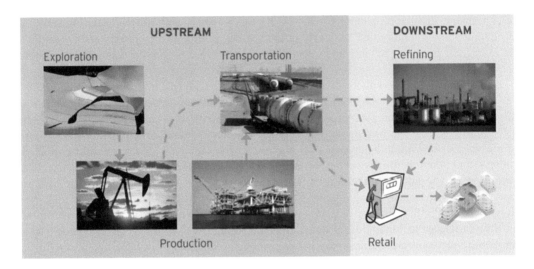

FIGURE 1.8 Petroleum Value Chain [Ovalles and Rechsteiner 2015].

the higher content of distillable products and lower level of contaminants (S, V, Ni, etc.) make the upgraded HO/B more valuable and less environmentally challenged. However, these "pre-processed" heavy crude oils and bitumens may bring new issues to Transportation and Downstream (Figure 1.8) due to the presence of unstable materials and hard-to-process compounds. These topics are covered in Chapters 2, 3, and 4.

As shown in Figure 1.9, five routes for underground crude oil upgrading have been reported in the open and patent literature. These concepts involve the following pathways and are discussed in detail in the following chapters:

- Chapter 5—Physical separation (e.g., downhole steam distillation and deasphalting)
- Chapter 6—Thermal conversion-carbon rejection (e.g., underground Visbreaking)
- Chapter 7—Thermal hydrogen addition (e.g., hydrogen gas, hydrogen precursor injection, or water)
- Chapter 8—Catalytic hydrogen addition (e.g., hydrogen gas or hydrogen donors)
- Chapter 9—*In-situ* combustion using air or oxygen
- Chapter 10—New concepts such as the use of electromagnetic and sonic energy

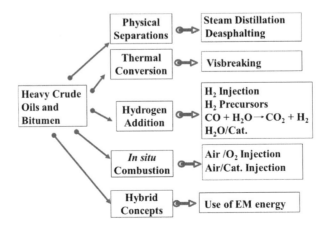

FIGURE 1.9 Routes for underground crude oil upgrading have been reported in the open and patent literature.

1.7.1 PHYSICAL SEPARATIONS

Except for the steam distillation, the main advantage of these processes is that they do not require external energy sources because they are operated at reservoir temperature [Pourabdollah and Mokhtari 2013] or in the presence of warm solvent [Nenninger and Nenninger 2005]. However, they all require a high solvent-to-crude ratio (minimum of 1:1 v/v) and ready availability of solvent on-site, and there is the potential of wellbore plugging due to asphaltene precipitation. Also, these processes need to recycle the solvent to improve the economic prospects; therefore, some surface facilities are required. This topic is discussed in Chapter 5.

1.7.2 THERMAL PROCESSES

In this route, the reservoir is heated to cracking temperature (> 230°C) so that chemical bonds (carbon–carbon, carbon–sulfur, etc.) within the HO/B are broken with the concomitant reduction in the viscosity of the produced oil and increase in distillable materials [Gray 2015]. These processes have the advantage of being relatively simple and in general, do not require complex surface installations [De Rouffignac *et al.* 2004]. However, thermal processes suffer several disadvantages such as lack of low-cost and abundant energy sources, the value of the upgraded products is limited by the lack of hydrogen, instability may exist due to asphaltene precipitation, formation of olefins, and operational problems may result from the potential for generation of "unusual" scales through dissolution/transport of formation minerals. Several energy sources have been reported to heat the reservoir at the desired temperature using steam, electrical heating, and electromagnetic energy. This route is discussed in Chapter 6.

1.7.3 THERMAL AND CATALYTIC HYDROGEN ADDITION

In this path, hydrogen gas or hydrogen precursors are used for the upgrading of heavy oils in the presence as an energy source such as steam or other electrical method [Kapadia *et al.* 2015, Muraza and Galadima 2015]. Most of the hydrocarbon conversion is thermal (> 230°C), and the presence of a hydrogen source increases the quality and stability of the upgraded products. However, these processes suffer several disadvantages such as poor mixing between the hydrogen source and heated heavy crude oil, and lack of availability of inexpensive and abundant of hydrogen and energy sources. This topic is discussed in Chapter 7.

Also, the uses of catalyst have been reported in the literature to enhance the hydrogenation reaction and further improve the properties of the produced oil. By injecting metal-containing catalysts downhole, hydrogen transfer is enhanced with the concomitant increase in the economic benefits [Guo *et al.* 2016]. However, the downhole mixing of the catalyst and its cost and maintaining sustained catalytic activity represent significant challenges. This catalytic route is discussed in Chapter 8.

1.7.4 IN-SITU COMBUSTION

In this route, air or oxygen gas is added downhole to generate a combustion front so that the heavy ends of the crude oil are burned inside the reservoir, and the lighter ends gravitate toward the producing well and flow to the surface [Greaves *et al.* 1999]. This subsurface upgrading process can be carried out in the absence (conventional *in-situ* combustion and Toe-to-Heel Air Injection) or presence of metal catalysts [Greaves and Xia 2001].

In-situ combustion processes have been studied for more than 60 years. They have the inherent advantage that no external energy sources are needed, and very high temperatures can be achieved for the conversion of the heavy crudes into lighter hydrocarbons. However, they suffer from significant drawbacks such as: partial oxidation of the distillable materials, lack of control of the burning front, significant safety concerns (valve failure, handling liquid O_2, high pressure, etc.), generation of toxic emissions to air (NH_3 and polynuclear aromatic hydrocarbons), large volumes of flue gas, lack of contact between catalyst and heated bitumens, and inactivation and/or fouling of catalyst. This route is discussed in Chapter 9.

1.7.5 HYBRID AND NEW CONCEPTS

Over the last few years, hybrid concepts have been reported in the literature that combine several elements of the previous four routes. For example, a consortium of several companies in Canada patented a process referred as Enhanced Solvent Extraction Incorporating Electromagnetic Heating ("ESEIH") for producing hydrocarbons from petroleum reservoirs. This technology involves preheating the mineral formation by using electromagnetic radiation (RF or MW) and injecting a solvent in a SAGD well configuration [Trautman and MacFarlane 2014].

Also, the use of sonic energy or ultrasound has been a very useful tool in enhancing the reaction rates for a variety of chemical systems [Sawarkar *et al.* 2009, Avvaru *et al.* 2018]. Thus, it is not surprising that the use of ultrasound for the HO/B subsurface upgrading had been proposed in the literature. More details are discussed in Chapter 10.

1.8 IMPACT ON SURFACE FACILITIES

Generally, the issues with surface facilities are centered on oil/water separation which, in turn, can be negatively affected due to the use of chemicals and/or catalysts downhole. These chemicals may stabilize water-in-oil emulsions so longer residence times and the use of demulsifiers may be necessary to break those stable emulsions and make the upgraded crude oils transportable and comply with pipeline requirements.

Due to the lower viscosity of the upgraded crude oil, SSU upgrading processes have the potential to reduce water usage for steam generation and handling of produced water. This situation translates into smaller surface facilities with lower carbon and areal footprints in comparison with conventional petroleum units.

Also, thermal, hydrogen addition, and *in-situ* combustion processes may suffer from corrosion and scaling issues. These processes may generate corrosive gases such as wet carbon dioxide, carbon monoxide, hydrogen sulfide, mercaptans, thiols, etc. Handling these issues can be a challenge but could be successfully addressed using high-temperature coatings or metallurgy.

Due to lack of infrastructure in the heavy crude oil areas (Orinoco, Northern Alberta, Russia, Colombia, and other remote locations), grass-root projects generally cost twice or three times as much as compared to the same facility located in the U.S. Gulf Coast. Additionally, a typical HO/B well pad completely lacks the facilities to install the surface units needed for underground upgrading. Therefore, new upstream installations should be designed to be compatible with current operations and with a future expansion toward SSU. In contrast, midstream HO/B units have medium complexity which could be potentially used for deploying the process units for subsurface upgrading.

1.9 ENVIRONMENTAL CONCERNS

HO/B production and upgrading technologies have a larger impact on the environment compared with those for medium and light crude oils. The energy consumed to lift and handle a barrel of heavy oil is as high as 20% or more of the oil produced [Smalley 2000]. This energy consumption is several times greater than that needed to produce a barrel of medium or light oil.

In general, there are three significant environmental risks associated with subsurface upgrading processes. Those are leakage to the surface, migration in the subsurface, and damage to company's reputation, even if there is no impact on the environment.

In the case a given contaminant finds its way to the surface, the major risks are an accumulation of gases in low-lying calm areas and poorly ventilated structures, and the possibility of asphyxiation. Also, there are chances of an explosion, human or ecological toxicity, contamination of surface waters with leaking gases or liquids, increased corrosion in facilities due to contact of acid gases with water, the release of greenhouse gases, and failure to comply with regulations.

In case the contaminant migrates underground, the following environmental risks can be envisioned such as contamination of groundwater resources with gases, hydrocarbon liquids, mobilization of heavy metals in groundwater resources (via acidification), and contamination of soil. Also, there is the possibility of contamination of hydrocarbon resources, loss of reservoir permeability by mineral precipitation, and degradation in geo-mechanical strength of the reservoir.

In general, hazard assessments are site and process specific. For people, the key hazards are exposure to high concentrations of CO_2, H_2S, and/or CO and the flammability of leaking hydrocarbon gases or liquids. Regarding the health risks, most of the hydrocarbons assessed have low health and reactivity hazards but pose significant flammability hazards.

For the atmosphere, the key hazards are the release of greenhouse gases and air pollutants, and for soil, the key hazards are an accumulation of gases that can impact vegetation and species viability or the accumulation of flammable hydrocarbons.

For groundwater, the key hazards are contamination with acid gases or products from the reaction of reduced pH water with formation minerals. Another hazard for groundwater is contamination with flammable gases or liquids that can be co-produced with the water. SO_x, NO_x, CO, and H_2S pose serious health risks, and acidic aqueous solutions of CO_2, NO_2, or H_2S can react with some aquifer formation minerals leading to the mobilization of heavy metals.

In the case of catalytic hydrogen addition processes, spent hydroprocessing contaminants are typically classified as hazardous wastes and must be removed from refinery streams and disposed in landfills according to relevant regulations. Produced spent catalysts would likely need to be handled in the same manner as spent catalysts from a refinery

Another environmental impact of SSU processes is that they require a significant areal footprint. Oil rates obtained by producing HO/B are relatively low, so more wells and ancillary infrastructure are needed to reach the same production as conventional crude oils. This small space between wells and the hazards associated with their exploitation amount to a significant impact from future heavy oil projects [Smalley 2000].

All of these environmental issues lead to the conclusion that high efficiency and lower greenhouse generation technologies are needed to reduce and mitigate the impact of HO/B exploitation to the environment. As discussed by Smalley, heavy crude oils and bitumens have a high carbon to hydrogen ratio, and when burned, they produce more carbon dioxide than conventional crude oils and natural gas. Thus, HO/B are much better suited to be used as feedstock to make petrochemical or asphalt products than as fuels [Smalley 2000].

REFERENCES

ASTM International, 2002, "*ASTM D 4052, Standard Test Method for Density and Relative Density of Liquids by Digital Density Meter*", ASTM International, West Conshohocken, PA.

ASTM International, 2006, *ASTM D 664, Standard Test Method for Acid Number of Petroleum Products by Potentiometric Titration*, ASTM International, West Conshohocken, PA.

Avvaru, B., Venkateswaran, N., Uppara, P., Iyengar, S. B., Katti, S. S., 2018, "Current Knowledge and Potential Applications of Cavitation Technologies for the Petroleum Industry", *Ultrason. Sonochem.*, 42, 493–507.

Boduszynski, M. M., 2015, "Petroleum Molecular Composition Continuity Model". In: *Analytical Methods in Petroleum Upstream Applications*, C. Ovalles, C. Rechsteiner, Ed., CRC Press, Boca Raton.

British Petroleum, 2016, *BP Statistical Review of World Energy*, 65th Ed., London, United Kingdom, downloaded on Aug. 22, 2018 from www.bp.com/content/dam/bp/pdf/energy-economics/statistical-review-2016/bp-statistical-review-of-world-energy-2016-full-report.pdf

Danyluk, M., 1982, "Addendum to UNITAR Proposal for a Definition of Heavy Crude and Sand", 2nd International Conference on Heavy Crude and Tar Sands, United Nations Institute for Training and Research (UNITAR), Caracas, Venezuela, February 7–17, Chapter 1.

De Rouffignac, E., Vinegar, H. J.,Wellington, S. L., Berchenho, I. E., Stegemeir, G. L., Zhang, E., Shahin, G. T., Fowler, T. D., Ryan, R. C., 2004, "In Situ Thermal Processing of Hydrocarbon Containing Formation to Pyrolyze Selected Percentage of Hydrocarbon Material", US Patent No. 6,732,795.

Donaldson, E. C., 1979, "Environmental Aspects of Enhanced Oil Recovery", *Energy Sources*, 4(3), 213–229.

Faergestad, I. M., 2016, *Defining Heavy Oil*, Schlumberger, www.slb.com/~/media/Files/resources/oilfield_review/defining_series/Defining-Heavy-Oil.ashx, downloaded 12/15/2016.

Gibson, B. J., 1982, "Method of Classifying Heavy Crude Oils Using the UNITAR Viscosity-Based Definition" 2nd International Conference on Heavy Crude and Tar Sands, United Nations Institute for Training and Research (UNITAR), Caracas, Venezuela, February 7–17, Chapter 1.

Gray, M. R., 2015 *Upgrading Oilsands Bitumen and Heavy Oil*, The University of Alberta Press, Edmonton, and references therein.

Greaves, M., Ren, S. R., Xia, T. X., 1999, "New Air Injection Technology for IOR Operations in Light and Heavy Oil Reservoirs", Paper SPE 57295 presented at the SPE Asia Improved Oil Recovery Conference, Kuala Lumpur, Malaysia Oct. 25–26.

Greaves, M., Xia, T., 2001, "CAPRI-Downhole Catalytic Process for Upgrading Heavy Oil: Produced Oil Properties and Composition", Paper 2001-023 presented at the Petroleum Society's Canadian International Petroleum Conference, Calgary, Alberta, Canada, June 12–14.

Guo, K., Li, H., Yu, Z., 2016, "In-Situ Heavy and Extra-Heavy Oil Recovery: A Review", *Fuel*, 185, 886–902.

Guzman, R., 2015 *Trends and Challenges in the Heavy Crude Oil Market*, 4th Heavy Oil Working Group, Bogota – September 22.

Huc, A.-Y., Ed., 2011, *Heavy Crude Oils. From Geology to Upgrading. An Overview*, Technip, Paris, France, and references therein.

International Energy Agency, 2016, *World Energy Outlook 2016. Executive Summary*, Paris, France (www.iea.org).

Kapadia, P. R., Kallos, M. S., Gates, I. D., 2015, "A Review of Pyrolysis, Aquathermolysis, and Oxidation of Athabasca Bitumen", *Fuel Process. Technol.*, 131, 270–289.

Luhning, R. W., Anand, A., Blackmore, T., Lawson, D. S., 2002, "Pipeline Transportation of Emerging Partially Upgraded Bitumen", Presented at the Petroleum Society's Canadian International Petroleum Conference, Calgary, Alberta, June 11–13, paper 202–205.

Meyer, R. F., Attanasi, E. D., Freeman, P. A., 2007, "Heavy Oil and Natural Bitumen Resources in Geological Basins of the World", U.S. Geological Survey, Open-File Report 2007-1084, available online at http://pubs.usgs.gov/of/2007/1084/.

Morrow, A. W., Mukhametshina, A., Aleksandrov, D., Hascakir, B., 2014, "Environmental Impact of Bitumen Extraction with Thermal Recovery", SPE Np. 170066, presented at SPE Heavy Oil Conference-Canada, 10–12 June, Calgary, Alberta, Canada.

Muñoz, J. A. D., Ancheyta, J., Castañeda, L. C., 2016, "Required Viscosity Values to Assure Proper Transportation of Crude Oil by Pipeline", *Energy & Fuels*, 30(11), 8850–8854.

Muraza, O., Galadima, A., 2015, "Aquathermolysis of Heavy Oil: A Review and Perspective on Catalyst Development", *Fuel*, 157, 219–231.

Nenninger, J., Nenninger, E., 2005, "Method and Apparatus for Stimulating Heavy Oil Production", US Patent 6,883,607.

Noguchi, T., Ed., 1991, *Heavy Oil Processing Handbook*, Research Association for Residual Oil Processing, Tokyo, Japan, 48–64.

Organization of the Petroleum Exporting Countries, 2015, *2015 World Oil Outlook*, OPEC Secretariat, October, Helferstorferstrasse 17, A-1010 Vienna, Austria (www.opec.org).

Ovalles, C., Rechsteiner, C., 2015, *Analytical Methods in Petroleum Upstream Applications*, CRC Press, Boca Raton, USA, and references therein.

Petroleum.co.uk, 2016, "The Future of Petroleum", www.petroleum.co.uk/the-future-of-petroleum downloaded July 1st.

Pourabdollah, K., Mokhtari, B., 2013, "The VAPEX Process, from Beginning up to Date", *Fuel*, 107, 1–33.

Ramirez-Corredores, M. M., 2017, *The Science and Technology of Unconventional Oils: Finding Refining Opportunities*, 1st Ed., Elsevier, London, Chapter 2, p 2 and references therein.

Saniere, A., Hénaut, I., Argillier, J. F., 2004, "Pipeline Transportation of Heavy Oils, a Strategic, Economic and Technological Challenge", *Oil Gas Sci. Technol. Rev. IFP*, 59(5), 455–466.

Sawarkar, A. N., Pandit, A. B., Shriniwas, D. S., Joshi, J. B., 2009, "Use of Ultrasound in Petroleum Residue Upgradation", *Can. J. Chem. Eng.*, 87(3), 329–342.

Schenk, C. J., Cook, T. A., Charpentier, R. R., Pollastro, R. M., Klett, T. R., Tennyson, M. E., Kirschbaum, M. A., Brownfield, M. E., Pitman, J. K., 2009, "An Estimate of Recoverable Heavy Oil Resources of the Orinoco Oil Belt, Venezuela", U.S. Geological Survey Fact Sheet 2009-3028. https://pubs.usgs.gov/fs/2009/3028/, downloaded Dec. 17, 2016.

Smalley, C., 2000, "Heavy Oil and Viscous Oil", In: *Modern Petroleum Technology*, R. A. Dawe, Ed., John Wiley and Sons Ltd., West Sussex, England, Volume 1, pp 409–435. ISBN: 978-0-470-85021-3.

Society of Petroleum Engineers (SPE), 1997, "Petroleum Reserves Definitions", www.spe.org/industry/petroleum-reserves-definitions.php, downloaded Dec. 26, 2016.

Speight, J. G., 2007, *The Chemistry and Technology of Petroleum*, 4th Ed., CRC Press, Boca Raton, USA, and references therein.

Tafneft Website, 2017, "Development of Natural Bitumen Fields" www.tatneft.ru/production-activity/technologies/development-of-natural-bitumen-fields/?lang=en, downloaded Jan. 15, 2017.

Trautman, M., MacFarlane, B., 2014, "Experimental and Numerical Simulation Results from RF Test in Native Oil Sands at the North Steepbank Mine" World Heavy Oil Conference, New Orleans LA, USE Paper No. WHOC14-301.

UNITAR, 1982, "Proposal for a Definition of Heavy Crude and Sand", 2nd International Conference on Heavy Crude and Tar Sands, United Nations Institute for Training and Research (UNITAR), Caracas, Venezuela, February 7–17, Chapter 1.

U.S. Geological Survey, 2006, "Natural Bitumen Resources of the United States", https://pubs.usgs.gov/fs/2006/3133/, downloaded Dec. 12, 2016.

2 Heavy Oil Reservoirs and Crude Oil Characterization

Before we start discussing the state-of-the-art in the area of subsurface upgrading (SSU), we must describe the reservoirs in which heavy crude oils and bitumens (HO/B) are found in nature. We start by discussing the general characteristics of heavy oil-bearing formations including fluid (aqueous and gas) and solid compositions and biogenesis. The effect of these components on SSU processes is described as well. Also, different aspects of the HO/B viscosity are presented, as this property is the key to effectively recover these valuable natural resources from underground formations [Kapadia *et al.* 2015].

Next, in this chapter, we present the physical and chemical aspects of heavy crude oils and bitumen composition. These topics include elemental analysis, SARA, simulated distillation (SimDis), the Boduszynski's continuum model of petroleum, and the relationships between viscosity and asphaltenes [Boduszynski 2015]. All of these aspects are very germane to the performance and success of a given subsurface upgrading process.

Once we understand the materials we are aiming to physically and chemically transform, we can effectively develop improved technologies to enhance oil production and increase recovery factors by using SSU. Also in this chapter, we define several concepts and measurements that are used for evaluation and selection of *in-situ* upgrading processes. As Dr. M. Boduszynski said *"Better understanding crude oil molecular composition and linking that knowledge to physical and chemical behavior is the key to better crude oil value assessments and predictions of the outcome of upstream and downstream operations"* [Boduszynski 2015].

2.1 GENERAL CHARACTERISTICS OF HEAVY OIL-BEARING FORMATIONS

As mentioned in Section 1.2, heavy crude oil and bitumen reservoirs can be found on all the continents except Antarctica. They are located in various basins and depositional settings, and most of the HO/B reservoirs are encountered at shallow or medium depths (< 2,000 m). Biodegradation, water washing, and loss of volatile components to the atmosphere are part of the mechanisms of formation and evolution of these critical raw materials [Eschard 2011]. At shallow depths, meteoric water invasion provided the necessary nutrients for biodegradation. Geologic periods of uplift and erosion favor further degradation of the oil.

Heavy oil-bearing formations (or reservoir rocks) are a combination of mineral matter, connate water, and heavy crude oil/bitumen coexisting in approximate 80(±10):10(±5):10 (±5) v/v/v ratios, respectively. The properties and composition of the reservoir rocks and the heavy oils and bitumens may exhibit significant variation with geographic and depth locations. Typically, there is almost no free gas zone, and the solution–gas content is relatively low compared to conventional oil reservoirs [Strausz and Lown 2003, Kapadia *et al.* 2015]. All of these variations on the physical and chemical properties are going to affect the subsurface upgrading processes significantly. Thus, detailed descriptions of those variables are necessary to assess their relative influence on the outcome of the process.

Gravity segregation is a concern as fluids in HO/B reservoirs may be more variable than those found in conventional crude oils. For example, Smalley reported that the produced fluids from an Alaska reservoir vary between 14 and 24°API and between 15 and 600 cP as measured at reservoir conditions [Smalley 2000]. These differences are significant since the crude oils with the lowest viscosity and highest API are attractive resources, but the ones with less favorable properties would need EOR or SSU processes to make their production economical. Also, gravity segregation may occur for the heavier components of the oil (e.g., resins and asphaltenes), so these fractions sink toward the base of the oil column. The end-results can be a significant accumulation of highly polar compounds at the bottom of the reservoir and near the oil-to-water interface. These changes of composition in function of the depth should be taken into consideration as they significantly affect the outcome of the subsurface conversion process [Smalley 2000, Mullins 2007].

The effect of heterogeneities is a critical factor when a fluid is injected downhole to heat, induce chemical transformation, or displace the heavy crude oils and bitumens. The reservoir facies of HO/B fields generally corresponds to the most porous and permeable sands. In consequence, facies with medium range porosity could act as a barrier to fluid flow [Eschard 2011]. The impact of reservoir heterogeneities on the subsurface upgrading of heavy oils is poorly understood and has not been studied in detail.

One additional complication is that HO/B cores and downhole oil samples are difficult to acquire at reservoir conditions even when using state-of-the-art coring techniques [Smalley 2000]. As we discuss later, HO/B are present in soft and/or unconsolidated sands. Their petrophysical properties such as porosity and permeability could be significantly altered by bringing them to the surface. Also, as the reservoir is put into production, unconsolidated materials, which have high compressibility, might lead to potential surface subsidence and compaction as well as permeability modifications. These characteristics affect the reservoir characterization, and thus the successful exploitation of HO/B fields should integrate a more detailed reservoir description than is usually carried out for conventional oils. The design and successful execution of subsurface upgrading processes may depend on a clear understanding of the reservoir geology and geophysics.

2.1.1 OVERALL COMPOSITION AND BIOGENESIS

The average petrophysical properties of reservoir rocks and API gravity of representative examples of heavy crude oils and bitumens are shown in Table 2.1. These examples, taken from the literature, were selected to illustrate the HO/B properties and to provide study cases in which a subsurface upgrading process could be potentially applied. Examples of Canada, Venezuela, Colombia, China, Kuwait/Saudi Arabia, and Russia are reported. As seen, the API gravity ranges from 6 to 19°API, depths from the surface to around 2,000 m. One exception is Boscan crude, from Western Venezuela, that is found between 1,500 and 2,700 m [Kumar *et al.* 2001].

In general, HO/B oil-containing sands are relatively thick with values ranging from 5 to 60 m (Table 2.1). As mentioned, heavy oil and bitumen reservoirs are relatively shallow, so their temperatures (8–70°C) and pressures (surface-10.3 MPa) are milder in comparison with the light and medium counterparts. In contrast, porosities (20–37%), permeabilities (1–16 D), and oil saturation (70–90%) are excellent which makes them attractive from the economic point of view. The other main fluid is the aqueous phase, which is discussed in the next section. Typically, there is almost no free gas zone, and the solution–gas content is relatively low (3 to 4 m^3 gas per m^3 of bitumen at reservoir conditions) than those found in lighter crude oil reservoirs [Strausz and Lown 2003, Kapadia *et al.* 2015].

One of the most studied HO/B reservoirs are the Alberta oilsands [Strausz 2003, Rottenfusser and Ranger 2004, Jiang *et al.* 2009, Ivory *et al.* 2008]. These materials are composed of bitumen and sand, which comprise up to 18 and 95% of the ore material, respectively. Other lesser components are found such as water, clay, silt, heavy minerals, gases, humic matter, and mineral-bound organic matter. Nevertheless, these minor components can profoundly affect the water flotation efficiency, the separability of the bitumen [Strausz 2003], as well as a possible subsurface upgrading process.

For the Canadian reservoirs, the sand grains are typically between 50 and 120 μm in size with pore sizes between sand grains typically equal to about 10 to 30 μm [Kapadia *et al.* 2015]. The sand is usually composed largely of quartz. A schematic diagram showing a structural model of

TABLE 2.1

Average Petrophysical Properties of Reservoir Rocks and API Gravity[a] of Heavy Crude Oils and Bitumens

Reservoir (Location)	Country	API Gravity[a]	Depth (m)[b]	Pay Zone (m)[c]	Temp (°C)[d]	Press. (MPa)[e]	Porosity (%)[f]	Perm. (Darcy)[g]	So (%)[h]	Sw (%)[i]	References
Athabasca (Alberta)	Canada	8	Surface–400	5–9	8–20	0.8–3.5	30	5–8	85	15	Strausz [2003] Rottenfusser [2004]
Cold Lake (Alberta)	Canada	6–12	275–500	6–12	8–20	0.8–3.5	33	1–5	70–85	20–30	Strausz [2003] Jiang et al. [2009]
Peace River (Alberta)	Canada	9	460–760	24	8–20	0.8–3.5	24		85	15	Strausz [2003] Ivory [2008]
Hamaca (Orinoco Belt)	Venezuela	8–9	600–1,200	9–30	54–65	8.2–9.7	36	8–16	80–90	10–20	Mirabal et al. [1996] Ovalles [2003]
Cerro Negro (Orinoco Belt)	Venezuela	8–9	450–1,200	10–35	60	6.2–9	32	9	90	10	Lugo [2001]
Junin/Zuata (Orinoco Belt)	Venezuela	9	300–550	10–20	43–54	4.8–8.3	20–32	8–20	80	20	Kopper et al. [2001] Cocco and Ernandez [2014]
Boscan (Western Maracaibo)	Venezuela	10.5	1,600–2,700	21	66–82	16.7–22	26	0.8–1.1	60–80	20–40	Kumar [2001]
Liaho (Guantao)	China	10	600	20–30	32	6	25–30	2–3	70–85	15–30	Bao et al. [2012]
Henan (Gucheng)	China	17	150–800	8–15	38	2–8	28–34	4	70	30	Hu [1998]
Llanos (Llanos Basin)	Colombia	8.5–9.5	–	12–55	68	–	35	1.6	70	30	Andarcia [2014]
Wafra 1st Eocene (Middle East)	Kuwait/Saudi Arabia	13–19	1,000–1,800	60	–	–	37	0.5	70–90	10–30	Rubin [2011]
Ashalchinskoye (Tartarstan)	Russia	18	85–250	25	8	0.44	34	1–3	40	60	Musin [2010] Ibatullin [2012]

[a] API gravity of the heavy crude oils and bitumens as measured using ASTM D 4052 [ASTM 2002].

[b] Reservoir average depth in meters.

[c] Average thickness of oil-bearing formation in meters.

[d] Reservoir average temperature in °C.

[e] Reservoir average pressure in MPa (1 MPa = 145 psi).

[f] Reservoir average porosity in % of porous space.

[g] Reservoir average permeability in Darcy.

[h] Reservoir average oil saturation as a percentage of porous space of the mineral formation occupied by oil phase.

[i] Reservoir average water saturation as a percentage of porous space of the mineral formation occupied by aqueous phase.

Athabasca oil sand is shown in Figure 2.1 [Takamura 1982]. The water in the oil sand appears in three forms: as pendular rings at grain-to-grain contact points, as a roughly 10 nm thick film which covers the sand surfaces (connate water), and as water retained in fines clusters. The remaining void space is occupied by bitumen.

Thus, a few percent of the connate water in all Alberta oil sands are present as a thin film covering the sand surfaces and separating the bitumen from the sand (see the insert in Figure 2.1). This significant structural feature renders the sand water-wet and the bitumen separable from the sand by water flotation. The water, clay, and bitumen contents in the ore are interrelated, and water and clay vary in parallel and inversely to bitumen [Strausz 2003].

As shown in Table 2.1, the porosity of the Alberta reservoir ranges from 24 to 33%, and the absolute permeability of the reservoir rock ranges from 1 to 8 Darcy depending on the porosity, shape of sand grains, and depth. In general, deeper pay zones imply greater overburden stress which, in turn, means lower porosity.

Due to the proximity to the surface, the initial temperature of the reservoir is typically between 8 and 20°C (Table 2.1), which means the viscosity of the bitumen is in the order of millions of mPa s. In general, it was found that this property does not depend strongly on pressure [Kapadia *et al.* 2015] but the initial reservoir pressure, which typically ranges from 0.8 MPa (116 psi) up to about 3.5 MPa (500 psi), depending on the depth.

In Table 2.1, four Venezuelan reservoirs are presented; three are located in the Orinoco Belt (Hamaca, Cerro Negro, and Junin/Zuata) and one in Eastern Venezuela (Boscan). As mentioned in Section 1.2, the Orinoco Belt is one of the most extensive accumulations of hydrocarbons in the world. The petrophysical information revealed the similarity of rock and fluid characteristics in most regions of this giant reservoir [Mirabal *et al.* 1996, Ovalles *et al.* 2003, Lugo *et al.* 2001, Kopper *et al.* 2001, Cocco and Ernandez 2014]. The data shown in Table 2.1 indicated an average porosity of 20–36%, permeabilities of 10–20 Darcy, and oil and water saturation of 80–90 and 10–20%, respectively. Orinoco Belt initial pressures and temperatures varied within the range of 5–10 MPa (700–1,450 psi) and 43–60°C (110–140°F), respectively. Oil gravity ranged from 6 to 10°API. Similar average properties have been reported by Vega-Riveros and Barrios during their evaluation of steam injection processes [Vega-Riveros and Barrios 2011].

In contrast, Boscan crude oil [Kumar *et al.* 2001] is found much deeper (1,500–2,700 m) and hence, at higher temperatures (66–82°C) and pressures (16–22 MPa) than the Orinoco counterparts (Table 2.1). Porosity is in the order of 26%, but permeability is around 1 Darcy. API gravity is slightly higher (10.5°API) than Orinoco heavy oils but has similar oil and water saturation (Table 2.1).

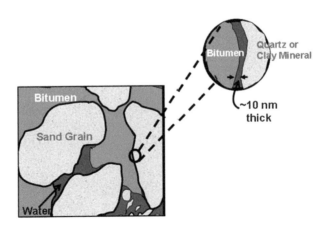

FIGURE 2.1 Schematic diagram showing a refined structural model of Athabasca oil sand. (Reprinted with permission [Takamura 1982].)

Other heavy crude oils around the world can be potentially produced by subsurface upgrading processes. Those are Liaohe and Henan from China [Bao *et al.* 2012, Hu *et al.* 1998]. As seen in Table 2.1, their petrophysical properties, i.e., porosity, permeability, and oil and water saturation, are not too different from those found in Canada and Venezuela. Also, they are found at depths as shallow as the Canadian oilsands (150–800 m). Other HO reservoirs are located in the Llanos Basin in Colombia [Andarcia *et al.* 2014], Wafra/1st Eocene in Kuwait/Saudi Arabia [Rubin 2011], and Ashalchinskoye in the Tatarstan/Russia [Musin *et al.* 2010, Ibatullin *et al.* 2012].

In conclusion, the above-described properties of heavy crude oils and bitumens are thought to be the consequence of water flows and microbial activities within a shallow depth (< 2 km) and relatively lower temperatures (80°C) as the oil accumulates over geological timescales [Faergestead 2016, Omajalia *et al.* 2017]. These microorganisms metabolize lighter hydrocarbons and produce methane leading to the enrichment of heavy components. In consequence, the *in-situ* biodegradation process increases oil density (lower API), viscosity, sulfur, nitrogen, and metal contents and lowers the hydrogen-to-carbon ratio, i.e., the crude oil becomes more aromatic. Other processes, such as preferential migration of light hydrocarbons, water washing, or evaporation also lead to oil degradation. Finally, HO/B are found in geologically young Pleistocene, Pliocene, and Miocene formations and the older Cretaceous, Mississippian, and Devonian formations [Faergestead 2016].

2.1.2 AQUEOUS PHASE COMPOSITION

Another critical component present in heavy oil-bearing formation is the aqueous phase. There are several reasons why this phase could significantly affect the performance of most of the subsurface upgrading process. Water could behave as a hydrogen donor (Aquathermolysis) [Muraza and Galadima 2015], oxidize metal-containing catalysts, be a poison, or lead to sintering of the catalytically active phases [Ovalles *et al.* 2015a] and act as heat sink, not allowing increasing the temperature beyond the steam-equilibrium conditions. As it is discussed in the following chapters, the water-containing phase can substantially alter the mechanism for chemical transformation of HO/B during SSU process.

In the Takamura's model (Figure 2.1) water in the oil sand appears in three forms: As pendular rings at grain-to-grain contact points, as a roughly 10 nm thick film which covers the sand surfaces (connate water), and as water retained in fines clusters [Takamura 1982]. A critical property of Alberta oilsands distinguishing them from most of the oilsands in the United States is their water-wet condition resulting from the presence of a thin film of connate water surrounding each mineral grain (Figure 2.1).

Reservoir aqueous phase is believed to have its ultimate origin in the ancient seawater in which the source sediments were deposited and migrated to the reservoir rock along with the oil [Strausz and Lown 2003]. For the Canadian oilsands, the total amount of water tied up in these surface films is quite small, only a few percent of the 2–10% connate water in the oil sand. The dominant ionic species in seawater and reservoir waters are the sodium and chloride ions, followed by the sulfate and magnesium ions (seawater). In reservoir waters, the minority components are the calcium and bicarbonate ions. The total dissolved solids (TDS) in the formation water is 46,400 ppm or mg/L whereas for the seawater, 35,100 mg/L [Strausz and Lown 2003].

The pH of the connate water can profoundly affect the solubility of ions and minerals and the behavior of the oil sand for a given subsurface upgrading technology. Consequently, the composition and the results of analyses are significantly dependent on the mode of sample preparation. The pH of most Canadian oil sands (connate water) is slightly basic with values of 8 ± 0.8 and with a considerable buffer capacity [Strausz and Lown 2003].

For the Orinoco case, the composition of the aqueous phase is highly variable, and the most noticeable trend is that the formation waters become fresher at shallower depths [Cocco and Ernandez 2014]. It has been reported the connate waters have high chlorides and sodium contents (> 1,000 ppm) with an Na/Cl < 1. The TDS is generally around 3,500 ppm for the Junin/Zuata area [Marcos *et al.* 2007].

In the case of the Henan oilfield in China, the formation water has a TDS of 4,735 ppm [Wang *et al.* 2014]. In conclusion, all these data indicate that the connate waters of heavy crude oil and bitumens have relatively high salinity with TDS values in the 3,500–40,000 ppm range. This characteristic should be taken into consideration during the development of SSU processes.

2.1.3 GAS PHASE COMPOSITION

As with the formation water, the gas phase is another critical component of HO/B formations. Even though much of the heavy oil reservoirs do not contain a free gas cap, in the case they do, this zone could behave like a thief zone for steam, solvents, catalysts, or hydrogen donors during SSU processes. Additionally, it has been reported that methane (the main component of the gas phase) is involved during upgrading reactions carried out in lab experiments under subsurface [Ovalles *et al.* 2003, Ovalles *et al.* 2015a] and surface [Sundaram 1987, Egiebor and Gray 1990, Ovalles *et al.* 1995, 1998] conditions.

Athabasca deposit is known to have thin gas zones near the top and zones that once contained gas but now are water-bearing [Strausz and Lown 2003]. For the Zuata reservoir located in the Orinoco Belt, the gas phase is present in minor quantities in isolated sands and silts. The gas zones are generally thin but are easily identified by conventional logging techniques [Kopper *et al.* 2001]. However, most of the gases contained in those zones ultimately escape through the overburden into the atmosphere.

A vast majority of the gas is present as a dissolved gas. This solution–gas content is typically relatively low compared to conventional oil reservoirs. For the Canadian oilsand, the gas-to-oil ratios (GOR) are generally lower than 3 to 4 m^3 gas per m^3 of bitumen (~2–3 SCF/BBL) at reservoir conditions [Kapadia *et al.* 2015]. For the Boscan reservoir located in Western Venezuela, the average GOR is higher, i.e., ~20 m^3 gas per m^3 of heavy oil (110 SCF/BBL) [Kumar *et al.* 2001]. For Hamaca crude from the Orinoco Belt, the average gas-to-oil ratio has similar values, i.e., ~19 m^3 gas per m^3 of crude (106 SCF/BBL) [Mirabal *et al.* 1996]. For Wafra-1st Eocene from Kuwait/Saudi Arabia, the reported GOR = ~11 m^3 gas per m^3 of heavy crude oil (62 SCF/STB) [Rubin 2011].

The composition of gases (and highly volatiles) from some of the Alberta oilsand reservoirs, namely the Athabasca, Cold Lake, and Peace River reservoirs, has been reported [Strausz and Lown 2003]. The compounds detected were C1–C7 hydrocarbons, carbon dioxide, carbon monoxide, carbonyl sulfide, carbon disulfide, hydrogen sulfide, sulfur dioxide, methanol, and acetaldehyde. The latter compound is believed to be produced by oxidation of the hydrocarbons.

For the Hamaca reservoir, the main component of the dissolved gas is methane (> 90 vol.%) followed by carbon dioxide and nitrogen. Small amounts of C2 to C6 hydrocarbons were also detected [Mirabal *et al.* 1996]. The presence of dissolved gas in HO/B reservoirs should be scrutinized due to the foamy behavior observed for some of these crude oils. Foamy oil is characterized by the entraining liberated solution gas. The gas bubbles may be entrained in the oil due to high viscosity, low diffusion coefficient, and high asphaltene content. This behavior has a profound impact on the recovery factor. Laboratory results showed the *in-situ* formation of non-aqueous oil foam with high gas retention, improving oil mobility and leading to high well productivity. An experimental recovery factor over 10% was obtained under primary production [Mirabal *et al.* 1996].

2.1.4 MINERAL FORMATIONS

The effect of the mineral formation on the SSU of heavy oils has not been studied in detail. It has been reported that the reservoir rock plays a significant role in the upgrading of HO/B in lab experiments under subsurface conditions, as discussed in Section 7.3.2 [Ovalles *et al.* 2003]. Two possible reasons were proposed to explain this beneficial effect. The first one attributes catalytic activity to the mineral matrix, even at low severity conditions (< 280°C) as suggested by several authors in the literature [Stapp 1989, Strausz *et al.* 1977, Phillips *et al.* 1985, Ovalles *et al.* 2003, Yoshiki and Phillips 1985]. Also, the mineral formation can be used as support material for metallic nano-catalysts as reported for the *in-situ* catalytic upgrading [Pereira-Almao *et al.* 2015].

Alternatively, the beneficial effect of the natural formation can be attributed to an improved heat transfer mechanism that diminishes the probability of the molecules to move away from the reaction zone. The extended contact period could increase the cracking reaction as reported in the literature [Phillips *et al.* 1985, Yoshiki and Phillips 1985]. Thus, the knowledge of the composition and morphology of the mineral formations can be useful to select potential reservoir candidates to apply SSU in the field.

As mentioned earlier, heavy oil-bearing formations (reservoir rocks) are a combination of mineral matter, connate water, and heavy crude oil/bitumen in which the solid phase represents around 80% of the volume. Generally, in many HO/B reservoirs, the primary component is sand (> 99 wt. % SiO_2), and the heavy oil acts like cement, and the removal of the oil causes the reservoir rock to collapse [Chopra 2010].

Specific drilling and completion issues are frequent in heavy oil containing unconsolidated sands. On the one hand, sand production may constitute a significant problem for producing HO/B reservoirs due to decreasing production, plugging the well botttomhole, and eroding subsurface and surface equipment. On the other hand, a process like Cold Heavy Oil Production with Sand (CHOPS) may benefit from the sand flow, with the sands being produced jointly with the oil, as discussed in Section 3.1.1. [Huc 2011].

Unconsolidated sands often have high compressibilities. This situation leads to a range of issues and concerns, including subsidence as the reservoir pressure is drawn down, permeability modification as the reservoir compresses, and the possibility of pressure support from reservoir compaction [Smalley 2000].

As mentioned, the primary component of all Alberta oilsands is sand, comprising mostly quartz (SiO_2) in the 60–90% range. The major groups of mineral constituents of several Alberta oil sand deposits are feldspar (small content with quartz), micas (muscovite, sericite, biotite), chlorite, and carbonates (siderite, dolomite, calcite, magnesite). Also, metal oxides (titanium-containing leucoxene and ilmenite and iron-containing hematite, magnetite, and limonite), silicates: tourmaline, zircon, epidote, and staurolite and sulfides (pyrite/marcasite and sphalerite) [Strausz and Lown 2003]. In the McMurray Formation, kaolinite and illite are the two major clay minerals, potash, feldspar, dolomite, and mica the major light minerals, and siderite and pyrite/marcasite are the main heavy minerals [Strausz and Lown 2003].

For the Orinoco Belt reservoirs, X-ray diffraction (XRD) showed that mineral composition of the Hamaca sand is 1% dolomite, 1% calcite, 4% feldspar, 8% clay, and 86% quartz (SiO_2) [Ovalles *et al.* 2003]. Reservoirs of the Oficina Formation within the Zuata Field are clean, unconsolidated fine- to coarse-grained (predominantly medium-grained) quartz-rich sands with porosity averaging 32% and permeability averaging 7,800 mD (Table 2.1). The producing sands have net-gross between 40–60% and are interconnected [Kopper *et al.* 2001].

Several heavy crude oils and bitumens are present in carbonates reservoirs such as dolomite and other minerals. Representative examples are the Grosmont formation in Alberta [Chopra 2010] and the Wafra/Eocene in Kuwait/Saudi Arabia [Rubin 2011]. It has been reported that the rock present in these mineral formations may be oil-wet and that the heavy oil is present only in the larger pore spaces [Chopra 2010, Huc 2011]. This situation makes the crude oil movement very challenging, and the pay zones must be fractured to obtain an economical oil production. Until now, the experience of heavy oil production in fractured carbonate reservoir is limited. As can be imagined, these reservoirs represent a challenge, not only for conventional thermal EOR processes but also for new subsurface upgrading concepts.

2.2   VISCOSITY OF HEAVY CRUDE OILS AND BITUMEN

The critical problem for enhancing the HO/B production is their high viscosity at reservoir and surface conditions. As seen in Table 2.2, heavy oils and bitumens have viscosities in the order of 10^5–10^6 cP (at 77°F); whereas light and medium oils have values in the 10–10^2 range. This data is the average taken from 50 basins and 305 deposits [Meyer et al. 2007]. For HO/B, viscosity can reach even higher values at temperatures lower than 20°C, which is typical for a Canadian reservoir (see Table 2.1). Thus, one of the primary objectives of SSU processes is aimed at permanently reducing oil viscosity, improving oil mobility at reservoir or surface conditions with the concomitant increases in oil rates and cumulative production.

2.2.1   VISCOSITY MEASUREMENTS AND EFFECT OF TEMPERATURE

In 1982, the petroleum experts gathered in Caracas, Venezuela, for the 2nd International Conference of Heavy Crude and Tar Sand [Gibson 1982] recognized the importance of performing representative and meaningful viscosity measurements to classify an oil as heavy oil or bitumen. Other end goals were to evaluate and select the efficient EOR process, to assess the commercial value, and to the design production and transport facilities. Heavy oils and bitumens are complex fluids, and

TABLE 2.2
Average Physical and Chemical Properties of Crude Oils

Property	Unit	Light Oil[a]	Medium Oil[b]	Heavy Oil[c]	Bitumen[d]
API gravity[e]	degrees	38.1	22.4	16.3	5.4
Depth	feet	5,140	3,280	3,250	1,224
Viscosity (77°F)	cP[f]	15.7	64.6	101,000	1,290,000
Viscosity (100°F)	cP[f]	13.7	34.8	641.7	198,000
Viscosity (130°F)	cP[f]	10.1	24	278.3	2,371
Carbon	wt. %[g]	85.3	83.2	85.1	82.1
Hydrogen	wt. %[g]	12.1	11.7	11.4	10.3
Molar ratio H/C	–	1.70	1.69	1.61	1.51
Nitrogen	wt. %[g]	0.1	0.2	0.4	0.6
Oxygen	wt. %[g]	1.2		1.6	2.5
Sulfur	wt. %[g]	0.4	1.6	2.9	4.4
Acid number	mgKOH/[h]	0.4	1.2	2	3
Vacuum residue	vol%[i]	22.1	39.8	52.8	62.2
Asphaltenes	wt. %[j]	2.5	6.5	12.7	26.1
Nickel	ppm[k]	8	33	59	89
Vanadium	ppm[k]	16	98	177	334

Meyer [2007].

[a] Light oils defined as API gravity greater than 26. Average property from 131 basins, 8,148 deposits.
[b] Medium oils defined as API gravity between 20 and 25. Average property from 74 basins, 774 deposits.
[c] Heavy oils defined as API gravity between 10 and 20. Average property from 127 basins, 1,199 deposits.
[d] Bitumen defined as API gravity lower than 10. Average property from 50 basins, 305 deposits.
[e] API gravity as measured using ASTM D 4052 [ASTM 2002].
[f] Centipoise 1 cP = 1 mPa·s.
[g] Percentage of weight.
[h] mgKOH/g = milligrams of potassium hydroxide per gram of sample.
[i] Volume percentage of 540°C vacuum residue.
[j] Weight percent of asphaltenes using pentane as precipitant.
[k] Parts per million (mg/Kg) of metal.

viscosity measurement is not a simple task. Thus, repeatability, reproducibility, and accuracy are not only affected by sample handling, storage, and cleaning procedures, but they are also affected by the selection of viscometers and the experimental methodology followed by different operators. All methods need careful calibration and attention to detail. This fact requires a certain level of skill and training in the operator of the equipment [Abivin *et al.* 2012].

Bitumens and heavy crude oils can have very different rheological properties ranging from a Newtonian and purely viscous character to non-Newtonian behavior, in the form of either stress-dependent viscosity (i.e., viscoelastic fluid) or time-dependent behavior (i.e., thixotropic fluid). This fact can give the perception of inconsistent viscosity measurement with routine laboratory techniques. For some oils, the viscoelastic character is linked to the presence of paraffinic wax crystals [Abivin *et al.* 2012]. For the others, viscoelastic oils, the elastic nature and macroscopic weak gel-like behavior seem to be related to their high amount of asphaltenes. This topic is discussed in Section 2.4.

Recently, Zhao *et al.* reviewed and summarized the best practices and suggested guidelines for heavy oil viscosity measurements [Zhao *et al.* 2016]. They carried out a systematic evaluation and comparisons of different viscometers typically used in HO/B viscosity measurements, provided references about viscometer selection, recommended measurement procedures for each viscometer, and generated a reliable viscosity database of dead and live heavy oils. The authors discussed three viscometers typically used in heavy oil systems: A capillary viscometer, an electromagnetic viscometer, and a rheometer. The authors concluded that, within testing conditions, each well-calibrated viscometer could reproduce reported values for viscosity standards, with a relative error of less than 5% [Zhao *et al.* 2016].

The effect of temperature on the viscosity of heavy crude oils and bitumens (dead oils) is presented in Figure 2.2 in the temperature range 10–120°C. As seen, the logarithm of viscosity in centipoise (cP = 1 mPa. s) is linearly correlated with the temperature (in °C) for crudes from Canada, Venezuela, Colombia, and China. These heavy crudes have viscosity values several orders of magnitude higher than those determined for medium or light oils (Figure 2.2).

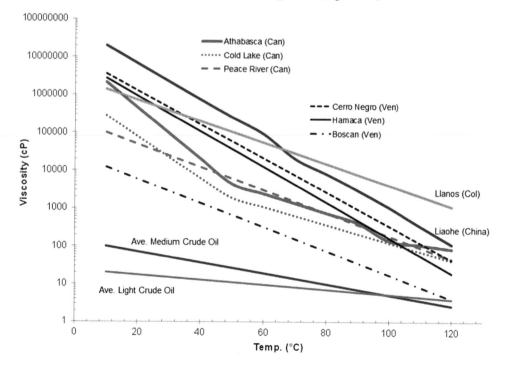

FIGURE 2.2 Effect of temperature on the viscosity of heavy crude oils and bitumens (dead oils).

In general, the viscosity of HO/B is fundamentally related to the motions and interactions of all the different molecules present within these materials. When heavy crude oils are cooled, the molecules are drawn more closely together. Because of the overall attractive intermolecular interactions, friction between the molecules thus increases, which, in turn, increases the overall apparent viscosity. In addition to friction, other interactions may occur, such as van der Waals, hydrogen bonding, or π-stacking. In consequence, due to the nature of these interactions, the viscosity depends strongly upon the temperature. The most commonly used model to predict the temperature dependence of the viscosity of HO/B is the Arrhenius equation (Equation 2.1).

$$\mu(T) = \mu_0 \exp\left[Ea/(RT)\right] \tag{2.1}$$

In this equation μ is the viscosity (in units of Pa s), T the temperature (in Kelvin), μ_0 a constant, Ea the activation energy (in units of J mol^{-1}) which is HO/B dependent, and R the universal gas constant (in units of J K^{-1} mol^{-1}).

It has been found that bitumens and some crude oils do not strictly follow an Arrhenius behavior but experience a more drastic increase in viscosity at low temperatures. Three examples can be seen in Figure 2.2 for Athabasca (continuous red line), Cold Lake (dotted red line), and Liaohe (purple line) crude oil. These deviations from the Arrhenius behavior can be taken into account by introducing a temperature-dependent activation energy [Henaut et al. 2001], which can be attributed to the chemical composition (asphaltenes) of the sample.

More recently, 13 heavy oils from different parts of the world were subjected to a full rheological characterization [Abivin et al. 2012]. The authors found that all heavy oils experienced a glass transition at low temperatures, and the Williams–Landel–Ferry (WLF) equation was used to successfully model the temperature dependence of the viscosity over a broad temperature range (−35°C to 200°C). Some heavy crude oils showed weak gel-like behavior which was attributed to the presence of paraffinic wax crystals, as quantified by Differential Scanning Calorimetry (DSC).

2.2.2 Effects of Diluent and Amount Needed for Transportation

Dilution is the most widely preferred technology for reducing the viscosity of HO/B and making them transportable by conventional methods such as pipelines, shipping, trains, etc. To avoid asphaltene precipitation problems, aromatic naphtha, kerosene, or diesel diluents are used in the 20–30% range. It is important to mention that as the aromaticity of the diluents decreased, the extent of asphaltene precipitation and the fouling rates increased [Alomair and Almusallam 2013].

Figure 2.3 shows the typical effect of the weight percentage of diluent on the viscosity of heavy crude oils and bitumens at 30°C. As seen, for Venezuelan and Canadian heavy oils the viscosity decreased one order of magnitude (from ~1000 to 70 cSt) by increasing the naphtha and diesel contents in the 20–30 wt. % range. In Figure 2.3, the points are the actual experimental data, and the continuous lines are the calculated values using the correlations shown in the figure.

Several methods for predicting the viscosity of hydrocarbon blends have been reported in the literature since 1947 [Al-Besharah et al. 1987]. Al-Besharah et al. developed a model and compared it with ASTM D341 and the Refutas index. The authors found that their methodology gave the best representation of experimental results with deviations in most cases less than 6% [Al-Besharah et al. 1987].

In 2011, Ancheyta and coworkers evaluated 17 mixing rules reported in the literature for predicting kinematic viscosity of petroleum and its fractions [Centeno et al. 2011]. The authors compared the calculated values with the experimental viscosities of four crude oils (21.31, 15.93, 12.42, and 9.89°API gravity) and their blends using diesel at several volumetric proportions. The tested mixing rules were classified as "pure mixing rules" (they require an experimental viscosity of components and composition), mixing rules with a viscosity blending index parameter, and mixing rules that need other extra parameters (they are usually obtained by mathematical methods)

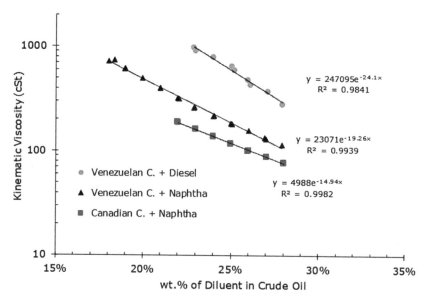

FIGURE 2.3 Effect of the weight percentage of diluent on the kinematic viscosity of heavy crude oils and bitumens (Dead oils) at 30°C. Points are the experimental data, and the continuous lines are the calculates values using the correlations shown.

[Centeno *et al.* 2011]. As the crude oil API gravity decreased, the authors found that all mixing rules failed to predict the viscosity of the HO blends. At higher temperature, the predictions were significantly improved. Nevertheless, no rule was capable of estimating viscosity for all the crude oils. This finding showed that predicting the viscosity of crude oil, and especially HO/B, is a very challenging task [Centeno *et al.* 2011].

One of the most common viscosity mixing rules is the one shown in Equation 2.2. This model is used for the numerical reservoir simulators and predicts with an acceptable degree of accuracy (< 10%) the viscosity of HO/B-diluent blends [CMG 2011].

$$\text{Log}\left(\text{Viscosity of blend}\right) = x_1 \text{Log}\left(\mu_{HO}\right) + y_1 \text{Log}\left(\mu_{Diluent}\right) \qquad (2.2)$$

Where x_1 and y_1 are the molar fractions, and μ_{HO} and $\mu_{Diluent}$ are the viscosities of the heavy crude oil or bitumen and diluent (naphtha or diesel), respectively.

In heavy oil production, the dilution step is one of the most expensive parts of the process. In general, the amount of added diluent is a crucial factor because of its higher cost compared with that of the HO/B alone. Thus, one of the primary objectives of subsurface upgrading processes is to minimize or eliminate the use of diluents.

Muñoz *et al.* [2016] evaluated the maximum viscosity values reported in the literature to define a range that allows for the crude oils to be fluidly transported. The authors found that the viscosity target varies according to the type of crude oil and the region where it is transported. For HO/B, they reported viscosity values in the range between 350 cSt at 7°C and 150 cSt at 50°C [Muñoz *et al.* 2016].

From the practical point of view, the viscosity requirements for heavy crude blends vary with the seasonal temperature. In Canada, pipelines are typically designed for a maximum viscosity of 350 cSt and a density of 940 kg/m³. The viscosity limit is at the pipeline reference temperature of 7.5°C during the winter and 18.5°C in the summer [CAPP 2012]. The pipelines are generally operated by keeping the HO/B content constant and by changing the diluent content throughout the year.

Based on the above discussions and the data presented in Figure 2.3, it can be calculated that the amounts of diluent needed to achieve the viscosity target of 350 cSt (see the dark black horizontal line) for the Venezuelan crude are for ~22 wt. % and ~27 wt. % for naphtha and diesel, respectively. For the Canadian crude, the amount needed is ~18 wt. % of naphtha. These examples illustrate the volumes of diluent required for pipeline transportation and the potential savings that could be achieved by using a subsurface upgrading process.

2.3 HEAVY CRUDE OIL COMPOSITION

This section discusses the current knowledge and understanding of the composition of petroleum and its fractions. The emphasis is placed on describing the properties that are germane to crude oil upgrading such as elemental composition, group-type components, and boiling point. Also, a comparison between the upstream versus downstream characterization is presented. Finally, the Boduszynski's continuum model of petroleum is discussed which describes petroleum composition on the basis of continuous changes as a function of the atmospheric equivalent boiling point (AEBP) [Boduszynski 2015]. The application of these concepts to the subsurface upgrading of heavy crude oils and bitumens is of paramount importance for the development of this technology in the next few decades.

2.3.1 ELEMENTAL COMPOSITION

Table 2.2 shows several of the average chemical and physical properties of conventional, medium, and heavy crude oils and natural bitumen [Meyer *et al.* 2007]. As mentioned previously (Section 1.1), the average API gravities of heavy crude oils (16.3°API) and bitumens (5.4°API) are lower than 20°API whereas light and medium oils (conventional oils) have average values of 38.1°API and 22.4°API, respectively.

As seen in Table 2.2, carbon contents are in the 82–85 wt. % range for all oils, but the hydrogen weight percentages decrease from light (12.1 wt. %) to bitumen (10.3 wt. %). By this way, the molar H/C ratios decrease from 1.7 to 1.5. Alternatively, the concentration of heteroatoms (nitrogen, oxygen, and sulfur) and the acid numbers also increase in the following order: light oil, medium oil, heavy oil, and bitumen.

On the other hand, the percentages of vacuum residue (non-distillable materials), asphaltenes (C5-insolubles), and nickel and vanadium contents are higher in the HO/B than those measured for the light and medium oils. These elevated levels of heteroatoms and relatively higher-molecular-weight materials found in the heavy oils and bitumens are one of the reasons for their relatively lower monetary value than the conventional counterparts and for the difficulties in processing at the refining centers. Thus, a goal for a subsurface upgrading process is to entirely or partially transform those heteroatom-containing/higher-molecular-weight compounds into higher value/easier to process hydrocarbon streams and at the same time, leave behind heavy metals and contaminants. This topic is discussed in Section 3.3.

2.3.2 COMPARISON BETWEEN UPSTREAM VS. DOWNSTREAM PETROLEUM CHARACTERIZATION

The *in-situ* upgrading of heavy oils brings together the integration of upstream production with downstream manufacturing. Both traditional operations are parts of the Petroleum Value Chain (Figure 1.8) and, as expected, have their priorities and challenges. Thus, it is not surprising that the requirements for petroleum characterization are significantly different.

As shown in Figure 2.4, in upstream, the composition of the crude oil has been defined as the content of the four group-type components of the SARA analysis, i.e., Saturates, Aromatics, Resins, and Asphaltenes. This analysis involves the precipitation of asphaltenes (alkane insolubles) using an n-paraffin solvent such as n-pentane (C5), n-hexane (C6), or n-heptane (C7). The content and composition of the asphaltenes depend not only on the precipitant used but the analytical procedure. The second step of the SARA analysis involves the use of high-performance liquid chromatography

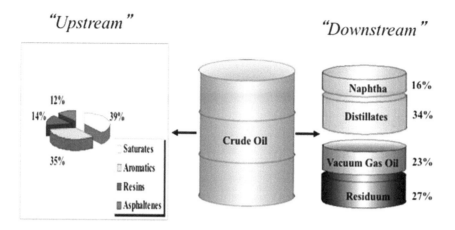

FIGURE 2.4   Upstream and downstream approaches to crude oil composition [Boduszynski 2015].

(HPLC) to separate maltenes (alkane solubles) into Saturates, Aromatics, and Resins (also known as polars). Different chromatographic methods produce different results. As discussed by Boduszynski, the use of similar terms to describe the group-type components produced by similar but significantly different methodologies introduces further ambiguity [Boduszynski 2015] so caution should be taken analyzing this type of data.

On the other hand, in the refining world (downstream) the crude oil composition is mainly defined in terms of the yield and quality of the distillable fractions obtained by simulated or conventional column distillation (see Figure 2.4). The main reason is to maximize fuel production in the boiling range of motor gasoline, diesel, jet, and bunker fuels. These virgin distillation cuts are then converted into useful fuel components by further conversion processes. In other to homogenize and standardize the distillation procedure an atmosphere-equivalent boiling point scale was developed, and it is described next.

2.3.3   Atmosphere Equivalent Boiling Point (AEBP)

The atmosphere equivalent boiling point scale covers the entire boiling range of crude oil, from atmospheric pressure, continuing with those accessible under vacuum, and finishing with the equivalent boiling ranges of non-distillable residuum solubility fractions [Boduszynski 2015]. By this way, the light, medium, and heavy crude oils can now be described regarding their various physical and chemical properties as they change with increasing AEBP.

As described by Boduszynski, Figure 2.5 shows a typical AEBP distribution curve for heavy crude oil [Boduszynski 2015]. The values for distillable fractions were derived from simulated distillation measurements whereas the values for the non-distillable residuum were calculated using vapor phase osmometry (VPO) average molecular weights [Boduszynski 2015].

It has been demonstrated that the development of the AEBP scale provides an organization for relating to the molecular structure/properties of an oil fraction and allows a comparison of HO/B on a shared and rational basis. For example, it is well known that the boiling point of a compound at a given pressure varies with the class and structure of molecules. As shown in Figure 2.6, compounds having similar molecular weights and carbon numbers cover a broad boiling point range and, alternatively, a narrow boiling point cut contains a wide molar mass range [Boduszynski 1987].

For a given homologous series of compounds, the boiling point increases with molecular weights as illustrated in Figure 2.6. This phenomenon is due to the increase of the weak, van der Waals attractive intermolecular forces as molecules become larger. However, compounds having fused aromatic rings or other types of polar interactions have additional attractive intermolecular

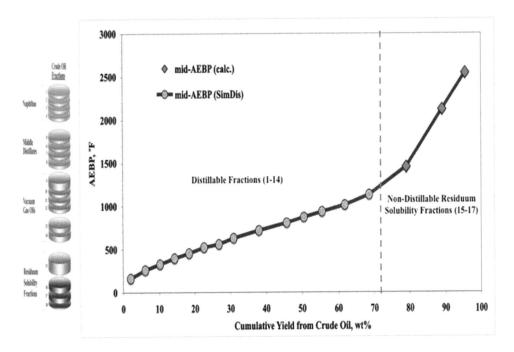

FIGURE 2.5   A typical example of an AEBP distillation curve of a heavy crude oil [Boduszynski 2015].

forces, and higher boiling points are expected. This observation is the main reason why polar compounds are shown to concentrate in the "heavy ends" or residues. For a complex mixture, the molecular weights range widens rapidly with increasing boiling point as illustrated in Figure 2.6 [Boduszynski 2015]. Thus, the AEBP scale allows for demonstrating the continuity of changing crude oil composition and properties and forms the basis of our current understanding of the petroleum composition.

2.3.4 BODUSZYNSKI'S CONTINUUM MODEL OF PETROLEUM

Dr. Mieczyslaw Boduszynski proposed the petroleum molecular composition continuity model some years ago, in series of papers [Altgelt and Boduszynski 1992, Boduszynski *et al.* 1980, 1987, 1988, Boduszynski and Altgelt 1992] and in the book entitled "Composition and Analysis of Heavy Petroleum Fractions" [Altgelt and Boduszynski 1994]. Recently, the author summarized his model in Chapter 1 of the monograph titled "Analytical Methods in Petroleum Upstream Applications" [Boduszynski 2015].

This model is based on the *continuous* variation of chemical composition and physical properties of petroleum and its fractions as a function of the Atmospheric Equivalent Boiling Point. This model applies to *all crudes*, and it is based on one central hypothesis, i.e., crude oil consists of relatively small molecules with ~95% of the species lower than 2,000 Da in molecular weight. Thus, crude oil composition increases gradually and continuously with regard to aromaticity, molecular weight, and heteroatom content, from the light distillates to non-distillable components (residuum). This model is known as the Boduszynski Continuum Model.

This concept has many applications throughout the petroleum value chain such as fluid-phase behavior, thermodynamics, fluid flow, chemical processes, and upgrading units. It has been successfully used in the developing of reservoir and process engineering numerical simulators [Boduszynski 2015].

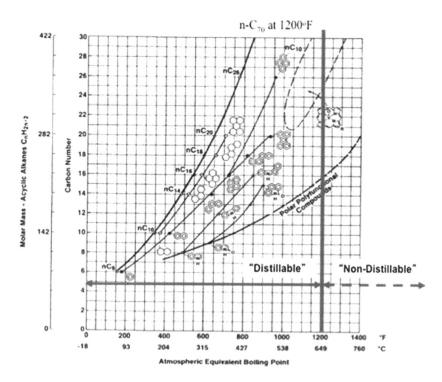

FIGURE 2.6 Effect of the molar mass (molecular weight) and carbon number on the Atmospheric Equivalents Boiling Point [Boduszynski 2015].

The continuity of changing molecular composition is especially useful when interpolating or extrapolating physical and chemical properties of crude oil fractions. An example of this model can be seen in Figure 2.7. In this figure, the changes in the micro-carbon residue (MCR) with the increasing AEBP are shown. MCR measures the propensity of a material to form coke during heating in subsurface and surface conversion and upgrading processes. As seen in Figure 2.7, the MCR increases with the AEBP, and very high micro-carbon residue values are found for non-distillable residuum solubility fraction [Boduszynski 2015].

This model has been validated in a series of papers by Marshall and Rodgers' group from Florida State University using Fourier Transform Ion Cyclotron Resonance Mass Spectrometry (FT-ICR-MS). In Part 1 of this series, the authors validated the Boduszynski model for middle distillates from high vacuum gasoil (HVGO) distillation fractions of Athabasca bitumen [McKenna et al. 2010a]. Part 2 reports the analysis of a Middle Eastern heavy crude oil distillation series by atmospheric pressure photoionization (APPI) FT-ICR-MS to extend the Boduszynski model to the limit of conventional distillation [McKenna et al. 2010b]. Parts 3 to 5 focused on the analysis of asphaltenes and maltenes by FT-ICR-MS [McKenna et al. 2013a, 2013b, Podgorski et al. 2013]. The authors found evidence of the existence of what it are now called "archipelago"- and "island"-type molecules in these fractions, giving further support to the Boduszynski Continuum model.

In 2018, the researchers from Florida State University reviewed their most relevant findings [Chacon-Patiño et al. 2018]. They tested the Continuum model for hundreds of thousands of species identified by FT-ICR-MS, from light distillates to asphaltenes, and found no deviations. In the process, they collected data that supported the low molecular weight of petroleum (< 2,000 Da), defined the maltene continuum, highlighted the effect of aggregation on FT-ICR-MS analysis, identified and overcame selective ionization, and confirmed that asphaltenes are composed of abundant island and archipelago structures [Chacon-Patiño et al. 2018].

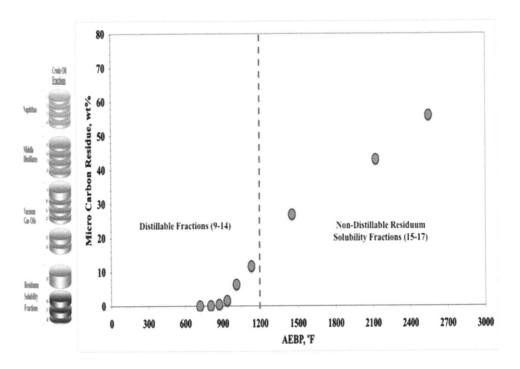

FIGURE 2.7   Micro-carbon residue (MCR) as a function of Atmospheric Equivalents Boiling Point (AEBP) [Boduszynski 2015].

2.4   RELATIONSHIPS BETWEEN ASPHALTENES AND VISCOSITY

Once we have described our current knowledge and understanding of the composition of petroleum and its fractions, we turned our attention to the asphaltenes. It is well known that this fraction is mainly responsible for the high viscosity of HO/B at room and reservoir temperatures [Henaut *et al.* 2001, Ovalles *et al.* 2011]. Also, they are considered to be the "bad actors" when heavy crude oils and bitumens are produced, transported, and upgraded. Asphaltene precipitation is a serious issue throughout the petroleum value chain (Figure 1.8). Besides the choking of the pipelines during transportation, asphaltenes can plug up well bores and decrease or stop oil production entirely. In downstream, asphaltenes are believed to be the source of sediment and coke during thermal upgrading processes, which in turn reduce and limit the yield of residue conversion. In catalytic upgrading processes, they contribute to catalyst poisoning by coke and metal deposition. Asphaltenes can also cause fouling in heat exchangers and other refinery units [Speight 2007, Ramirez-Corredores 2017].

From the reasons mentioned above, some subsurface upgrading routes focus on the removal or conversion of asphaltenes to reduce the HO/B viscosity and increase oil production and crude oil value. Removing the asphaltenes out of the petroleum also leads to a higher API and higher amount of distillable materials and lower level of contaminants.

2.4.1   WHAT ARE ASPHALTENES?

Asphaltenes are a solubility class of compounds and are arbitrarily defined as the fraction of petroleum that is insoluble in light alkanes (C3–C7) but soluble in aromatic (benzene, toluene, etc.) and chlorinated (dichloromethane, chloroform, carbon tetrachloride, etc.) solvents. In general, they are composed of polyaromatic rings substituted with alkyl groups and contain a disproportionate number of heteroatoms (S, N, and O) and trace metals (V, Ni, and Fe). C5-precipitated asphaltenes are dark brown solids (Figure 2.8a) whereas C6- and C7- materials are black dry powers (Figure 2.8b) at ambient conditions.

The amount, chemical constituency, and physical properties of precipitated asphaltenes vary with the nature of the crude precipitant, pressure, temperature, and solvent-to-oil ratio [Speight 2007, Ancheyta *et al.* 2009, Mullins *et al.* 2007]. In Table 2.2 the average weight percentages of C7-asphaltenes are reported for light, medium, and heavy crude oils and bitumens. In Table 2.3, various properties of heavy crude oils from different parts of the world are shown. As seen, the C7-asphaltene values range from 0.01 wt. % for a North Sea crude to ~14 wt. % for an Eastern Venezuelan Crude. These asphaltene contents not only vary with respect to the oil source but also with the composition [Speight 2007, Ramirez-Corredores 2017].

Table 2.4 shows the typical properties of asphaltenes (heptane insolubles) from different parts of the world. As seen, these materials have very high Microcarbon Residue (MCRT), H/C molar ratio between 1.04 and 1.15, relatively high heteroatom (N, S, and O) and metal (V and Ni) contents, and aromaticities in the 0.5–0.6 range. All those properties make the asphaltenes the most challenging petroleum fraction to handle and process during HO/B production, transporting, and refining.

The effect of the nature and the number of carbon atoms in the solvent on the percentage of insolubles (asphaltenes) is shown in Figure 2.9 for Orinoco and Athabasca crudes. For the latter, the data was reported by Mitchell and Speight [1973]. As seen, as the number of carbon atoms increases, the weight percentages of asphaltenes decrease for paraffins, cycloparaffins, and methylcycloparaffins. Thus, the amount of asphaltenes increases in the following order:

$$\text{Cycloparaffins}\,(\text{Naphthenes}) > \text{methylcycloparaffins} > \text{n-paraffins} \qquad (2.3)$$

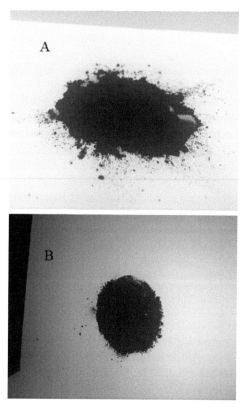

FIGURE 2.8 Photograph of (A) Pentane- and (B) n-Heptane-precipitated asphaltenes.

TABLE 2.3

Typical Properties of Heavy Crude Oils from Different Parts of the World

Heavy Crude Oil	API[a]	Viscosity at 40°C (cSt)[b]	Pour Point (°C)[c]	% Asphaltenes[d]
Canadian 1	7.4	47,648	20	9.3
Canadian 2	8.1	193.6		7.1
Californian 1	13.6	1,254	−3	1.6
Californian 2	12.7	1,566	−17	6.9
Californian 3	10.9	9,398	10	4.7
North Sea	19.5	129	−28	0.01
Eastern Venezuelan	9.4	18,925	13	13.6
Orinoco	7.7	65,689	28	8.7

[a] API gravity as measured using ASTM D 4052 [ASTM 2002].

[b] Viscosity in cSt at 40°C.

[c] Pour point in °C.

[d] Weight percentage of n-heptane asphaltenes as measured using ASTM 6560 [ASTM 2005].

Carbognani-Ortega *et al.* used a variety of analytical techniques (density, solubility parameter profiling, viscosity, and fluorescence spectroscopy) to determine the effect of different alkane precipitants on asphaltene properties [Carbognani-Ortega *et al.* 2015]. The authors found that denser, more polar, higher molecular weight, more viscous, red-shifted fluorescence asphaltenes follow the order:

$$\text{Solvent extracted asphaltenes} > \text{C7 - (unwashed) asphaltenes} > \text{C5 - (unwashed) asphaltenes} \qquad (2.4)$$

The effect of pressure on asphaltene precipitation was reported by Mehrotra and coworkers [Nielsen *et al.* 1994] using C5 as a precipitating agent. The authors studied four crude oils, ranging from an asphaltic condensate to a tar sand bitumen at 0–150°C and 0–5.6 MPa of pressure. They found that the mean asphaltene particle size increases from 266 to 495 pm with the pressure of the system [Nielsen *et al.* 1994]. Consistent with this data, Pasadakis *et al.* investigated the effect of reservoir pressure on the concentration of dissolved asphaltenes in heavy crude oil [Pasadakis *et al.* 2001]. As shown in Figure 2.10, the amount of the dissolved asphaltenes in oil decreases as pressure falls from the initial reservoir pressure (~5,000 psi, 34.5 MPa) to bubble point pressure (~900 psi, 6.2 MPa) and subsequently increases as the pressure is reduced further to 1 atm [Pasadakis *et al.* 2001].

The effect of temperature on asphaltene precipitation was studied by Andersen [Andersen 1995]. The author separated n-heptane asphaltenes from Boscan and Kuwait crudes in the 0–80°C range [Andersen 1995]. He observed a decrease in gravimetric yield of asphaltenes as the temperature increased. For Boscan crude the percentages of asphaltenes decreased from 24 wt. % at 0°C to 17 wt. % at 80°C (29% decrease). A similar phenomenon was observed for Kuwait C. These results agree with the known behavior of ideal and regular solutions and with the reduction of intermolecular association upon heating [Andersen 1995]. Similarly, Figure 2.11 shows the changes in weight percentages of C7-asphaltene fraction versus temperature for a preparatively separated sample from a Mexican VR. In this work, the asphaltene content was determined by the on-column precipitation method [Ovalles *et al.* 2015b]. As shown, the values decrease from ~98 wt. % at 35°C to 78 wt. % at 200°C. These changes correspond to a 25% decrease in the asphaltene content. Other authors have observed similar behavior, as reported by Ovalles, Rogel, Moir, and Morazan [2015b].

The effect of solvent-to-oil ratio on asphaltene precipitation was studied by Alboudwarej and coworkers using Athabasca bitumen at 23°C, at 24 h of contact time [Alboudwarej *et al.* 2002]. As

TABLE 2.4

Typical Properties of Asphaltenes (*Heptane Insolubles*) from Different Parts of the World

Origin[a]	MCRT (wt. %)[b]	% C[c]	% H[d]	H/C molar Ratio[e]	% N[f]	% S[g]	% O (by dif.)[h]	V (ppm)[i]	Ni (ppm)[j]	Total Metals[k]	Aromaticity (fa)[l]
Orinoco	52.80	83.33	7.70	1.11	1.92	3.84	2.11	1,907	427	2,334	0.59
California	54.50	86.36	7.97	1.11	2.08	2.39	0.09	910	495	1,405	0.61
SDA Tar	60.60	83.04	7.17	1.04	1.11	6.25	1.40	576	261	838	0.62
Mexican	58.10	80.84	7.66	1.14	1.47	7.34	1.56	1,875	363	2,238	0.66
Eastern Venezuelan	50.29	81.22	7.76	1.15	2.01	6.35	1.50	5,014	405	5,419	0.51
Canadian	51.20	79.82	7.67	1.15	1.36	8.01	2.00	1,066	387	1,453	0.60
Russian	58.10	84.56	7.45	1.06	1.52	3.54	1.88	1,038	330	1,368	0.60
Middle East	55.50	81.92	7.46	1.09	1.08	7.52	0.93	633	197	830	0.60

a Geographic region of origin of the Asphaltenes.
b Micro-carbon residue.
c Weight percentage of carbon.
d Weight percentage of hydrogen.
e H/C molar ratio.
f Weight percentage of nitrogen.
g Weight percentage of sulfur.
h Weight percentage of oxygen calculated by difference.
i Vanadium content in mg/Kg.
j Nickel content in mg/Kg.
k Metal content of V + Ni in mg/Kg. l Aromaticity as determined by ^{13}C-NMR.

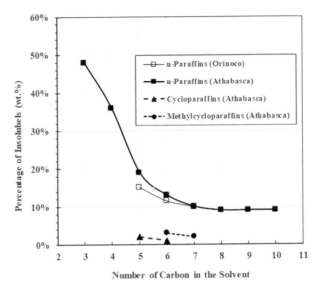

FIGURE 2.9 Effect of the nature and the number of carbon atoms in the solvent on the percentage of insolubles (asphaltenes) in crude oils. Athabasca data taken from Mitchell [1973].

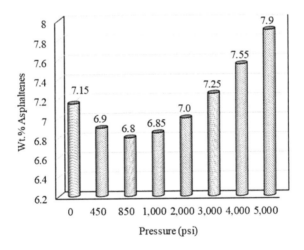

FIGURE 2.10 Effect of pressure on the amount of precipitated asphaltenes. Reprinted with permission. (Data taken from Pasadakis [2001].)

shown in Figure 2.12, the asphaltene yield increases with the heptane-to-bitumen ratio [Alboudwarej *et al.* 2002]. A similar effect has been reported by other authors [Speight 2007].

In the last 50 years, many different research groups from all over the world have dedicated significant efforts trying to unravel the molecular structure of the asphaltenes. Based on spectroscopic analytical techniques and elemental analysis measurements, average structures have been proposed for virgin and converted asphaltenes. Even though the use of these representations to understand asphaltene chemistry has been considered useful, the complexity and molecular diversity associated with this solubility-defined material has limited the progress and only given a high-level view of their chemical structure.

Recently, researchers at the IBM Zurich combined non-contact atomic force microscopy (AFM) and molecular orbital imaging using scanning tunneling microscopy to study more than ~300

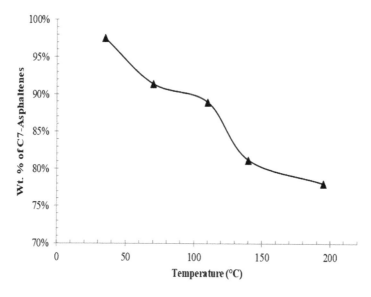

FIGURE 2.11 Weight percentage of C7-asphaltenes vs. temperature for preparative separated asphaltenes from a Mexican VR. Asphaltenes were determined by the On-Column Precipitation method.

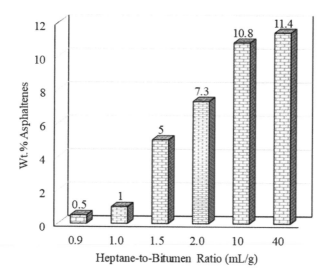

FIGURE 2.12 Effect of heptane-to-bitumen ratio on the asphaltene yield from Athabasca bitumen No. 2 at 23°C, at 24 h of contact time. (Data taken from Alboudwarej [2002].)

asphaltene molecules and other heavy oil fractions from different geographic/geologic origin and processing steps [Schuler *et al.* 2015, 2017]. The collected AFM data of individual molecules provided information about the molecular geometry, aromaticity, hydrogen deficiency, locations of heteroatoms, and occurrence, length, and connectivity of alkyl side chains.

Figure 2.13 shows three examples of AFM images and structure proposals from heavy oil VR asphaltenes [Schuler *et al.* 2017]. As seen, molecules B1.1 and B2.2 (Figure 2.13) have multi-aromatic ring systems and thus, substantial hydrogen deficiency. B1.4 shows a long-chain paraffinic structure with a high uncertainty in the group R. From the analysis of the ~300 individual molecules, the authors found that the "island"-type architecture, i.e., a central aromatic core with attached side chains, is predominant over "archipelago"-type molecules (that were also detected but

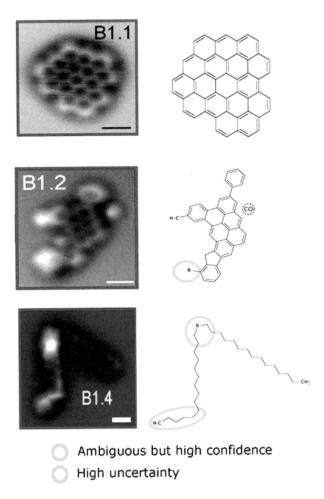

Ambiguous but high confidence

High uncertainty

FIGURE 2.13 Examples of AFM images and structure proposals from heavy oil VR asphaltenes. Regions where a full structure assignment was not possible or was uncertain are highlighted with red and blue circles, respectively. Scale bars: 5 Å. (Reproduced with permission [Schuler 2017]. Copyright 2017 American Chemical Society.)

only with occurrences of less than 10% in all samples) [Schuler *et al.* 2017]. These results clearly demonstrated the extremely high complexity of asphaltene fractions and may have implications for upstream oil production and downstream oil processing and in particular for the subsurface upgrading of HO/B.

2.4.2 EFFECT ON VISCOSITY

Asphaltenes play a critical role in the rheological behavior of heavy crude oils and bitumens as a result of their particular physical and self-associating properties. In Figure 2.14, the viscosity of a heavy crude oil and its deasphalted fraction (Maltene F.) is shown as a function of the temperature. As seen, by removing the asphaltenes from the heavy crude (darker trace), a reduction of two to three orders of magnitude on the viscosity of the maltenes (lighter trace) is obtained over all the temperature range [Ovalles *et al.* 2011].

Asphaltenes interact extensively within themselves and with other similar molecules within the crude oil forming extended networks. Whereas the viscosity of polymer-like compounds increases proportionally with their amount in solution, the asphaltene viscosity shows an exponential growth

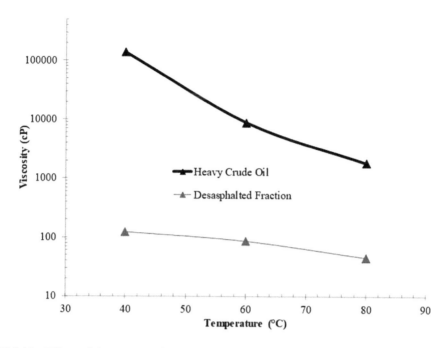

FIGURE 2.14 Effects of the removal of asphaltenes on the viscosity of a heavy crude oil.

with their concentration [Altgelt and Harle 1975, Luo and Gu 2007]. These results are attributed to the formation of molecular aggregates in the asphaltene solution. When the asphaltenes content is high enough, the system can form a gel [Altgelt and Harle 1975] and a fractal network [Abivin *et al.* 2012]. Also, the tendency of asphaltenes to form aggregates or colloidal structures is associated with the formation of stable emulsions and also to the fouling of operating equipment which results in increased operating costs throughout the petroleum value chain [Luo and Gu 2007].

When asphaltenes are added to the maltenes, it is generally observed that the relative viscosity of the fluid increases exponentially with the asphaltene content [Abivin *et al.* 2012]. Ghanavati and coworkers precipitated the asphaltenes from a heavy crude oil, and then ten well-defined reconstituted heavy oil samples were made by dispersing the asphaltenes into the maltene [Ghanavati *et al.* 2013]. Viscosity measurements were carried out at temperatures from 25 to 85°C. The results indicated that the viscosity values of the reconstituted heavy oil samples increase exponentially as the asphaltene content increases at a constant temperature [Ghanavati *et al.* 2013].

Henaut *et al.* identified two domains in the viscosity vs. asphaltene concentration curve [Henaut *et al.* 2001]. The first area concerns the dilute samples for which the relative viscosity increases linearly with the weight fraction of asphaltenes. In this domain, the aggregates of asphaltenes stay independent from each other and have approximatively the same radius. For the more concentrated samples, the viscosity increases dramatically because of the aggregates entanglement as observed by small-angle X-ray scattering (SAXS). The results revealed an analogy between heavy crude oils and concentrated colloidal systems [Henaut *et al.* 2001].

According to Muñoz *et al.*, the mechanism for asphaltene aggregation involves the $\pi-\pi$ interactions between aromatic sheets [Yen *et al.* 1961], hydrogen bonding between functional groups, and charge transfer interactions between the molecular species [Muñoz *et al.* 2016]. Taborda and coworkers attributed the increase in viscosity of asphaltene-containing fractions to the formation of a viscoelastic network of interacting asphaltenes nanoaggregates [Taborda *et al.* 2017].

To further demonstrate the profound effect of asphaltenes on the HO/B viscosity, the use of additives that can act as modifiers of the aggregation behavior of asphaltenes was studied [Ovalles *et al.* 2011]. Firstly, a solution of the heavy crude oil was prepared near the flocculation point of the

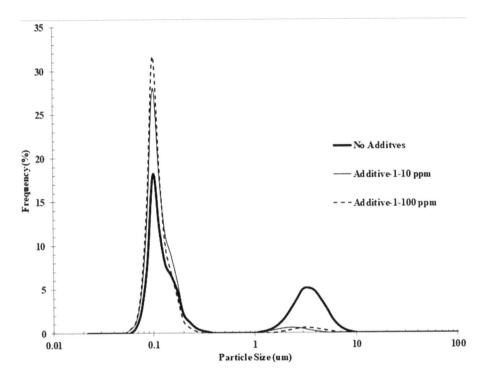

FIGURE 2.15 Effect of the asphaltene dispersants on the asphaltene particle size distributions of a heavy crude oil solution. (Additive concentration 10 ppm.)

asphaltenes (450 ppm of HO in 38% toluene in n-heptane) to measure particle size distribution. By this way, the efficiency of the additive to interact with the asphaltenes was studied. As shown in Figure 2.15, there are three regions of asphaltene particles in the heavy crude oil solution. The zone for sizes less than 0.1 μm is called "stable asphaltenes," whereas in size range of 0.1–1 μm it is called "stabilized asphaltenes" or "colloidal asphaltenes." The asphaltenes with a size greater than 1 μm are called "flocculated asphaltenes" [Kraiwattanawong et al. 2009].

The effect of an additive on the asphaltene particle size distributions was studied next. Additive 1 at 10 ppm (Figure 2.15, lighter trace) led to a visible reduction in the flocculated asphaltenes peak (> 1 μm) with the concomitant increase in the stabilized asphaltenes at ~0.1 μm. An increase of the concentration of the additive to 100 ppm (Figure 2.15 dashed trace) yielded a further enhancement in the stabilized asphaltenes (~0.1 μm) and a reduction in the concentration of the flocculated counterparts (> 1 μm). The net effect of changing the aggregation behavior of the asphaltene is a decrease in the viscosity of heavy crude oil/diluent blends (80/20 wt. %). As can be seen in Figure 2.16, reductions of the viscosity in the 8–9% range were observed using additive 1 [Ovalles et al. 2011].

Taborda and coworkers also observed viscosity reduction of the heavy oil and extra-heavy crude oils but in the presence of nanoparticles of different chemical natures (SiO_2, Fe_3O_4, and Al_2O_3), particle size, surface acidity, and concentration at low-volume fractions [Taborda et al. 2017]. A maximum viscosity reduction of ~52% was obtained at a concentration of 1,000 mg/L with 7 nm SiO_2 nanoparticles at shear rates below 10 s⁻¹. These results were attributed to a hampering interaction of the nanoparticles to the asphaltene aggregates present in crude oil. The authors also reported that, as the particle size increases, the effect on the reduction of viscosity decreases [Taborda et al. 2017].

Finally, it is important to mention that the influence of the asphaltenes on the relative viscosity of crude oil can be useful to predict viscosity variations resulting from asphaltene gradients within

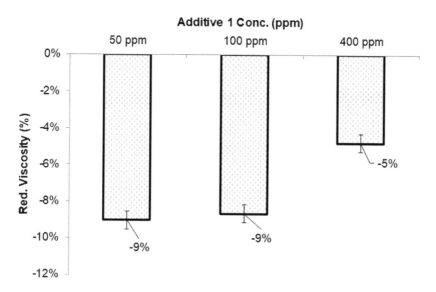

FIGURE 2.16 Reduction of the viscosity of heavy crude oil/diluent (80/20 wt. %) blends using additive 1. Error bars 0.5%.

a petroleum-containing reservoir [Abivin *et al.* 2012]. Asphaltene compositional grading is indeed an essential aspect of the reservoir management strategy [Mullins *et al.* 2003] because it can lead to substantial viscosity variations within a few meters in a pay zone. In turn, this phenomenon may affect the development of a subsurface upgrading process. The models used for calculating the fluid viscosity changes during reservoir simulations are discussed in Section 5.3.5.

REFERENCES

Abivin, P., Taylor, S. D., Freed, D., 2012, "Thermal Behavior and Viscoelasticity of Heavy Oils", *Energy & Fuels*, 26(6), 3448–3461 and references herein.

Al-Besharah, J. M., Salman, O. A., Akashah, S. A., 1987, "Viscosity of Crude Oil Blends", *Ind. Eng. Chem. Res.*, 26(12), 2445–2449.

Alboudwarej, H., Beck, J., Svrcek, W. Y., Yarranton, H. W., Akbarzadeh, K., 2002, "Sensitivity of Asphaltene Properties to Separation Techniques", *Energy & Fuels*, 16(2), 462–469.

Alomair, O. A., Almusallam, A. S., 2013, "Heavy Crude Oil Viscosity Reduction and the Impact of Asphaltene Precipitation", *Energy & Fuels*, 27(12), 7267–7276.

Altgelt, K. H., Boduszynski, M. M., 1992, "Composition of Heavy Petroleums. 3. An Improved Boiling Point–Molecular Weight Relation", *Energy & Fuels*, 6(1), 68–72.

Altgelt, K. H., Boduszynski, M. M., 1994, *Composition and Analysis of Heavy Petroleum Fractions*, Marcel Dekker Inc., New York, 495.

Altgelt, K. H., Harle, O. L., 1975, "The Effect of Asphaltenes in the Viscosity of Asphalts", *Ind. Eng. Chem. Prod. Res. Dev.* 14, 242.

Ancheyta, J., Trejo, F., Rana, M. M., 2009, *Asphaltenes. Chemical Transformation during Hydroprocessing of Heavy Oils*, CRC Press, Boca Raton, FL, Chapter 1, and references therein.

Andarcia, L., Bermudez, J. M., Reyes, Y., Caycedo, H., Suarez, A. F., 2014, "Potential of Steam Solvent Hybrid Processes in Llanos Basin, Colombia", SPE No. 171049, presented at SPE Heavy and Extra Heavy Oil Conference: Latin America, Medellín, Colombia, 24–26 September.

Andersen, S. I., 1995, "Effect of Precipitation Temperature on the Composition of n-Heptane Asphaltenes, Part 2", *Fuel Sci. Technol. Int.*, 13(5), 579 and references therein.

ASTM International, 2002, *ASTM D 4052, Standard Test Method for Density and Relative Density of Liquids by Digital Density Meter*, ASTM International, West Conshohocken, PA.

ASTM International, 2005, *ASTM D6560, Standard Test Method for Determination of Asphaltenes (Heptane Insolubles) in Crude Petroleum and Petroleum Products*, ASTM International, West Conshohocken, PA.

Bao, Y., Wang, J. Y. J., Gates, I. D., 2012, "History Match of the Liaohe Oil Field SAGD Operation - A Vertical-Horizontal Well Reservoir Production Machine" SPE No. 157810 presented at SPE Heavy Oil Conference Canada, Calgary, Alberta, Canada, 12–14 June.

Boduszynski, M. M., 1987, "Composition of Heavy Petroleums. 1. Molecular Weight, Hydrogen Deficiency, and Heteroatom Concentration as a Function of Atmospheric Equivalent Boiling Point Up to 1400 OF (760°C)", *Energy & Fuels*, 1(1), 2–11.

Boduszynski, M. M., 1988, "Composition of Heavy Petroleums: Molecular Characterization," *Energy & Fuels*, 2, 597–613.

Boduszynski, M. M., 2015, "Petroleum Molecular Composition Continuity Model". In: *Analytical Methods in Petroleum Upstream Applications*, C. Ovalles, and C. Rechsteiner, Ed., CRC Press, Boca Raton, FL, Chapter 1, and references therein.

Boduszynski, M. M., Altgelt, K. H., 1992, "Composition of Heavy Petroleums. 4. Significance of the Extended Atmospheric Equivalent Boiling Point (AEBP) Scale", *Energy & Fuels*, 6(1), 72–76.

Boduszynski, M. M., McKay, J. F., Latham, D. R., 1980, "Asphaltenes, Where Are You?", *Proceedings of the Assessment Asphalt Paving Technology*, Louisville, KY, February 18–20, 49, 123–143.

CAPP- Canadian Association of Petroleum Producers, 2012, "Canada's Oil Sands Overview and Bitumen Blending Primer", presented before the US National Academy of Science, 23 October, downloaded from http-onlinepubs.trb.org-onlinepubs-dilbit-Segato102312.pdf on Feb. 13, 2017.

Carbognani Ortega, L., Rogel, E., Vien, J., Ovalles, C., Guzman, H., Lopez-Linares, F., Pereira-Almao, P., 2015, "Effect of Precipitating Conditions on Asphaltene Properties and Aggregation", *Energy & Fuels*, 29(6), 3664–3674.

Centeno, G., Sánchez-Reyna, G., Ancheyta, J., Muñoz, J. A. D., Cardona, N., 2011, "Testing Various Mixing Rules for Calculation of Viscosity of Petroleum Blends", *Fuel*, 90(12), 3561–3570.

Chacón-Patiño, M. L., Rowland, S. M., Rodgers, R. P., 2018, "The Compositional and Structural Continuum of Petroleum from Light Distillates to Asphaltenes: The Boduszynski Continuum Theory as Revealed by FT-ICR Mass Spectrometry". In: *The Boduszynski Continuum: Contributions to the Understanding of the Molecular Composition of Petroleum*, C. Ovalles, and M. E. Moir, Ed., ACS Press, Chapter 6, and references therein.

Chopra, S., Lines, L., Schmitt, D. R., Batzle, M., 2010, "Heavy-Oil Reservoirs: Their Characterization and Production". In *Heavy Oils: Reservoir Characterization and Production Monitoring, Society of Exploration Geophysicists, Geophysical Developments Series*, Vol. 13, 1–69.

Cocco, M. J., Ernandez, J. E., 2014, "Reservoir Characterization of Junín Area, Orinoco Oil Belt Region, Venezuela", SPE 171136, presented at SPE Heavy and Extra Heavy Oil Conference: Latin America, Medellín, Colombia, 24–26 September.

Computer Modelling Group (CMG), STARS, 2011, "*User's Manual*", CMG, Calgary.

Egiebor, N. O., Gray, M. R., 1990, "Evidence for Methane Activation during Coal Pyrolysis and Liquefaction", *Fuel*, 69(10), 1276.

Eschard, R., 2011, "Reservoir Geology". In: *Heavy Crude Oils. From Geology to Upgrading. An Overview*, A.-Y. Huc, Ed., Technip, Paris, France, Chapter 5.

Faergestad, I. M., 2016, "Defining Heavy Oil", Schlumberger, downloaded from www.slb.com/~/media/Files/resources/oilfield_review/defining_series/Defining-Heavy-Oil.ashx on 15 Decemeber 2016.

Ghanavati, M., Shojaei, M.-J., Ramazani, A. S. A., 2013, "Effects of Asphaltene Content and Temperature on Viscosity of Iranian Heavy Crude Oil: Experimental and Modeling Study", *Energy & Fuels*, 27(12), 7217–7232.

Gibson, B. J., 1982, "Methods of Classifying Heavy Crude Oils Using the UNITAR Viscosity-Base Definition", 2nd International Conference on Heavy Crude and Tar Sands, United Nations Institute for Training and Research (UNITAR), Caracas, Venezuela, 7–17 February, Chapter 1.

Henaut, I., Barre, L., Argillier, J.-F., Brucy, F., 2001, "Rheological and Structural Properties of Heavy Crude Oils in Relation with Their Asphaltenes Content", SPE 65020 presented at SPE Int. Symp. on Oilfield Chemistry, Houston, TX, 13–16 February.

Hu, C., Liu, X., Wang, J., Song, Z., Fan, Z., Yang, F., 1998, "Cold Production of Thin-Bedded Heavy Oil Reservoir in Henan Oilfield", PSE No. 50885, presented at SPE International Conference and Exhibition in China held in Beijing, 2–6 November.

Huc, A.-Y., Ed., 2011, *Heavy Crude Oils. From Geology to Upgrading. An Overview*, Technip, Paris, France, and references therein.

Ibatullin, R. R., Ibragimov, N. G., Khisamov, R. S., Zaripov, A. T., 2012, "Problems and Decisions for Development Shallow Fields of Heavy Oil", SPE No. 191998, presented at SPE Russian Oil and Gas Exploration and Production Technical Conference and Exhibition, Moscow, Russia, 16–18 October.

Ivory, J., Rajan, R. V., Chang, J., Bjorndalen, N., London, M., Akinlade, O., 2008, "Handbook of Canadian Heavy Oil and Oil Sand Properties for Reservoir Simulation", AERI/ARC Core Industry Research Program, Report #0708-22, March.

Jiang, Q., Thornton, B., Houston, J. R., Spence, S., 2009, "Review of Thermal Recovery Technologies for the Clearwater and Lower Grand Rapids Formations in the Cold Lake Area in Alberta", presented at the Canadian International Petroleum Conference (CIPC) 2009, Calgary, Alberta, Canada, 16–18 June.

Kapadia, P. R., Kallos, M. S., Gates, I. D., 2015, "A Review of Pyrolysis, Aquathermolysis, and Oxidation of Athabasca Bitumen", *Fuel Proc. Technol.*, 131, 270–289.

Kopper, R., Kupecz, J., Curtis, C., Cole, T., Dorn-López, D., Copley, J., Muñoz, A., Caicedo, V., 2001, "Reservoir Characterization of the Orinoco Heavy Oil Belt: Miocene Oficina Formation, Zuata Field, Eastern Venezuela Basin", SPE No. 69697 presented at the 2001 SPE International Thermal Operations and Heavy Oil Symposium held in Porlamar, Margarita Island, Venezuela, 12–14 March.

Kraiwattanawong, K., Fogler, H. S., Gharfeh, S. G., Singh, P., Thomason, W. H., Chavadej, S., 2009, "Effect of Asphaltene Dispersants on Aggregate Size Distribution and Growth", *Energy & Fuels*, 23(3), 1575.

Kumar, M., Akshay, S., Alvarez, J. M., Heny, C., Vaca, P., Hoadley, S. F., Portillo, M., 2001, "Evaluation of IOR Methods for the Boscán Field", presented at SPE International Thermal Operations and Heavy Oil Symposium, Porlamar, Margarita Island, Venezuela, 12–14 March.

Lugo, R. G., Eggenschwiler, M., Uebel, T., 2001, "How Fluid and Rock Properties Affect Production Rates in A Heavy-Oil Reservoir Cerro Negro, Venezuela" presented at SPE Int. Thermal Operations and Heavy Oil Symposium, Porlamar, Margarita Island, Venezuela, 12–14 March.

Luo, P., Gu, Y., 2007, "Effects of Asphaltene Content on the Heavy Oil Viscosity at Different Temperatures", *Fuel*, 86(7–8), 1069.

Marcos, D. J., Pardo, E. M., Casas, J., Delgado, D. G., Rondon, M. A., Exposito, M. A., Zerpa, L. B., Ichbia, J.-M., Bellorini, J.-P., 2007, "Static and Dynamic Models of Formation Water in Orinoco Belt, Venezuela", SPE No. 107378, presented at Latin American & Caribbean Petroleum Engineering Conference, Buenos Aires, Argentina, 15–18 April.

McKenna, A. M., Blakney, G. T., Xian, F., Glaser, P. B., Rodgers, R. P., Marshall, A. G., 2010b, "Heavy Petroleum Composition. 2. Progression of the Boduszynski Model to the Limit of Distillation by Ultrahigh-Resolution FT-ICR Mass Spectrometry", *Energy & Fuels*, 24(5), 2939–2946.

McKenna, A. M., Donald, L. J., Fitzsimmons, J. E., Juyal, P., Spicer, V., Standing, K. G., Marshall, A. G., Rodgers, R. P., 2013a, "Heavy Petroleum Composition. 3. Asphaltene Aggregation", *Energy & Fuels*, 27(3), 1246–1256.

McKenna, A. M., Marshall, A. G., Rodgers, R. P., 2013b, "Heavy Petroleum Composition. 4. Asphaltene Compositional Space", *Energy & Fuels*, 27(3), 1257–1267.

McKenna, A. M., Purcell, J. M., Rodgers, R. P., Marshall, A. G., 2010a, "Heavy Petroleum Composition. 1. Exhaustive Compositional Analysis of Athabasca Bitumen HVGO Distillates by Fourier Transform Ion Cyclotron Resonance Mass Spectrometry: A Definitive Test of the Boduszynski Model", *Energy & Fuels*, 24(5), 2929–2938.

Meyer, R. F., Attanasi, E. D., Freeman, P. A., 2007, "Heavy Oil and Natural Bitumen Resources in Geological Basins of the World", U.S. Geological Survey Open-File Report 2007-1084, available online at http://pubs.usgs.gov/of/2007/1084/.

Mirabal, M., Gordillo, R., Rojas, G., Rodriguez, H., Huerta, M., 1996, "Impact of Foamy Oil Mechanism on the Hamaca Oil Reserves, Orinoco Belt-Venezuela", SPE No. 36140, presented at SPE Latin America/Caribbean Petroleum Engineering Conference, Port-of-Spain, Trinidad, 23–26 April.

Mitchell, D. L., Speight, J. G., 1973, "The Solubility of Asphaltenes in Hydrocarbon Solvents", *Fuel*, 52(2), 149.

Mullins, O. C., Bethancourt, S., Cribbs, M. E., Creek, J. L., Dubost, F. X., Andrews, A. N., Venkataramanan, L., 2007a, Asphaltene Gravitational Gradient in a Deepwater Reservoir as Determined by Downhole Fluid Analysis, presented at SPE Int. Symp. on Oilfield Chemistry held in Houston, TX, 28 February–2 March.

Mullins, O. C., Elshahawi, H., Flannery, M., O'Keefe, M., Vannuffelen, S., 2003, "The Impact of Reservoir Fluid Compositional Variation and Valid Sample Acquisition on Flow Assurance Evaluation SS-FA", presented at the Offshore Technology Conference, Houston, TX, 5–8 May.

Mullins, O. C., Sheu, E. Y., Hammani, A., Marshall, A. G., 2007, *Asphaltenes, Heavy Oils and Petroleomics*, Springer, New York, Chapters 1 and 3, and references therein.

Muñoz, J. A. D., Ancheyta, J., Castañeda, L., 2016, "Required Viscosity Values to Ensure Proper Transportation of Crude Oil by Pipeline", *Energy & Fuels*, 30(11), 8850–8854.

Muraza, O., Galadima, A., 2015, "Aquathermolysis of Heavy Oil: A Review and Perspective on Catalyst Development", *Fuel*, 157, 219–231.

Musin, M., Khisamov, R., Dinmuhamedov, R., 2010, "Solution to Problem of Evaluation of Unconsolidated Heavy Oil Reservoirs in Tatarstan", SPE No. 136389, presented at SPE Russian Oil and Gas Conference and Exhibition, Moscow, Russia, 26–28 October.

Nielsen, B. B., WilSvrcek, W. Y., Mehrotra, A. K., 1994, "Effects of Temperature and Pressure on Asphaltene Particle Size Distributions in Crude Oils Diluted with n-Pentane", *Ind. Eng. Chem. Res.*, 33(5), 1324–1330.

Omajalia, J. B., Hartb, A., Walkerc, M., Joseph Wood, J., Macaskie, L. E., 2017, "In-Situ Catalytic Upgrading of Heavy Oil Using Dispersed Bionanoparticles Supported on Gram-Positive and Gram-Negative Bacteria", *Appl. Catal. B: Environ.*, 203, 807–819.

Ovalles, C., Filgueiras, E., Rojas, I., Morales, A., de Jesus, J. C., Berrios, I., 1998, "Use of Dispersed Molybdenum Catalyst and Mechanistic Studies for Upgrading Extra-Heavy Crude Oil Using Methane as Source of Hydrogen", *Energy & Fuels*, 12(2), 379–385.

Ovalles, C., Hamana, A., Rojas, I., Bolívar, R. A., 1995, "Upgrading of Extra-Heavy Crude Oil by Direct Use of Methane in the Presence of Water. Deuterium Labeled Experiments and Mechanistic Consideration", *Fuel*, 74(8), 1162.

Ovalles, C., Rivero, V., Salazar, A., 2015a, "Downhole Upgrading of Orinoco Basin Extra-Heavy Crude Oil Using Hydrogen Donors under Steam Injection Conditions. Effect of the Presence of Iron Nanocatalysts", *Catalysts*, 5(1), 286–297.

Ovalles, C., Rogel, E., Moir, M. E., Morazan, H., 2015b, "Effect of Temperature on the Analysis of Asphaltenes by the On-Column Filtration/Redissolution Method", *Fuel*, 146, 20–27 and references therein.

Ovalles, C., Rogel, E., Segerstrom, J., 2011, "Improvement of Flow Properties of Heavy Oils Using Asphaltene Modifiers", SPE-146775, presented at SPE Annual Technical Conference and Exhibition, Denver, CO, 30 October–2 November.

Ovalles, C., Vallejos, C., Vasquez, T., Rojas, I., Ehrman, U., Benitez, J. L., Martinez, R., 2003, "Downhole Upgrading of Extra-Heavy Crude Oil Using Hydrogen Donor and Methane under Steam Injection Conditions", *Pet. Sci. Technol.*, 21(1–2), 255.

Pasadakis, N., Varotsis, N., Kallithrakas, N., 2001, "The Influence of Pressure on the Asphaltenes Content and Composition in Oils", *Pet. Sci. Technol.*, 19(9–10), 1219–1227.

Pereira-Almao, P., Scott, C., Carbognani, L., Maini, B., Chen, J., 2015, "Advancing the Unconventional Upgrading of Heavy Oils", WHOC15-174, presented at World Heavy Oil Congress, Edmonton, Alberta, Canada.

Phillips, C. R., Haidar, N. I., Poon, Y. C., 1985, Kinetic Models for the Thermal Cracking of Athabasca Bitumen-the Effect of the Sand Matrix, *Fuel*, 64(5), 678.

Podgorski, D. C., Corillo, Y. E., Nyadong, L., Lobodin, V. V., Bythel, B. J., Robbins, W. K., McKenna, A. M., Marshall, A. G., Rodgers, R. P., 2013, "Heavy Petroleum Composition. 5. Compositional and Structural Continuum of Petroleum Revealed", *Energy & Fuels*, 27(3), 1268–1276.

Ramirez-Corredores, M. M., 2017, *The Science and Technology of Unconventional Oils: Finding Refining Opportunities*, 1st Ed., Elsevier, London, Chapter 2, p 41 and references therein.

Rottenfusser, R., Ranger, M., 2004, "A Geological Comparison of Six Projects in the Athabasca Oil Sands", presented at CSPG/CSEG 2004 GeoConvention, Calgary, AB, Canada, May 31–June 4.

Rubin, E., 2011, "Full Field Modeling of Wafra First Eocene Reservoir 56-year Production History", SPE No. 150575, presented at SPE Heavy Oil Conference and Exhibition, Kuwait City, Kuwait, 12–14 December.

Schuler, B., Fatayer, S., Meyer, G., Rogel, E., Moir, M., Zhang, Y., Harper, M. R., Pomerantz, A. E., Bake, K. D., Witt, M., Peña, D., Kushnerick, J. D., Mullins, O. C., Ovalles, C., van den Berg, F. G. A., Grossy, L., 2017, "Heavy Oil Based Mixtures of Different Origins and Treatments Studied by AFM", *Energy & Fuels*, 31(7), 6856–6861.

Schuler, B., Meyer, G., Peña, D., Mullins, O. C., Gross, L., 2015, "Unraveling the Molecular Structures of Asphaltenes by Atomic Force Microscopy", *J. Am. Chem. Soc.*, 137(31), 9870–9876.

Smalley, C., 2000, "Heavy Oil and Viscous Oil", In: *Modern Petroleum Technology*, R. A. Dawe, Ed., John Wiley and Sons Ltd., West Sussex, England, Volume 1, p 414.

Speight, J. G., 2007, *The Chemistry and Technology of Petroleum*, 4th Ed., CRC Press, Boca Raton, FL, and references therein.

Stapp, P. R., 1989, "In Situ Hydrogenation", Report NIPER-434, Bartlesville, Oklahoma, December and references therein.

Strausz, O. P., Jha, K. N., Montgomery, D. S., 1977, "Chemical Composition of Gases in Athabasca Bitumen and in Low-Temperature Thermolysis of Oil Sand, Asphaltene and Maltene", *Fuel*, 56(2), 114.

Strausz, O. P., Lown, E. M., 2003, *The Chemistry of Alberta Oil Sands, Bitumens and Heavy Oils*, Alberta Energy Research Inst., Calgary, Alberta, Canada.

Sundaram, M., 1987, "Direct Use of Methane in Coal Liquefaction", U.S. Patent 4,687,570.

Taborda, E. A., Franco, C. A., Ruiz, M. A., Alvarado, V. B., Cortés, F. N., 2017, "Experimental and Theoretical Study of Viscosity Reduction in Heavy Crude Oils by Addition of Nanoparticles", *Energy & Fuels*, 31(2), 1329–1338.

Takamura, M., 1982, "Microscopic Structure of Athabasca Oil Sand", *Can. J. Chem. Eng.* 60(Aug.), 538.

Vega-Riveros, G. L., Barrios, H., 2011, "Steam Injection Experiences in Heavy and Extra-Heavy Oil Fields, Venezuela", SPE No. 150283, presented at SPE Heavy Oil Conference and Exhibition, Kuwait City, Kuwait, 12–14 December.

Wang, J., Wang, T., Yu, Y., Chen, Z., Niu, Z., Yang, C., 2014, "Nitrogen Foam Anti-Edge Water-Incursion Technique for Steam Huff-Puff Wells of Heavy Oil Reservoir With Edge Water", SPE No. 170044 presented at SPE Heavy Oil Conference-Canada, Calgary, Alberta, Canada, 10–12 June.

Yen, T. F., Erdman, J. G., Pollack, S. S., 1961, "Investigation of the Structure of Petroleum Asphaltenes by X-Ray Diffraction", *Anal. Chem.*, 33(11), 1587.

Yoshiki, K. S., Phillips, C. R., 1985, "Kinetics of the Thermo-Oxidative and Thermal Cracking Reactions of Athabasca Bitumen", *Fuel*, 64(11), 1591–1596.

Zhang, C., Zhao, H., Hu, M., Xiao, Q., Li, J., Cai, C., 2007, A Simple Correlation for the Viscosity of Heavy Oils From Liaohe Basin, NE China, *J. Can. Pet. Tech.*, 46(4), 8–11.

Zhao, H., Memon, A., Gao, J., Taylor, S. D., Sieben, D., Ratulowski, J., Alboudwarej, H., Pappas, J., Creek, J., 2016, "Heavy Oil Viscosity Measurements: Best Practices and Guidelines", *Energy & Fuels*, 30(7), 5277–5290.

3 Fundamentals of Heavy Oil Recovery and Production

Before we start discussing the state-of-the-art in subsurface upgrading (SSU), the basic concepts associated with heavy oil production and transportation are presented. This chapter is addressed to our downstream and refining colleagues. It represents an introduction to the heavy crude oils and bitumens (HO/B) production. The idea is that by reading these concepts, the understanding of the SSU processes can be made more accessible and straightforward. If more information is needed, please check the references included at the end of this chapter.

Firstly, the primary processes used to recover HO/B are summarized but only the ones closely related to SSU, i.e., thermal and downhole solvent injection processes as well as *in-situ* combustion. These recovery methods have been modified or enhanced to achieve subsurface upgrading of HO/B either by solvent deasphalting, thermal or catalytic conversion, reaction with air or oxygen, or by using unconventional forms of heating. These topics are covered in Chapters 5 through 10.

Next, several HO/B production topics are presented such as well issues, oil as gas separation, and heavy oil transportation. These subjects could potentially impact subsurface upgrading processes and need to be appropriately addressed to enhance the technical and economic feasibility of a given SSU project.

3.1 RECOVERY PROCESSES

The primary processes used for HO/B production are cold production using vertical and horizontal wells, Cold Heavy Oil Production with Sand (CHOPS), and surface mining. To date, none of these recovery methods have been used in conjunction with subsurface upgrading processes. However, these technologies could affect the performance of SSU. Thus, a short description of each is presented.

3.1.1 COLD PRODUCTION

The cold production using vertical and horizontal wells is the preferred primary process for the Orinoco Belt reservoir [Dusseault 2001], but it is also used in other countries such as China, Russia, Canada, etc. Due to the complexity associated with heavy oil reservoirs, multilateral wells are used, in which several wellbore branches are built around a center borehole to increase well productivity and economics [Pasicznyk 2001]. For the Orinoco case, the estimated recovery factor is 10–15% of the original oil in place [Schenk *et al.* 2009].

CHOPS involves the deliberate sand production throughout most of the productive life of the well and the implementation of how to separate the sand from the oil for disposal. No sand control devices (screens, liners, gravel packs, etc.) are used. The sand is produced along with oil, water, and gas and separated before upgrading to a synthetic crude. This process can generate environmental concerns over disposal of the oily sand at the surface as well as the damage caused to the artificial lift system. It has been used almost exclusively in the Canadian heavy oil-sands and in shallow (< 800 m), low-production-rate wells (up to 100 to 125 m³/d). However, a few examples can be found in China and Alaska as well [Istchenko and Gates 2012].

For HO/B reservoirs located very close to the surface (< 200 m), mining is the primary process used and especially in the province of Alberta, Canada. The term mining is applied to the surface or subsurface excavation of petroleum-bearing formations for subsequent removal of the oil by washing, flotation, or retorting treatments [Strausz and Lown 2003].

3.1.2 THERMAL RECOVERY

Once the primary production is exhausted, there is a need to maintain high economic oil rates and achieve a high recovery factor. Thus, the subsequent use of an Enhanced Oil Recovery Processes (EOR) is essential. Due to the high viscosity of HO/B (Section 2.2), thermal recovery methods are commonly employed [Prats 1982]. Mainly, three types of thermal processes have been used. Those are: cyclic steam stimulation (CSS), steamflood, and steam assisted gravity drainage (SAGD).

3.1.2.1 Cyclic Steam Stimulation

Figure 3.1 illustrates the CSS process, also known as "Huff and Puff" [NIPER 1986]. This process injects steam, usually between 5,000–20,000 bbl of cold water equivalent (CWE), into a well over a period of several days or weeks ("huff"). Following steam injection, the well is shut in for 3 to 14 days to maximize heat transfer to the reservoir and at the same time minimize heat loss to the over- and under-burden [Hong 1994]. After the steam soak period, the well is allowed into production (Figure 3.1) for a period of weeks to months ("puff").

It has been found that the oil production response to the first cycle is usually higher than subsequent cycles [Hong 1994]. The initial period of oil production can recover over 30–50% of the stimulated oil production. A typical cumulative oil versus time plot for a horizontal well is shown in Figure 3.2 [Perez-Perez 2002]. Due to the steam injection step at the beginning of the cycle, CSS cumulative oil production (continuous trace) starts later than that for cold production (dashed line). At the end of the period, CSS cumulative oil is ~20% higher than that at the end of the 1st cycle [Perez-Perez 2002].

When oil production declines to a point where it is no longer produced at an economical rate (see Figure 3.3, 1st cycle), the whole CSS process is repeated. A producing well can have many steam–soak–production cycles depending on the amount of oil recovered per each cycle. Usually, three or more are used in a single well [Rivas and Boccasdo 1994]. The oil recovery per cycle varies

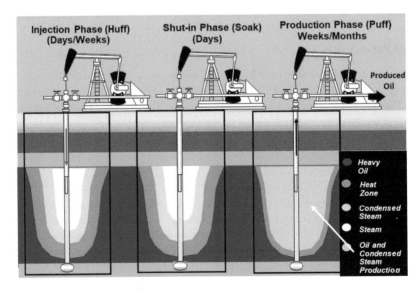

FIGURE 3.1 Schematic diagram showing the three phases of Cyclic Steam Stimulation (CSS also known as Huff and Puff), i.e., injection, soaking, and production [NIPER 1986]. Acknowledgment to the U.S. Department of Energy and National Energy Technology Lab.

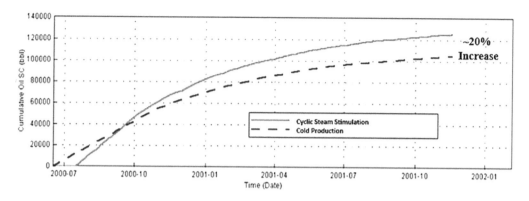

FIGURE 3.2 Comparison of the cumulative oil production for the cyclic steam stimulation of a horizontal well versus cold production.

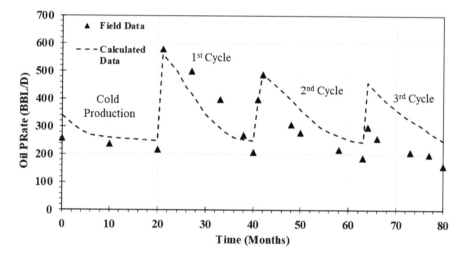

FIGURE 3.3 Actual and predicted production behavior of a South Monagas well. Data taken from Rivas and Boccasdo [1994].

with the volume of steam, formation thickness, reservoir pressure, OOIP, and the number of pre-ceding "Huff and Puff" steps [Hong 1994]. With each cycle, the water cut increases, oil production declines, and the CSS cycles become longer (Figure 3.3, 2nd and 3rd cycles). After a few Huff and Puff cycles, the typical recovery factor is in the 20–30% range of the OOIP, and then, the well is converted to a continuous injection (steamflood).

The primary production mechanism for CSS is HO/B viscosity reduction, but other physical phenomena may be occurring at the same time and are responsible for the enhanced oil production. Those phenomena are: density decreases due to increase temperature, gravity drainage for thick reservoirs, and pressure reduction for the removal of all fluids [Hong 1994]. CCS is commonly used in several parts of the world, especially in Venezuela, California, and Canada. Several examples of coupling subsurface upgrading with CSS are reported and are discussed in the following chapters.

3.1.2.2 Steamflooding

Steamflooding, also known as Continuous Steam Injection or "Steam Drive," is a multi-well process in which steam is injected into injector wells in a variety of location patterns and with different spacing (Figure 3.4). Due to its lower density, steam rises from the injector well until it reaches the overburden (impermeable barrier). Then, it spreads out toward the producing wells until it breaks

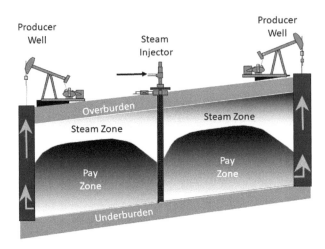

FIGURE 3.4   Schematic diagram of steamflooding.

through. Then, the steam extends downward as oil is produced by gravity drainage from the producer well [Prats 1982].

A more realistic description of steamflooding is shown in Figure 3.5 [Kopper *et al.* 2002]. Reservoir and wellbore complexities allow steam to travel along different pathways. These complexities include discontinuous impermeable barriers, allowing pay zone contacts, and loss of steam containing. Inadequate cement jobs and incomplete zonal isolation require high steam volumes to heat and contact with air-filled sands, resulting in high heat losses.

Figure 3.6 shows the relative contribution of the different production mechanisms to the overall oil recovery by steamflooding of heavy oil. The principal driver is the reduction of the viscosity with temperature as discussed in Section 2.2. Oil recovery has been in the 50–60% range vs. the OOIP [Hong 1994]. Other mechanisms are also contributing especially if the temperature increases beyond ~250°F (~120°C), and those are thermal expansion, steam distillation, solution gas drive, and miscible emulsion drive (Figure 3.6). Recovery factors up to 85% have been found for steamflooding. Examples of steamflooding are Duri field in Indonesia, Kern River field in California, and Pikes Peak Lloydminster in Alberta, Canada.

During the steam drive, the oil produced immediately before the steam breakthrough has been found to be lighter than the oil produced afterward [Prats 1982, Hong 1994]. This phenomenon occurs due to the lowering of the boiling temperature by the presence of steam (i.e., steam distillation) so that the oil produced initially is enriched in light hydrocarbons. In turn, this crude oil has higher API and lower viscosity than the oil present in the reservoir [Hong 1994, Ovalles *et al.* 2002]. This mechanism has been used for the subsurface upgrading of heavy crude oils [Sharpe *et al.* 1995] and is discussed in Section 5.1.

3.1.2.3   Steam Assistance Gravity Drainage

Steam Assisted Gravity Drainage (SAGD) was invented by Roger Faergestad in the 1970s [Butler 1981, 1985]. A pair of parallel horizontal wells is drilled with a spacing of 16 ft. (5 m) to 23 ft. (7 m) between them (see Figure 3.7). Steam injected into the top well rises and heats the tar sand, reducing the HO/B viscosity. Gravity causes the mobilized oil to flow down toward the lower horizontal well. Initially, the communication between the injector and producer is established by cyclic or continuous steam injection. Then, as steam is injected, the steam chamber (see insert on the right-hand side of Figure 3.7) expands until it reaches the overburden [Zhao *et al.* 2005]. This technology is particularly useful for thick reservoirs (30 m) with high horizontal and vertical homogeneity and good permeability [Guo *et al.* 2016]. The presence of a bottom aquifer or gas cap reduces the efficiency of

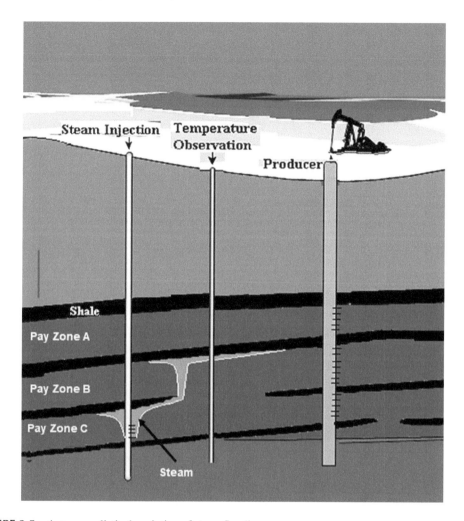

FIGURE 3.5 A more realistic description of steamflooding.

the SAGD process. Also, the process requires a relatively large amount of steam so that the steam-to-oil (v/v) ratio becomes uneconomical around a value of four [Gates and Chakrabarty 2013].

The left-hand side of Figure 3.7 (SAGD-Cross Section) shows the primary production mechanisms occurring during SAGD [Albahlani 2008]. As steam rises in the steam chamber, different dynamic processes are happening at the same time such as steam fingering, co-/counter current flow, thermal conduction and convection, w/o emulsification, imbibition, interfacial tension changes, wettability, and geomechanics effects [Albahlani 2008]. However, this process is very efficient in recovering the HO/B. The estimated recovery factor is between 50–70% of the OOIP. SAGD is currently used in several fields in Canada including Christina Lake and MacKay River. It has also been used in the eastern Venezuela heavy oil fields [Guinand *et al.* 2011].

Since large volumes of water are required during SAGD operations, there is a need for water treatment for steam generation and recycle from the production wells, which could negatively affect the economics of the process. Additionally, the use of natural gas as fuels leads to a considerable amount of greenhouse gas (GHG) emissions. The co-injection of low-cost solvents has been explored to control those issues. These processes are known as Expanding Solvent SAGD (ES-SAGD), have been reported in the literature [Govind *et al.* 2008, Nasr 2001], and are described in the next section.

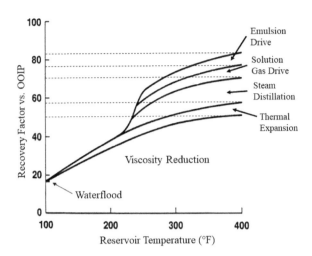

FIGURE 3.6   Relative contribution of mechanisms to the overall oil recovery by steamflooding of a heavy oil.

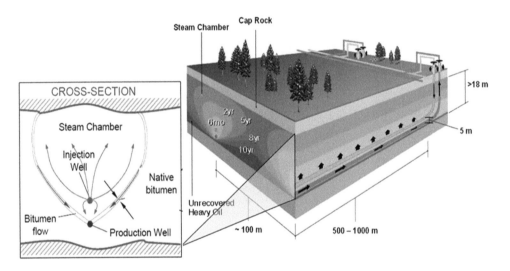

FIGURE 3.7   Schematic diagram of steam assistance gravity drainage (SAGD).

3.1.3   Steam and Solvent Co-Injection

There are several technical issues associated with the thermal-based methods that have affected their commercial development. Those issues are high energy and water consumptions, substantial heat losses and expensive water treatment, and significant greenhouse gas emissions [Lin *et al.* 2014]. Three types of thermal processes, cyclic steam stimulation (CSS), steamflooding, and steam assisted gravity drainage (SAGD), have been employed in conjunction with solvent co-injection to address all these issues, [Ardali 2012, Nasr and Isaacs 2001, Nasr 2001, Guo *et al.* 2016]. Since 1980 [Redford and McKay 1980], these processes have been extensively studied in the literature and have been field tested.

The main disadvantage of the solvent–steam co-injection processes is the higher cost of the solvents used (C5, naphtha, etc.) in comparison with the produced crude oil. Similarly, the recovery and possible recycling of the solvent is of paramount importance to increase the economic benefits. Also, the handling and storage of these materials during steam injection represents an added burden that could make field operation more complicated.

3.1.3.1 Cyclic Steam Stimulation with Solvent Co-Injection

Escobar and coworkers reported their experiences of CSS using solvent-based processes in the Bachaquero reservoir, Maracaibo Lake, Venezuela [Escobar *et al.* 1997]. These authors use of slug of solvent (6–7 wt. %) before steaming in two wells and a mixture of steam and solvent (3–7 wt. %) in the other nine wells. The results showed an average increase of 86% in extra oil production in comparison to the previous cycle.

Figure 3.8 shows the stages of the liquid addition to steam for enhancing recovery (LASER) process as applied to Clearwater and Cold Lake reservoirs in Alberta, Canada [Leaute 2002, Leaute and Carey 2007, Stark 2013]. This process was implemented and patented by Imperial Oil as a late-life technology after CSS. In Stage 1, steam (1,400–1,900 bbl/d of cold water equivalent, CWE, containing 5–6% v/v of solvent) is injected into vertical wells at pressures in the 1,450–1,750 psi range. In Stage 2, the wells are soaked to allow the injected heat to penetrate the formation. In Stage 3, the wells are put into production. The process is repeated over multiple cycles with each well experiencing from eight to ten periods of CSS [Stark 2013].

As seen in Figure 3.8, the solvent is added to the vapor phase and rises along with the steam. Stark reported that the production mechanism for the LASER Process is via transportation of the diluent (C5+ condensate) into the deeper parts of the reservoir until steam condensation occurs along the colder boundaries of the formation [Stark 2013]. The solvent is mixed with the bitumen, which in turn leads to a viscosity reduction. The results have shown significant oil increases (30–60%) over the conventional thermal project. However, solvent recovery has been reported in the 60–80% range

3.1.3.2 Solvent Co-Injection Steamflooding

The co-injection of solvent to steam during steamflooding has been implemented by Shell in a process known as Vertical-well Steam Drive (VSD). Core flooding experiments were carried out using a slug of 15% solvent (C4–C20) [Hedden and Verlaan 2014]. Numerical simulations have shown that, in the beginning, bitumen rates are higher in the presence of the solvent (Figure 3.9, darker trace) than those calculated for the no-solvent case (Figure 3.9, lighter case). Recovery improvement

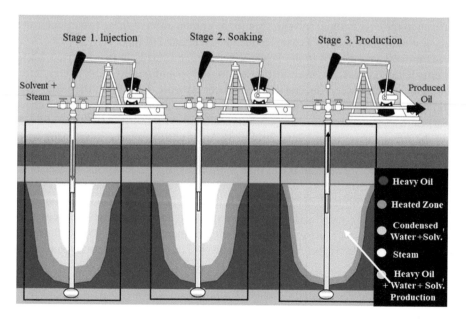

FIGURE 3.8 Stages of the liquid addition to steam for enhancing recovery (LASER) process. (Adapted from NIPER 1986.)

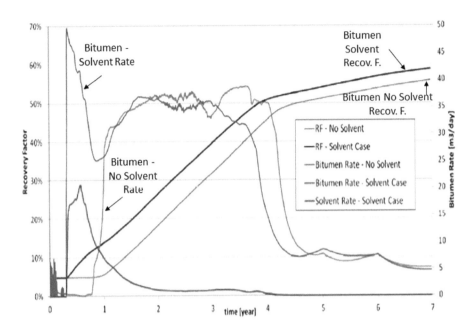

FIGURE 3.9 Recovery factors, bitumen solvent rates vs. time for the numerical simulation of the solvent/ steam co-injection and the comparison with the no-solvent case. [Hedden and Verlaan 2014]. Reprinted with permission of Society of Petroleum Engineers.

in the order of 3–6% of the OOIP (Figure 3.9, darker trace) was obtained in comparison with the control experiment (no-solvent, lighter curve).

The simulations showed that the dominant recovery mechanism is the formation of a solvent bank that mixes the bitumen ahead of the steam zone. Due to this mixing, the viscosity of the latter is significantly reduced and thus accelerated oil rates are obtained [Hedden and Verlaan 2014, Hedden *et al.* 2014b]. Also, the solvent bank displaces the heavy crude oil and reduces its saturation in comparison to the case without solvent. Due to the combination of these two effects, an increase in recovery is obtained for the VSD case [Hedden and Verlaan 2014].

The VSD process was piloted in the Peace River area in Canada in two inverted 5-spot 5-acre drive patterns [Castellanos-Diaz *et al.* 2016]. Before the steamflooding, CSS cycles were used to create communication between injectors and producers. Figure 3.10 shows the instantaneous oil/ steam ratio (iOSR) versus time for the numerical simulation of the solvent/steam co-injection and the comparison with the no-solvent case [Castellanos-Diaz *et al.* 2016]. After several months from the start of the stream drive (from 1 June 2014 to 31 August 2014), a large solvent slug was injected (15% for six months), subsequently followed by a steam drive. As seen, higher iOSR were obtained in the presence of the solvent (green trace) than those found for the steam-only case (black curve) [Castellanos-Diaz *et al.* 2016].

After the initial peak, oil rates declined but sustained higher values than those found for steam-only patterns for about six months (Figure 3.10). After a data uncertainty management plan and aided by reservoir simulation, the results showed oil rate increases of 27–45% with respect to the no-solvent case. The authors reported 51–59% of solvent recovery after the trial [Castellanos-Diaz *et al.* 2016].

3.1.3.3   Solvent/Steam Assistance Gravity Drainage

In this concept, a hydrocarbon solvent is co-injected at low concentration (< 15%) with steam in the SAGD process. The hydrocarbon solvent is selected so that it evaporates and condenses at the same conditions as the water phase at the boundary of the steam chamber. This process is known as

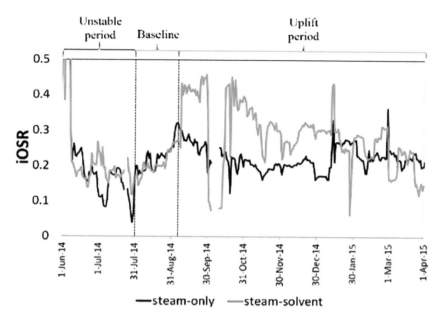

FIGURE 3.10 Instantaneous oil/steam ratio (iOSR) vs. time for the numerical simulation of the solvent/steam co-injection and the comparison with the no-solvent case. *Field Test at Peace River* [Castellanos-Diaz 2016]. Reprinted with permission of Society of Petroleum Engineers.

Expanded Solvent SAGD (ES-SAGD) because the solvent is injected with steam in a vapor phase. This phenomenon leads to a condensation around the interface of the steam chamber, dilutes the oil, and in conjunction with heat, reduces its viscosity [Ayodele *et al.* 2009, Bayestehparvin 2018, Nasr *et al.* 2009].

In effect, lab experiments have shown that adding solvent to steam led to a higher oil rate and reduced the steam-to-oil ratio which, in turn, increased the ultimate recovery [Ayodele 2009, Nasr *et al.* 2009]. Figure 3.11 shows a mechanistic diagram of the ES-SAGD process and the different regions of the steam chamber during solvent co-injection [Jha 2012]. As the temperature drops near the chamber boundary, steam starts condensing first because its mole fraction and partial pressure are higher than those for the solvent. As temperature decreases toward the chamber boundary, the

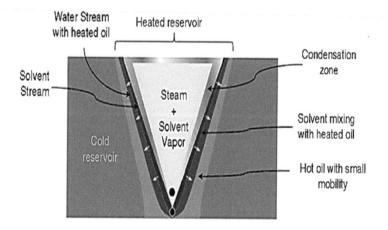

FIGURE 3.11 Mechanistic diagram of the ES-SAGD process showing the different regions of the steam chamber during solvent co-injection [Jha 2012]. Reprinted with permission of Society of Petroleum Engineers.

vapor phase becomes increasingly richer in solvent, and this material and steam condense simultaneously. Thus, contrary to steam-only injection, where condensation occurs at the injected steam temperature, condensation of the steam/solvent mixture is accompanied by a reduction in temperature in the condensation zone (Figure 3.11) and the farther regions [Jha 2012].

The condensed steam/solvent mixture drains outside the chamber, leading to the formation of a mobile liquid stream (drainage region) where heated oil, condensed solvent, and water flow together to the production well (blue and red are in Figure 3.11). The condensed solvent mixes with the heated oil and further reduces its viscosity. This additional viscosity reduction more than offsets the effect of reduced temperature near the chamber boundary [Jha 2012]. As the steam chamber expands laterally because of continued injection and temperature increase, the condensed solvent mixes with oil, which in turn lowers the residual oil saturation (ROS) in the steam chamber. Therefore, ultimate oil recovery with the steam/solvent-co-injection process is higher than that in the steam-only counterpart [Jha 2012].

Table 3.1 presents a summary of several field tests of solvent co-injection with SAGD [Gupta 2005 [Gupta 2006, Orr 2009, Alberta 2014, 2015, 2016a, 2016b]]. For the Senlac, Christina Lake, Long Lake, Surmont, and Algar projects, increases in the oil rate and reductions of SOR were observed. However, this observation is not general, and in some cases, the evidence is inconclusive.

Similarly, solvent recovery has not been definitively proven (Table 3.1). The percentage of solvent recovery ranges from 20–80%. Several reasons have been proposed such as that the projects are in the early phase, difficulties in sampling operations, and solvent composition overlapping with the bitumen making quantification difficult. Finally, increases in the API of the produced oil have been observed for two of the seven cases studied (Table 3.1). The observed upgrading has been attributed to the solvent deasphalting (both Senlac and Christina Lake projects used C4, an asphaltene precipitant solvent as discussed in Section 2.4). This topic is discussed in Section 4.1.1.

3.1.4 SOLVENT-ONLY PROCESSES

In these processes, also known as solvent flooding, the injection of hydrocarbon solvents is carried out cyclically or continuously into the reservoir to dissolve the oil without the need of thermal energy [Guo *et al.* 2016, Bayestehparvin 2018]. Chang *et al.* numerically and experimentally simulated the performance of cyclic steam injection (CSI) in horizontal wells as a follow-up process for CHOPS in the Cold Lake and Lloydminster reservoirs [Chang *et al.* 2009]. After primary production, six cycles of 28% C3 and 72% CO_2 was used to recover ~50% of OOIP. The numerical model matched well the experimental results [Chang *et al.* 2009].

Chang *et al.* carried out the numerical simulation of the $40 million 2006–2010 Joint Implementation Vapor Extraction Program in support of the Husky CSI field test [Chang *et al.* 2015]. As reported by Worth and Quale the solvent injection process could be economically viable for depleted heavy oil reservoirs depending on the oil and solvent prices, production response, and fiscal regime [Chang *et al.* 2015, Worth and Quale 2010].

Other field tests have been reported in the literature using downhole solvent injection (VAPEX and N-Solv). Those examples are discussed in Chapter 5.

3.1.5 *IN-SITU* COMBUSTION

In-situ combustion (ISC), also known as fireflooding, is the oldest of the thermal production methods. In this multi-well process, air or oxygen is injected downhole to create a combustion front that propagates from the injector well to the producing well (Figure 3.12). By doing so, some of the oil is burned, and the generated heat reduces the remaining HO/B viscosity to increase oil rate and total

TABLE 3.1

Summary of Several Field Tests of Solvent Co-injection with Steam Assistance Gravity Drainage (SAGD)[a]

Project (Year)	Operator	Solvent Used[b] (% v/v vs. Steam)	Oil Rate (B/d)[c]	Steam-to-Oil Ratio[d]	Solvent Recovery[e]	Upgrading[f]	References
Senlac (2002)	Encana	Butane (15 wt. %)	From 1,900 B/d to 3,000 B/d (~58% Inc.)	From 2.6 to 1.6	~70%	1°API increase	Gupta and Gittins [2005]
Christina Lake (2004)	Encana	Butane (15 wt. %)	From 630 B/d to 1,900 B/d (~58% Inc.)	From 5 to 1	N-A	1°API increase	Gupta and Gittins [2006]
Long Lake (2006)	Nexen	Jet B (C7–C12) (5%)	Same as SAGD	Same as SAGD	N-A	N-A	Orr [2009]
Jackfish (2013)	Devon	C6	Inconclusive	Inconclusive	N-A	N-A	Alberta [2014]
Surmont (2012–2013)	Conoco-Phillips	C3–C4–C6 (20%)	Increase of 20–30%	Reduction of 10–30%	50% (Feb 2016)	N-A	Alberta [2015]
Algar (2011–2015)	Connacher	C5–C6 (Ave. 6%)	Increase of 20–30%	Reduction of 28–32%	89% (April 2015)	N-A	Alberta [2016]
Leismer (2013–2015)	Statoil	C5–C6 (Ave. 5%)	Non-responsive	From 1.5 to 1.0	20% (Q2 2015)	N-A	Alberta [2016b]

[a] N-A = Not available or not reported.

[b] Percentage by v/v of solvent vs. cold water equivalent of steam.

[c] Oil rate produced in barrels per day. The number in brackets indicates the percentage of increase vs. SAGD.

[d] Cold water equivalent of steam vs. oil produced (v/v).

[e] Percentage of solvent recovered vs. volume injected. Numbers in bracket indicate the time of measurement.

[f] Produced oil upgrading as measured by the increase in the API gravity in the tank.

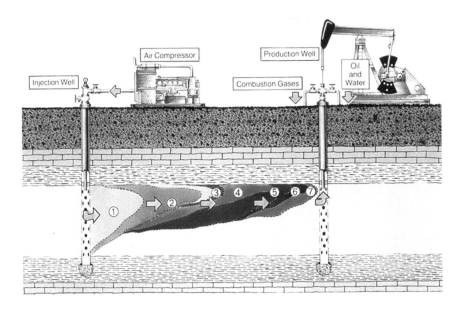

FIGURE 3.12 Schematic diagram showing the phases of *in-situ* combustion [NIPER 1986]. (1) Burned out zone, (2) air or vaporized water zone, (3) combustion zone (315–650°C), (4) steam or vaporizing zone (~200°C), (5) condensing or hot water zone (< 100°C), (6) original oil at reservoir temperature, and (7) cold combustion gases. Acknowledgment to the U.S. Department of Energy and National Energy Technology Lab.

recovery. A burn oil or residue is left behind. The most widely used form of ISC is simple air injection and is referred to as "dry combustion." In another embodiment, water is co-injected with air, and it is known as "wet combustion." In principle, ISC is more efficient and economical than steam injection processes. The only energy required is for the use of air compressors and is much lower than a steam generator. However, ISC suffers from other disadvantages such as high corrosivity of the products, a high level of contaminants, and safety considerations [Huc 2011, Speight 2007, Guo *et al.* 2016].

In ISC, ignition is the first step and occurs in reservoirs where the temperature is equal to or greater than 60°C. Otherwise, the combustion is initiated by artificially heating the injection well by using an electric heater, burner hot-fluid, or a highly oxidable chemical. Figure 3.12 shows the phases of *in-situ* combustion [NIPER 1986]. In zone 1 (Figure 3.12), the burning has already taken place and injected air and water are present. In region 2, air or vaporized water move from the injector to the producer. Area 3 is called the burning or combustion zone and is where the oxygen is used up to generate heat. The temperature is in the 315–650°C range. In zone 4, steam is moving ahead of the combustion toward the producing well. The temperature is approximately 200°C. In region 5, vapor generated in zone 4 condenses, and mostly hot water is observed. Part 6 is the area of the reservoir in which the original oil is present at reservoir temperature. Finally, in region 7, the cold combustion gases escape via the producing well [NIPER 1986, Huc 2011, Speight 2007].

ISC is typically used in high permeability and homogeneous sandstones and shallow on-shore formations. Although detailed geological characterization is always carried out, there is no guarantee of success due to the difficulties in controlling and maintaining the combustion front (zone 3, Figure 3.12). For these reasons, ISC is usually the last option among all thermal recovery processes. However, more than 200 combustion processes have been performed over the years with different degrees of failure and success. Some tests were economically and technically successful, while others were only technical, and many were failures. Current operations are in Romania, India, and the US [Huc 2011]. The general mechanism and chemical pathways of *in-situ* combustion are described in Section 9.3.1.

3.2 PRODUCTION ISSUES

During HO/B production, several challenges could potentially impact subsurface upgrading processes and need to be adequately addressed to enhance the technical and economic feasibility. These topics are well issues, oil and gas separation, and HO/B transportation. It is important to mention that these challenges are interlinked, so they need to be analyzed as a complex system going from the mineral formation, completion, well design, lift process, fluid separation, to storage and transportation [Smalley 2000]. Only by achieving an integrated view of these elements and their interactions can the optimal economic benefits be obtained.

3.2.1 WELL ISSUES

Three basic types of well exist, i.e., production, injection, and observation wells. In heavy oil and bitumen production, the designs of the injector and producer wells can differ from those used in more conventional light and medium crude oils. Due to the high temperature and pressures in thermal recovery processes, specialized technologies and care are needed to provide proper design and operations of downhole equipment. Similarly, some HO/B wells and wellheads are designed for the use of diluents, most of the time have artificial lift, and require sand management. All these topics are discussed next.

3.2.1.1 Well Completions

The typical completions used for HO/B producing well are shown in Figure 3.13. These are (A) slotted liner, (B) perforated casing, and (C) gravel pack. The first two are used where the pay zone requires sand control [Hong 1994]. If zonal segregation is important, jet perforated of the casing, which has been previously cemented in place, can be used. These completions can be used during thermal recovery processes [Mahmoudi *et al.* 2016]. Due to the high viscosity of HO/B, in many areas such as Orinoco or Alberta, most of the wells are prudently completed to be used with thermal recovery processes (Section 3.1.2), even though they are initially used for cold production (Section 3.1.1). This situation leads to an initial high cost per well but, in the long run, it saves money and maintains oil production [Robles 2001]. A proper selection of the completion scheme is of paramount importance for a successful subsurface upgrading process.

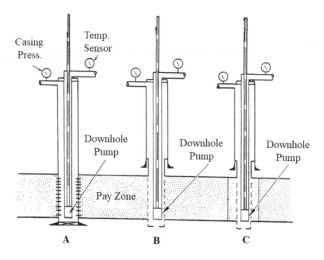

FIGURE 3.13 Most commonly used thermal completions for producing well slotted liner. (**A**) Perforated casing, (**B**) gravel pack, and (**C**) open hole with gravel pack.

As mentioned in Section 2.2.2., solvent dilution is the most widely preferred technology for reducing the viscosity of HO/B and making them transportable by conventional methods from downhole to the surface and processing centers. Generally, the diluent is injected into the annular space by using progressive cavity pumps. This approach, in some cases, does not guarantee proper dilution at the entrance of the pump because of the gas accumulation at the annular section. In turn, this situation causes lifting problems, such as high torque, high loads in the rod strings, damages in the rods, and operational difficulties in surface lifting units and pumps working at maximum discharge pressure limits [Rojas 2001]. To achieve reliable operation, the HO/B should be diluted before entering the pump by using a capillary tubing anchor at pump intake [Rojas 2001].

Injector wells are completed with casing cemented throughout the oil-bearing formations, and then jet- or bullet perforated in the sands in which steam or additives are to be injected. In relatively thick and uniform pay zones, the perforations may be shot at the bottom of the interval to help to reduce the steam gravity override [Hong 1994].

3.2.1.2 Artificial Lift

Many crude oil wells require artificial lift to produce oil to the surface. Sucker rod pumping (SRP), electrical submersible pumping (ESP), jet hydraulic pumps, and gas-lift are the various modes of artificial lift currently used. However, when heavy oils and bitumens are encountered, these methods of artificial lift fail to provide the necessary pressure for the crude to reach the surface [Awarval et al. 2015]. In some fields, electrical submersible (or submergible) pumps or ESPs are favored because of the significant drawdown they can produce. ESPs comprise a series of rotating turbine type blades that have the drawback of being very sensitive to even small amounts of sand production. Pump failures can form the primary component of operating expenditures in such fields [Smalley 2000].

In the last 20 years, progressive cavity pumps (PCP) have found great utility under conditions of heavy crude production [Robles 2001]. PCPs have a rotating spiral shaft that is designed to produce a low-pressure zone on the underside and draw liquid upwards. PCPs are suited to lifting viscous oil and very tolerant to sand at low sand contents. PCPs are favored in cold production projects such as Canada and Orinoco [Smalley 2000]. These pumps are mainly surface driven PCPs where the motors are located on the surface and the downhole pump operated by a long rotating rod. Such devices are relatively easy to unseat, clean out, retrieve, and the motor is readily available for service [Robles 2011, Awarval 2015].

For deeper reservoirs like Boscan in Western Venezuela or the curved trajectories, downhole PCP may be highly favored. However, the possibility of rod wear and failure is higher with the concomitant increase in cost [Smalley 2000].

3.2.1.3 Sand Management

Many unconsolidated HO/B reservoirs are prone to sand production, and a technically sound sand management strategy is of paramount importance. For the Duri field in Sumatra, Indonesia, as much as 1,000,000 lbs of sand per day are produced. The co-production of this amount of solids leads to additional costs in well re-completions, safe oil-contaminated sand disposal, and the effect on the surface facilities such as stabilization of emulsions.

Sand can be controlled downhole by mechanical sand exclusion methods such as a screen, gravel packing, and "frac packing." Frac packing involved the simultaneous hydraulic fracturing of a reservoir and the placement of a gravel pack [Mahmoudi et al. 2016, Kalgaonkar et al. 2017]. However, these mechanical devices add cost to the well and may introduce additional flow restrictions that reduce the oil flow rate. Several limitations such as screen plugging and expenses associated with deploying expandable screens make the current mechanical sand control practices more expensive. Sand production can also be minimized by using horizontal wells with small drawdowns, by using slotted liners with a slot size designed to exclude sand or induce fractures filled with resin coated proppant [Smalley 2000].

Chemical sand consolidation could be a cost-effective alternative for sand management in HO/B wells. These technologies would also allow the wellbore to be free of tools, screens, and pack sands. New developments of chemical systems are based on resins with a curing agent which hardens under reservoir temperature. A permeability enhancing additive is also incorporated into the system to achieve higher and controllable permeability of the consolidated sand pack [Marfo *et al.* 2015, Kalgaonkar *et al.* 2017].

In conclusion, sand management is an important variable that should be taken into consideration during the development of a subsurface upgrading process. All well issues presented in this section (completion, artificial lift, and sand handling) should be analyzed with an integrated view, so the optimal alternative is chosen to maximize economic benefits.

3.2.2　Oil, Water, and Gas Separation

After taking the oil-containing fluid to the surface, oil, water, and gas should be separated before they can be transported to the refining centers. The main issue with HO/B is that the density difference with water is small which makes the dehydration/dewatering processes difficult. In contrast, the difference in density of the liquid and gaseous hydrocarbons allows acceptable segregation in a conventional oil and gas separator [Smith 1987].

Another issue with heavy oils and bitumens is their high viscosity (Section 2.2) which makes the separation process slower and time-consuming. If one adds the even higher viscosity of water-in-oil (w/o) emulsions, the dehydration of HO/B becomes a significant challenge. In general, these w/o emulsions are formed by the mixing and shearing during artificial lift by ESP, PCP, or gas-lift [Correra *et al.* 2016]. The presence of solid particles and natural surfactants (asphaltenes) leads to further emulsion stabilization [Kokal 2005, Kilpatrick 2012, Delgado-Linares *et al.* 2016].

As mentioned in Section 1.8, subsurface upgrading processes may affect the oil/water separation processes. For example, in the case of solvent deasphalting, the lower asphaltene concentration of the upgraded crude oil may lead to a faster and easy to operate dehydration process. However, during *in-situ* combustion, an increase in oxidation compounds in the treated oil may result in w/o emulsion stabilization with the concomitant loss of the efficiency of the water removal systems. For HO/B, three types of oil and water separation processes are typically used. Those are by adding diluent and heat to dehydration process and by destabilization of emulsions. These methods can be used together or separated [Smalley 2000].

3.2.2.1　Use of Diluent and Heat

As mentioned in Section 2.2.2, dilution is the most widely preferred technology for reducing the viscosity of HO/B and making them transportable by conventional methods. Three additional advantages can be envisioned by doing so, related to the oil and water separation. First, an increase in the efficiency of the process can be seen due to the lower viscosity of the HO/B diluent blend. Second, by adding diluents, the dehydration process becomes more manageable due to the density reduction of the hydrocarbon mixture. This reduction makes a greater density difference with water with a concomitant increase in gravity segregation. Lastly, the presence of the diluent helps to destabilize w/o emulsions and thus, increasing the rate of water separation.

The use of heat in conjunction with diluents further enhance the efficiency of the dehydration process. The main advantage of heating is the more significant reduction of the HO/B viscosity with temperature (see Figure 2.2) in comparison with water. In effect, water viscosity slightly changes in the 20–80°C temperature range in which most of the separators operate. On the contrary, crude oil and water densities vary only 3–4% in the same temperature range [Boberg 1988].

As mentioned in Section 1.8, oil/water separation can be negatively affected due to the use of chemicals and/or catalysts downhole during SSU processes. It is known that these chemicals may stabilize w/o emulsions so longer residence times and the use of demulsifiers may be necessary to

break those stable emulsions and make the upgraded crude oils transportable and comply with pipelines requirements. These topics are discussed next.

3.2.2.2 Destabilization of Emulsions

To attain the necessary oil/water separation and adhere to the water specification (less than 1% of
basic sediment and water or BS&W), w/o emulsions found in HO/B are destabilized by increasing
temperature and residence time, removal of solids, and by addition of chemical demulsifiers [Kokal
2005, Kelland 2009]. The mechanisms involved in demulsification are flocculation, aggregation,
sedimentation, creaming, and coalescence and are focused on the destabilization of the w/o interfacial films [Kilpatrick 2012].

It has been reported that the stabilization of the interfacial films involves the presence of
asphaltenes as lipophilic surfactants adsorbed at the interface. Also, asphaltenes contribute to the
formation of a viscous or rigid film around the water drops that slows down the inter-drop film drainage and retards or inhibits the coalescence [Alvarez 2009, Delgado-Linares et al. 2016]. HO/B are
known for their relatively higher asphaltene concentration than the L/M counterparts (see Tables 2.2
and 2.3). Thus, the role of the demulsifier, which is a hydrophilic surfactant, is to counteract these
stabilization mechanisms so effective water separation can be achieved.

There are many classes of w/o demulsifiers [Kelland 2009]. The most commonly used are polyalkoxylate block copolymers and their ester derivatives, alkylphenol-aldehyde resin alkoxylates,
polyalkoxylate of polyols or glycidyl esters, and polyurethanes (carbamates) and polyalkoxylates
derivatives. These demulsifiers are oil-soluble and are deployed as a solution in hydrocarbons solvents (diesel). Additionally, alcohol (butanol) is used as co-solvent to improve the transportation of
the amphiphile to the water/oil interface, especially for high viscosity crudes such as HO/B. Bottle
testing is the most accepted method to evaluate demulsifier performance, but more advanced analytical techniques have been used such as interfacial tension, dielectric constant, and electric field
measurements [Kelland 2009].

3.2.3 Transportation

As mentioned in Section 2.2, one of the main issues associated with the commercialization of heavy
oils and bitumens is their high viscosity at the reservoir and ambient conditions, which makes difficult their transportation using conventional pipelines. Several routes have been investigated to
overcome this issue such as diluents (see Section 2.2. and [Argillier et al. 2005]), heated pipelines
[Escojido 1991, Hart 2014], emulsions [Rimmer et al. 1992, Nuñez et al. 1995 Salager et al. 2001],
train and trucking [Smalley 2000], annular flow [Bannwart 2001, Saniere et al. 2004], asphaltene
slurries [Brito et al. 2013, Dieckmann et al. 2015], and additives that modify the aggregation behavior of asphaltenes (see Section 2.4.2 and [Ovalles et al. 2011]).

One of the primary objectives of subsurface upgrading is reducing the viscosity of HO/B, and the
efficiency and economics should be similar or better than the conventional routes mentioned in the
previous paragraph. Among the latter, dilution, heating, and emulsions are the most commercially
used processes, so they are described in the next sections.

3.2.3.1 Using Diluents

The effect of the diluent on the HO/B viscosity and the amount needed for transportation were
outlined in Section 2.2.2. Figure 2.3 shows a decrease of one order of magnitude (from ~1,000 to 70
cSt at 30°C) of the viscosity of Venezuelan and Canadian heavy crude oils by increasing the naphtha
and diesel contents in the 20–30 wt. % range. Similar results have been reported by other authors
[Argillier et al. 2005].

Dilution can be used in two different ways, depending on whether the diluent is recycled or not.
Figure 3.14 shows two Orinoco Belt heavy oil projects (Eastern Venezuela) in which recycled diluent is used for transporting the crude oil from the production facilities to the upgrader unit in Jose

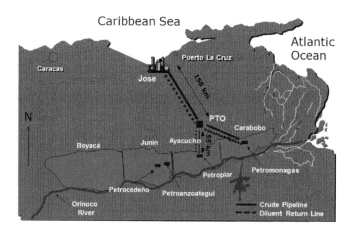

FIGURE 3.14 Map of the Orinoco Belt heavy oil projects showing the diluted crude oil (solid lines) and diluent return (red dashed lines) pipelines.

[Burgos *et al.* 2014]. As seen, Petropiar and Petromonagas receive diluent using the diluent return pipeline (dashed lines). The diluted heavy oil is produced from the Ayacucho and Carabobo blocks, respectively. Then it is pumped (solid lines) to the intermediate storage facility PTO ("Patio de Tanques de Oficina") located ~55 km of distance. From there, it is transported to Jose (~158 km) to be upgraded and shipped abroad. In the last facility, the diluent is separated and returned to the production field. Similar schemes are used for the other two projects, Petrocedeño and Petroanzoategui located in the Junín block [Burgos 2014].

In Alberta, Canada, Albian Sands Energy Inc. runs the oil-sands mining operation at the Muskeg River Mine located 75 kilometers north of Fort McMurray. In oil-sands mining, a mix of oil and sand is removed from just below the surface using shovels and trucks. This material is mixed with warm water to separate the oil from the sand. The produced bitumen is first diluted, then transported via the Corridor Pipeline System to a regional upgrader in Scotford where it is transformed into synthetic crude. As with the Orinoco case, the Corridor Pipeline System is a dual pipeline transporting diluted bitumen to the Scotford upgrader and the diluent back to the mine [Saniere *et al.* 2004].

Alternately, several projects in Canada export diluted bitumen in which the diluent is not recycled. The diluent is usually natural gas condensate, naphtha, or a mix of other light hydrocarbons and constitutes 24–50% of the bitumen blend [Saniere *et al.* 2004]. Typical examples are Bow River Heavy, Western Canadian Select, Cold Lake Blend, and Wainwright–Kinsella.

In all these dilution processes (recycling or not), diluent separation, construction of a dedicated return pipeline for the diluent, and operating a separate pipeline are costly steps. Also, light hydrocarbons suitable for dilution are more expensive than the bitumen, and natural gas condensate is not readily available in large quantities. These shortcomings represent opportunities for subsurface upgrading processes to reduce capital and operating expenses with the concomitant increase in commercial revenues.

3.2.3.2 Heated Pipelines

As mentioned in Section 2.2 and Figure 2.2, the viscosity of HO/B decreases several orders of magnitude as the temperature increases in the 15–120°C range. Thus, it is not surprising that this method is widely used in several parts of the world, especially in Venezuela, where it has been employed since 1955 [Escojido *et al.* 1991]. In Canada, bitumen producers who want to ship their products to a regional upgrader or major pipeline terminal without diluent require a heated/insulated pipeline. For example, Enbridge Pipeline operates a pipe system to ship bitumen from the MacKay River production site to Fort MacMurray where it is blended before being exported to the

North American market [Saniere *et al.* 2004]. Another example is the Chad–Cameroon pipeline in central Africa, where six pump-heating stations are used to transport 500 MBPD along a 640-mile distance between Palogue and Port Sudan [Dunia and Edgar 2012].

However, heating to increase the temperature of the HO/B involves a considerable amount of energy which in turn, translates into higher operating costs. Issues such as corrosion, flow instability, loss of heat to the surroundings, and sudden expansion and contraction of the pipeline have led to increasing capital and operating expenditures over the long distance from the oil field to the final storage or refinery [Hart 2014]. These challenges represent opportunities to the subsurface upgrading processes to generate upgraded HO/B that does not require heated pipeline for their transportation.

3.2.3.3   Use of Emulsions

The use of oil-in-water (o/w) emulsion for the transportation of heavy crude oils and bitumens is a successful commercial process. A typical example is a technology developed by PDVSA-Intevep called Orimulsion® [Nuñez 1995, Salager 2001]. In this process, the HO/B is suspended by mixing and shearing in water (the external phase) in the form of microspheres stabilized by chemical additives. The final oil and water concentrations are 70% and 30% v/v, respectively. By this way, a reduction of the apparent viscosity of the fluid is achieved from ~100,000 mPa s to less than ~100 mPa s [Rivas and Colmenares 1999, Rivas *et al.* 2003]. No heat or diluent is needed. As shown in Figure 3.15, Orimulsion® was transported from the manufacturing center located in Morichal to the Jose terminal on the Caribbean Sea via a dedicated pipeline. From there, it was exported to several places around the world.

As reported by Rivas *et al.* [2003], the most important factors controlling the process of emulsification of highly viscous hydrocarbons in water are the concentration of the nonylphenol ethoxylated surfactant, the presence or absence of electrolytes in water, temperature, speed and time of mixing, and shear rate. By following the changes in mean droplet diameters and viscosity, it was found that the emulsion stability decreases as a function of the storage time. This "aging" process can be controlled by adjusting the type and concentration of the surfactant and the electrolytes dissolved in water [Rivas *et al.* 2003].

Orimulsion® is directly used as a feedstock for power generation in thermoelectrical plants. Another strategy for crude oil-in-water emulsions could be to separate the crude and the water after transportation, at the refinery entry, and subsequently to upgrade the crude.

Besides Orimulsion®, the technical viability of pipeline transportation of HO/B oil as o/w emulsions that contain high fractions of oil has been demonstrated in an Indonesian pipeline in 1963 and a 13-mile long, 8-in diameter pipeline in California [Langevin *et al.* 2004].

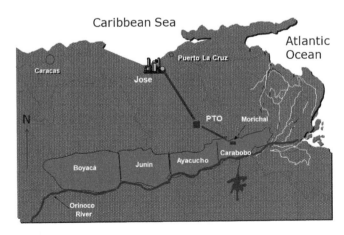

FIGURE 3.15   Map showing the pipeline used for the transportation of Orimulsion® in Eastern Venezuela.

As with dilution and heating, the costs associated with emulsion preparation can be significantly high. First, the water used should be relatively clean and free of non-desirable electrolytes that may cause excessive "aging." Next, the heavy crude oil or bitumen must have low water and sediment concentrations to avoid the formation of multiple emulsion [Nuñez 1995]. Also, the surfactant dosage should be kept to a minimum because its cost can be as high as $3 per barrel of processed crude. Finally, high standards of quality control should be established to guarantee a stable product from the production center to its destination. Thus, there is considerable room for improvement in which a subsurface upgrading process could be a more economically attractive option.

REFERENCES

Albahlani, A. M., Babadagli, T., 2008, "A Critical Review of the Status of SAGD: Where Are We and What Is Next?", SPE No. 113283, presented at SPE Western Regional and Pacific Section AAPG Join Meeting held in Bakersfield, California, U.S.A., March 31–April 2 and references therein.

Alberta Energy Regulator, 2014, "Devon Subsurface Performance Presentation", Oct. 2014, p 84. www.aer.ca/documents/oilsands/insitu-presentations/2014AthabascaDevonJackfishSAGD10097.pdf. Downloaded on June 12, 2017.

Alberta Energy Regulator, 2015, "Annual Surmont SAGD Performance Review", www.aer.ca/documents/oilsands/insitu-presentations/2015AthabascaConocoSurmontES-SAGD11596.pdf. Downloaded on June 11, 2017.

Alberta Energy Regulator, 2016a, "Connacher Performance Presentation", p 35. www.aer.ca/documents/oilsands/insitu-presentations/2016ConnacherGreatDivide10587.pdf. Downloaded on June 12, 2017.

Alberta Energy Regulator, 2016b, "Statoil Annual Performance Presentation", March 2016, p 57. www.aer.ca/documents/oilsands/insitu-presentations/2016AthabascaStatoilLeismerSAGD10935-Presentations.zip. Downloaded on June 12, 2017.

Alvarez, G., Poteau, S., Argillier, J. F., Langevin, D., Salager, J. L., 2009, "Heavy Oil–Water Interfacial Properties and Emulsion Stability: Influence of Dilution", *Energy Fuels*, 23(1), 294–299.

Ardali, M., Barrufet, M., Mamora, D. D., Qiu, F., 2012, "A Critical Review of Hybrid Steam/Solvent Processes for the Recovery of Heavy Oil and Bitumen", SPE No. 159257, presented at the SPE Annual Technical Conference and Exhibition, San Antonio, TX, 8–10 October, and references therein.

Argillier, J.-F., Henaut, I., Gateau, P., Heraud, J.-P., Glenat, P., 2005, "Heavy Oil Dilution", SPE No. 97763, presented at the SPE International Thermal Operations and Heavy Oil Symposium, 1–3 November, Calgary, Alberta, Canada.

Awarval, S., Aggarwal, A., Kumar, A., 2015, "PCPs Coming of Age as a Viable Artificial Lift Solution for Low API Crude Fields", SPE No. 172686 presented at SPE Middle East Oil & Gas Show and Conference, 8–11 March, Manama, Bahrain and references therein.

Ayodele, O. R., Nasr, T. N., Beaulieu, G., Heck, G., 2009, "Laboratory Experimental Testing and Development of an Efficient Low Pressure ES-SAGD Process", *J. Can. Pet. Soc.*, 48(9) Sept., 54.

Bannwart, A. C., 2001, "Modeling Aspects of Oil–Water Core–Annular Flows", *J. Pet. Sci. Eng.*, 32(2–4), 127–143.

Bayestehparvin, B., Farouq Ali, S. M., Abedi, J., 2018, "Solvent-Based and Solvent-Assisted Recovery Processes: State of the Art", SPE No. 179829, SPE Res. Eval. Eng., August, 1–21, and references therein.

Boberg, T. C., 1988, *Thermal Methods of Oil Recovery*, Wiley, New York, p 358.

Brito, A., Cabello, R., Mendoza, L., Salazar, H., Trujillo, J., 2013, "Heavy Oil Transportation as a Slurry", BHR-2013-C1, presented at 16th International Conference on Multiphase Production Technology, 12–14 June, Cannes, France.

Burgos, E. C., Peñaranda, J., Gonzalez, K., Trejo, E., Meneses, Rosales, A., Martinez, J., 2014, "Shallow Horizontal Drilling Meets with Very Extended Reach Drilling in the Venezuelan Faja", SPE No. 169365, presented at the SPE Latin America and Caribbean Petroleum Engineering Conference, 21–23 May, Maracaibo, Venezuela.

Butler, R. M., 1985, "A New Approach to the Modeling of Steam-Assisted Gravity, Drainage", *J. Can. Pet. Technol.*, 24(3), 42–51.

Butler, R. M., Stephens, D. J., 1981, "The Gravity Drainage of Steam-Heated Heavy Oil to Parallel Horizontal Wells", *J. Can. Pet. Tech.*, 20(2), 90–96.

Castellanos-Diaz, O., Verlaan, M. L., Hedden, R., 2016, "Solvent Enhanced Steam Drive: Results from the First Field Pilot in Canada", SPE No. 179815, presented at the SPE EOR Conf. Oil & Gas west Asia, Muscat, Oman, 21–23 March.

Chang, J., Ivory, J., London, M., 2015, "History Matches and Interpretation of CHOPS Performance for CSI Field Pilot", presented at SPE Can. Heavy Oil Tech. Conf., Calgary, Alberta, Canada, 9–11 June.

Chang, J., Ivory, J., Rajan, R. S. V., 2009, "Cyclic Steam-Solvent Stimulation Using Horizontal Wells", PETSOC-2009-175, presented at the Can. Intern. Pet. Conf., Calgary, Alberta, Canada, 16–18 June.

Correra, S., Iovane, M., Pinneri, S., 2016, "Role of Electrical Submerged Pumps in Enabling Asphaltene-Stabilized Emulsions", *Energy Fuels*, 30(5), 3622–3629.

Delgado-Linares, J. G., Pereira, J. C., Rondon, M., Bullon, J., Salager, J. L., 2016, "Breaking of Water-In-Crude Oil Emulsions. 6. Estimating the Demulsifier Performance at Optimum Formulation from Both the Required Dose and the Attained Instability", *Energy Fuels*, 30(7), 5483–5491.

Dieckmann, G. H., Segerstrom, J., Ovalles, C., Rogel, E., Kuehne, D. L., Subramani, H. J., O'Rear, D. J., 2015, "Method and System for Processing Viscous Liquids Crude Hydrocarbons", US Patent No. 9,028,680, 12 May.

Dunia, R., Edgar, T. F., 2012, "Study of Heavy Crude Oil Flows in Pipelines with Electromagnetic Heaters", *Energy Fuels*, 26(7), 4426–4437.

Dusseault, M. B., 2001, "Comparing Venezuelan and Canadian Heavy Oil and Tar Sands", presented at Can. Int. Pet. Conf. held in Calgary, Alberta, Canada, 12–14 June, Paper 2001-061.

Escobar, M. A., Valera, C. A., Perez, R. E., 1997, "A Large Heavy Oil Reservoir in Lake Maracaibo Basin: Cyclic Steam Injection Experiences," SPE 37551, presented at the Int. Therm. Oper. Symp., held in Bakersfield, California, 10–12 February.

Escojido, D. M., Urribarri, O., Gonzalez, J., 1991, "Part 1: Transportation of Heavy Crude Oil and Natural Bitumen," presented at 13th World Petroleum Congress, 20–25 October, Buenos Aires, Argentina, and references therein.

Gates, I. D., Chakrabarty, N., 2013, "Optimization of Steam Assisted Gravity Drainage in McMurray Reservoir", *J. Can. Pet. Technol.*, 45(9), 54–62.

Govind, P. A., Das, S. K., Srinivasan, S., Wheeler, T. J., 2008, "Expanding Solvent SAGD in Heavy Oil Reservoirs", presented at Int. Thermal Operations and Heavy Oil Symp. held in Calgary, Alberta, Canada, 20–23 October.

Guinand, P., Ruiz, F., Mago, R., Hernandez, A., Ospina, R., Soto, R., Mendez, O., 2011, "Drilling the First SAGD Wells in the Orinoco Oil-Belt Bare Field: A Case History", presented at SPE Western Venezuela Section South American Oil and Gas Congress held in Maracaibo, Venezuela 18–21 October.

Guo, K., Li, H., Yu, Z., 2016, "In-Situ Heavy and Extra-Heavy Oil Recovery: A Review", *Fuel*, 185, 886–902, and references therein.

Gupta, S. C., Gittins, S. D., 2006, "Christina Lake Solvent Aided Process Pilot", *J. Can. Pet. Technol.*, 45(9), 15–18.

Gupta, S., Gittins, S., Picherack, P., 2005, "Field Implementation of Solvent Aided Process", *J. Can. Pet. Tech.*, 44(11), 8–13.

Hart, A., 2014, "A Review of Technologies for Transporting Heavy Crude Oil and Bitumen via Pipelines", *J. Pet. Exp. Prod. Technol.*, 4(3), 327–336.

Hedden, R., Verlaan, M., 2014, "Recovery Improvement by Solvent Co-Injection in a Vertical Well Steam Drive", presented at the World Heavy Oil Congress WHOC14-106, Calgary, Alberta, Canada, 7–9 September.

Hedden, R., Verlaan, M., Lastovka, V., 2014, "Solvent Enhanced Steam Drive", SPE No. 169070 presented at SPE Improved Oil Recovery Symposium, Tulsa, OK, 12–16 April.

Hong, K. C., 1994, *Steamflood Reservoir Management*, Pennwell Books, Tulsa, Oklahoma, p 20, and references therein.

Huc, A.-Y., Ed., 2011, "Heavy Crude Oils: From Geology to Upgrading: An Overview", Technip, Paris, France, p 133.

Istchenko, C., Gates, I. D., 2012, "The Well-Wormhole Model of CHOPS: History Match and Validation", presented at the SPE Heavy oil Conf. Canada held in Calgary, Alberta, Canada, 12–14 June.

Jha, R. K., Kumar, M., Benson, I., Hanzlik, E., 2012, "New Insights into Steam-Solvent Co-injection Process Mechanism", SPE No. 159277, presented at the SPE Annual Technical Conf. and Exhibition held in San Antonio, Texas, USA, 8–10 October.

Kalgaonkar, R., Chang, F., Ballan, A. N. A., Abadi, A., Tan, X., 2017, "New Advancements in Mitigating Sand Production in Unconsolidated Formations", SPE No. 188043, presented at SPE Kingdom of Saudi Arabia Annual Technical Symposium and Exhibition, Dammam, Saudi Arabia, 24–27 April.

Kelland, M. A., 2009, *Production Chemicals for the Oil and Gas Industry*, CRC Press, Boca Raton, FL, pp 291–312 and references therein.

Kilpatrick, P. K., 2012, "Water-in-Crude Oil Emulsion Stabilization: Review and Unanswered Questions", *Energy Fuels*, 26(7), 4017–4026.

Kokal, S., 2005, "Crude Oil Emulsion:, A State-of-the-Art Review", SPE No. 77497, SPE Prod. Facilities, February, pp 5–13.

Kopper, C. C. R., Decoster, E., Guzmán-Garcia, A., Huggins, C., Knauer, L., Minner, M., Kupsch, N., Linares, L. M., Rough, H., Waite, M., 2002, "Heavy-Oil Reservoirs", *Oilfield Rev.*, Autumn, 30–51.

Langevin, D., Poteau, S., Hénaut, I., Argillier, J. F., 2004, "Crude Oil Emulsion Properties and Their Application to Heavy Oil Transportation", *Oil Gas Sci. Technol. Rev. IFP*, 59(5), 511–521.

Leaute, R. P., 2002, "Liquid Addition to Steam for Enhancing Recovery (LASER) of Bitumen with CSS: Evolution of Technology from Research Concept to a Field Pilot at Cold Lake", CIM/CHOA 79011, Presented at Int. Therm. Oper. Symp., held in Calgary Alberta, Canada, 4–7 November.

Leaute, R. P., Carey, B. S., 2007, "Liquid Addition to Steam for Enhancing Recovery (Laser) of Bitumen with CSS: Results from the First Pilot Cycle", *J. Can. Pet. Technol.*, 46(9), 22–30.

Lin, L., Ma, H., Zeng, F., Gu, Y., 2014, "A Critical Review of the Solvent-Based Heavy Oil Recovery Methods", SPE No. 170098 presented at the SPE Heavy Oil Conf. Canada, Calgary, Alberta, Canada, 10–12-June, and references therein.

Mahmoudi, M., Fattahpour, V., Nouri, A., Yao, T., Baudet, B. A., Leitch, M., Fermaniuk, B., 2016, "New Criteria for Slotted Liner Design for Heavy Oil Thermal Production", SPE No. 182511, presented at SPE Thermal Well Integrity and Design Symposium, Banff, Alberta, Canada, 28 November–1 December.

Marfo, S. A., Appah, D., Joel, O. F., Ofori-Sarpong, G., 2015, "Sand Consolidation Operations, Challenges, and Remedy", SPE No. 178306, presented at SPE Nigeria Annual International Conference and Exhibition, Lagos, Nigeria, 4–6 August.

Nasr, T. N., Beaulieu, G., Golbeck, H., Heck, G., 2009, "Novel Expanding Solvent-SAGD Process ES-SAGD", *J. Can. Pet. Technol.*, 42(1), 13–16.

Nasr, T. N., Isaacs, E. E., 2001, "Process for Enhancing Hydrocarbon Mobility Using a Steam Additive", US Patent 6,230,814.

NIPER, 1986, *Enhanced Oil Recovery Information Booklet as Part of the U.S. Department of Energy (DOE) EOR Research Program*, downloaded from www.netl.doe.gov/research/oil-and-gas/enhanced-oil-recovery/eor-process-drawings on 29 March 2017.

Nuñez, G. A., Rivas, H. J., Rodriguez, D. J., Layrisse, I. A., 1995, "Development of a New Technology: Profiting from Temporary Setbacks During Scale-Up", *J. Pet. Technol.*, 47(5), 400, and references therein.

Orr, B., 2009, "ES-SAGD; Past, Present, and Future. Paper", SPE 129518 presented at SPE Annual Tech, Conf. Exhib. New Orleans, Louisiana, 4–7 October.

Ovalles, C., Rico, A., Pérez-Pérez, A., Hernández, M., Guzman, N., Anselmi, L., Manrique, E., 2002, "Physical and Numerical Simulations of Steamflooding in a Medium Crude Oil Reservoir, Lake Maracaibo, Venezuela", presented in SPE/DOE 13th Symposium on Improved Oil Recovery, Tulsa, Oklahoma, 13–17 April.

Ovalles, C., Rogel, E., Segerstrom, J., 2011, "Improvement of Flow Properties of Heavy Oils Using Asphaltene Modifiers", SPE-146775, presented at SPE Annual Technical Conference and Exhibition, Denver, CO, 30 October–2 November.

Pasicznyk, A., 2001, "Evolution Toward Simpler, Less Risky Multilateral Wells", SPE No. 67825, presented at the SPE/IADC Drilling Conference held in Amsterdam, the Netherlands, 27 February–1 March.

Pérez-Pérez, A., Ovalles, Alvarez, C. E., Salas, K., 2002, "Simulación Numérica de Inyección Alternada de Vapor en un Pozo Horizontal en un Yacimiento de Crudos Pesados Del Lago de *Maracaibo*", presented at the IV Congreso Latinoamericano de Ingeniería del Petróleo INGEPET, Lima, Perú, 5–8 November.

Prats, M., 1982, *Thermal Recovery, SPE Monograph Series*, Chapter 2, Society of Petroleum Engineers, New York, pp 6–14.

Redford, D. A., McKay, A. S., 1980, "Hydrocarbon-Steam Processes for Recovery of Bitumen from Oil Sands", SPE No. 8823 presented at SPE/DOE Enhanced Oil Recovery Symposium, 20–23 April, Tulsa, Oklahoma.

Rimmer, D. P., Gregori, A. A., Hamshar, J. A., Yildirim, E., 1992, "Pipeline Emulsion Transportation for Heavy Crude Oils", *Adv. Chem.*, 231, Chapter 8, pp 295–312 and references therein.

Rivas, H., Colmenares, T., 1999, "Orimulsion®: Nuevas Generaciones y Perspectivas Futuras", *Vision Technologica*, Edicion Especial, 49–60.

Rivas, H., Gutierrez, X., Silva, F., Chirinos, M., 2003, "Sobre Emulsiones de Bitumen en Agua", *Acta Cient. Venez.*, 54(3), 216–234.

Rivas, O. R., Boccasdo, G., 1994, "Transient Analytical Modelling of Cyclic Steam Injection", SPE No. 27060, presented at the 3rd Latin American/Caribbean Petroleum Engineering Conference held in Buenos Aires, Argentina, 27–29 April.

Robles, J., 2001, "Application of Advanced Heavy-Oil-Production Technologies in the Orinoco Heavy-Oil-Belt", SPE No. 69848 presented at SPE International Thermal Operations and Heavy Oil Symposium, Porlamar, Margarita Island, Venezuela, 12–14 March.

Robles, J., Perez, M., Bettenson, J., Nobel, E., 2011, "Design and Application of Charge PCP Systems in High GVF Heavy Oil Wells", SPE No. 153038.

Rojas, A. R., 2001, "Orinoco Belt, Cerro Negro Area: Development of Downhole Diluent Injection Completions", SPE No. 69433, presented at SPE Latin American and Caribbean Petroleum Engineering Conference, Buenos Aires, Argentina, 25–28 March.

Salager, J. L., Briceño, M. I., Bracho, C. L., 2001, "Heavy Hydrocarbon Emulsions - Making Use of the State of the Art in Formulation Engineering", In: *Encyclopedic Handbook of Emulsion Technology*, J. Sjöblom, Ed., Chapter 20, Marcel Dekker, New York, pp 455–495.

Saniere, A., Hénaut, I., Argillier, J.-F., 2004, "Pipeline Transportation of Heavy Oils, a Strategic, Economic and Technological Challenge", *Oil Gas Sci. Tech. Rev. IFP*, 59(5), 455–466.

Schenk, C. J., Cook, T. A., Charpentier, R. R., Pollastro, R. M., Klett, T. R., Tennyson, M. E., Kirschbaum, M. A., Brownfield, M. E., Pitman, J. K., 2009, "An Estimate of Recoverable Heavy Oil Resources of the Orinoco Oil Belt, Venezuela", U.S. Geological Survey Fact Sheet 2009-3028, downloaded from https://pubs.usgs.gov/fs/2009/3028/on 17 December 2016.

Sharpe, H. N., Richardson, W. C., Lolley, C. S., 1995, *Representation of Steam Distillation and In-Situ Upgrading Processes in a Heavy Oil Simulation*, SPE 30301 presented at the SPE Int. Heavy Oil Symp., Calgary, AB, Canada, 19–21 June.

Smalley, C., 2000, "Heavy Oil and Viscous Oil", In: *Modern Petroleum Technology*, R. A. Dawe, Ed., John Wiley and Sons Ltd., West Sussex, England, Volume 1, pp 409–435.

Smith, V. H., 1987, *Petroleum Engineering Handbook*, Bradley, H. B., Ed. Society of Petroleum Engineers, Richardson, TX, 1992, pp 12-1–12-44.

Speight, J. G., 2007, *The Chemistry and Technology of Petroleum*, 4th Ed., CRC Press, Boca Raton, FL, p 188.

Stark, S. D., 2013, "Cold Lake Commercialization of the Liquid Addition to Steam for Enhancing Recovery (LASER) Process", IPTC 16795, presented at the Int. Pet. Tech. Conf., Beijing, China, 26–28 March.

Strausz, O. P., Lown, E. M., 2003, *The Chemistry of Alberta Oil Sands, Bitumens and Heavy Oils*, Alberta Energy Research Inst., Calgary, Alberta, Canada, pp 21–26.

Worth, K., Quale, H., 2010, "Joint Implementation of Vapour Extraction – Advances in Enhanced Oil Recovery", Paper 2012-249, presented at the World Heavy oil Cong., Aberdeen, Scotland, 10–13 September.

Zhao, L., Law, D. H.-S., Nasr, T. N., Coates, R., Golbeck, H., Beaulieu, G., Heck, G., 2005, "SAGD Wind-Down: Lab Test and Simulation", *J. Can. Pet. Tech.*, 44(1), 49.

4 Fundamentals of Heavy Oil Upgrading

Before we start discussing the state-of-the-art in subsurface upgrading (SSU) of heavy oils and bitumens (HO/B), the fundamental elements of heavy oil upgrading are presented. This chapter is addressed to our upstream and midstream colleagues. It represents an introduction to the HO/B upgrading. The idea is that by reading these concepts, the understanding of the chemistry and processes of subsurface upgrading can be made more accessible and straightforward. If more information is needed, please check the references included at the end of this chapter. The definition of upgrading and reasons to upgrade HO/B were discussed in Section 1.3. As in the preceding chapter, only the processes closely related to SSU are described.

Many of HO/B upgrading technologies have been successfully applied in refineries across the globe and some upstream facilities in Canada and Venezuela. In this chapter, the chemistry and mechanistic pathways involved in the HO upgrading processes are discussed. Also, the process flow diagrams, catalyst compositions, and catalytic pathways are described as well. Comparison between the different upgrading technologies in terms of conversion, quality of products, and cost are presented. This chapter discusses the stability of the upgraded products, the analytical methodologies used to measure it, and the factors that affect it. The last section shows the residue conversion and stability needed for transportation of heavy crude oils and bitumens after the subsurface upgrading process.

Figure 4.1 shows a general scheme of the routes and a few examples used for the upgrading of heavy oils and bitumens that have been investigated for potential application at subsurface conditions. Due to their relatively low molar hydrogen-to-carbon ratio (H/C = 1.5–1.6, Table 2.2), there are two logical pathways to increase their hydrogen content and improve the properties of HO/B, i.e., Carbon Rejection and Hydrogen Addition [Noguchi 1991, Huc 2011, Gray 1994, 2015, Colyar 2009 Solari and Baumeister 2007, Speight 2007, Ramirez-Corredores 2017]. As previously discussed in Figure 1.9 and Section 1.7, the first pathway involves a step in which an upgraded product with an H/C ratio greater than 1.6 is obtained along with a residue, coke, or asphaltenes with H/C in the 0.8–1.4 range. The generation of the later material with higher carbon content than the starting material (HO/B) is the reason why this route is called Carbon Rejection. Several examples can be found in the literature (Figure 4.1) that involve physical separation (steam distillation, solvent deasphalting, etc.) and thermal conversion (Visbreaking, Delayed and Fluid Coking) technologies. Even though the physical separation processes do not involve chemical transformation *per se*, they continue to be very attractive due to their simplicity and relatively lower cost. These technologies are discussed in Section 4.1.

Alternatively, the HO/B can be upgraded by hydrogen addition technologies to yield a product with an H/C ratio higher than 1.6 using a thermal (no catalyst) or catalytic route (Figure 4.1). A few examples of these pathways are hydrovisbreaking, hydrocracking, hydrotreating, hydroprocessing, and the use of slurry catalysts. These technologies are discussed in Section 4.2.

4.1 CARBON REJECTION

As mentioned, there are two types of carbon rejection routes for the upgrading of heavy oils and bitumens, i.e., Physical Separation and Thermal Conversion (see Figure 4.1). These concepts have been used for the subsurface of upgrading of HO/B and are described in Chapters 5, 6, and 7.

As mentioned, the distillation route does not involve chemical transformations of the crude oil. However, it was the first method by which petroleum was refined since the early days of the industrial era. All refineries have different types of distillation units, and its use is widespread throughout

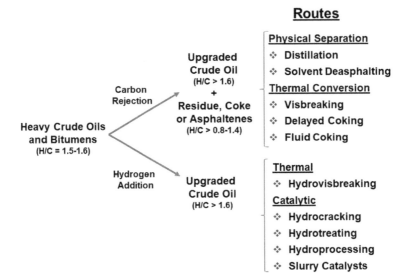

FIGURE 4.1   Routes used and a few examples for the upgrading of heavy crude oils and bitumens.

the world. There are several types of distillation, and they depend on the cut-off point of the products, i.e., atmospheric and vacuum distillation. More information on this process can be found elsewhere [Speight 2007, Ramirez-Corredores 2017].

4.1.1   Solvent Deasphalting

As in distillation, the solvent deasphalting route (SDA) does not involve chemical transformation. It is based on the precipitation of asphaltenes by the addition of a paraffinic solvent ("anti-solvent" or "precipitant solvent"). In this route, the solvent-to-oil ratio is generally from 4:1 to 10:1 v/v, and the paraffins used are from propane to hexane. Higher solvent/oil ratios give cleaner separation of the more valuable products such as the deasphalted oil (DAO).

Solvent deasphalting has the advantage of being a relatively low-cost process with flexibility over a wide range of feeds and products. The process has excellent selectivity to generate a relatively easy to process stream (DAO) and a fraction rich in paraffin-insoluble and coke precursor compounds (asphaltenes). However, this process is less selective to sulfur and nitrogen [Speight 2017].

Consistent with Figure 2.9, as the number of carbon atoms of the paraffinic solvent increases, the weight percentages of insolubles (asphaltenes) decrease [Huc 2011, Ramirez-Corredores 2017]. Similarly, the asphaltene yield increases as the solvent-to-bitumen ratio increases. As seen in Figure 2.12, the yield of C7-insolubles increases with the heptane-to-bitumen ratio [Alboudwarej *et al.* 2002]. A similar effect has been reported by other authors [Speight 2007]. As shown in Figure 2.11, the percentages of asphaltenes decreases as the temperature increases which is consistent with the known behavior of ideal and regular solutions and with the reduction of intermolecular association upon heating [Andersen 1995, Ovalles *et al.* 2015a].

In the SDA process, typical temperature range is from 50 to 235°C with pressures of 20–40 bar. In general, the primary utility cost is solvent recovery as the other separation units are relatively inexpensive and easy to operate [Noguchi 1991]. For this reason, the separation of feed and solvent is carried out above the critical temperature of the later material, which in turn, minimizes the energy costs [Huc 2011].

Solvent deasphalted processes are offered commercially by four licensors [Ramirez-Corredores 2017]. Those are Residuum Oil Supercritical Extraction (ROSE) process licensed by Halliburton-KBR, Demex process licensed by UOP, Solvahl process by Axens/Institut Francais Du Petrol (IFP), and Advanced Supercritical Solvent Deasphalting Process (PASD). These commercially available

SDA processes differ in the used solvent, the mixing conditions, and the method for recovery of the separated fractions [Ramirez-Corredores 2017].

One of the best known and commercially proven technologies is the ROSE process. Figure 4.2 shows the block diagram and some of the feed and product characteristics [Noguchi 1991]. The full characterization of the feed and products is reported in Table 4.1 [Noguchi 1991].

In Figure 4.2 the residue is mixed with a solvent in a volumetric ratio of eight or nine to one and passed into the asphaltene separator [Noguchi 1991]. The precipitated asphaltenes are separated from the bottom and steam stripped to remove small quantities of dissolved solvent (not shown in Figure 4.2). The solvent-free asphaltenes (SDA Tar) are pumped to fuel oil blending or further processing due to their low specific gravity (–7.8°API), and high metal and sulfur contents. The main

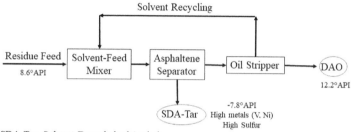

SDA-Tar: Solvent Deasphalted Asphaltenes
DAO: Deasphalted Oil

FIGURE 4.2 Block diagram for the Residuum Oil Supercritical Extraction (ROSE) process and some of the feed and product characteristics.

TABLE 4.1
Typical Properties of Feed and Products for the ROSE Process

Analysis	Feed Resid[a]	SDA TarAsphaltenes[b]	DAO[c] Resins	Oils
Yield[d]	100	18	30	52
Sp. gravity[e]	1.01	1.14	1.09	0.924
API[f]	8.6	–7.4	–1.7	21.6
H/C molar ratio[g]	1.135	1.216	1.329	1.602
Wt. % S[h]	1.5	2.2	1.8	1
C. Conradson[i]	19.1	25.4	16.8	3
Visc. (Pa s)[j]	6.36 (99°C)	2.3 (243°C)	1.4 (149°C)	636 (54°C)[k]
	0.21 (149°C)		0.03 (249°C)	65 (99°C)[k]

Noguchi [1991]
[a] Feed (See Figure 4.1).
[b] Solvent deasphalted (See Figure 4.1).
[c] Deasphalted Oil (See Figure 4.1).
[d] Weight percentage.
[e] Specific gravity measured at 15.5°C.
[f] API gravity calculated as (141.5/Sp. Gravity) – 131.5.
[g] Hydrogen-to-carbon molar ratio.
[h] Weight percentage of sulfur.
[i] Carbon Conradson as measured by ASTM D189-05.
[j] Viscosity in Pa s. Numbers in brackets are the temperatures at which the viscosity was measured in °C.
[k] Viscosity in cSt.

flow of solvent and extracted oil passes from the asphaltene separator to the oil stripper in which the deasphalted oil (DAO, 12.2°API) is obtained. Then the solvent is recycled to mix with the feed.

As shown in Table 4.1, the ROSE process leads to increases in the API of ~4°API (from 8.6 to 12.2 for the whole DAO) and in the H/C molar ratio (from 1.135 to 1.50), as well as reductions in the percentage of sulfur, carbon Conradson, and viscosity. In 2008, it was reported that 48 units were licensed or designed for a total capacity of 900 MBD, and 25 units were operating [Ramirez-Corredores 2017].

Finally, the economics of SDA rely heavily on the ability of the refinery to further upgrade the DAO and the differential values between the cutter stocks and the high-sulfur residual oil. Also, the disposal of the SDA Tar should be made as inexpensive as possible. For the latter process, several solutions have been used such as dilution with the residual oil, cogeneration of steam and power, and gasification to hydrogen production [Speight 2017].

4.1.2 Thermal Cracking and Visbreaking

One of the first processes used for the upgrading of HO/B is thermal cracking. It involves the breaking of chemical bonds to generate lighter hydrocarbons via decomposition at high temperatures or pyrolysis. This phenomenon can occur at temperatures as low as 200°C. However, at these low thermal conditions, the conversion of the petroleum-containing compounds may take extended periods of time from months and years to thousands or millions of years [Huc 2011]. At refineries, temperatures in the order of 380–500°C are typically used to achieve the necessary conversion levels in a reasonable amount of time (from minutes to hours). Examples of these processes are Visbreaking and Delayed Coking.

The chemistry of low-, medium-, and high-temperature cracking is fundamentally the same. The reaction mechanism involves the generation, propagation, and termination of free radicals. A free radical is an atom or group of atoms possessing an unpaired electron. These species are very reactive, and their chemistry has been intensely studied since their discovery in the 1930s. In the next section, a description of the thermal cracking reactions is presented as an introduction to the subsurface upgrading processes that utilize this route.

4.1.2.1 Thermal Cracking Reactions

Hydrocarbon thermal cracking leads to the formation of free radicals. These reactions are endothermic and involve three fundamental steps. Those are Initiation, Propagation, and Termination.

In the Initiation step, the homolytic cleave or homolysis of a chemical bond takes place with the subsequent generation of two free radicals as shown in Equation 4.1.

$$A : B \xrightarrow[\text{Cleavage}]{\text{Homolytic}} A^\bullet + B^\bullet \qquad (4.1)$$

During the homolytic cleave, the two bonding electrons of the molecule A:B are evenly allocated to each of the combined atoms with the simultaneous generation of two free radicals (A$^\bullet$ and B$^\bullet$). The main bonds of interest during upgrading of HO/B are the carbon–hydrogen, carbon–carbon, and carbon–sulfur bonds. Other bonds of interest are carbon–nitrogen and carbon–oxygen. Table 4.2 shows the bond dissociation energy at 289K for the homolytic cleavage reactions of selected compounds [Dean 1999]. These values give indications of the difficulty of breakage of each bond and serve to help in understanding the chemistry of thermal cracking and Visbreaking.

First, the homolysis of C–H (Equation 4.2), i.e., the homolytic cleavage of a C–H bond, depends strongly on the type of carbon involved:

$$R_1 - R_2 \rightarrow R_1^\circ + R_2^\circ \qquad (4.2)$$

Where R$_1$–R$_2$ is the parent hydrocarbon and R$^\bullet$ and R$_2$$^\bullet$ are hydrocarbon free radicals.

As shown in Table 4.2, the dissociation energy increase and thus, the ease of bond breakage decrease in the order shown in Equation 4.3:

$$\text{Benzylic C.} > \text{Phenyl C.} > \text{Tertiary C.} > \text{Secondary C.} > \text{Primary C.} \qquad (4.3)$$

For the cracking of the C–C bond, the data showed in Table 4.2 indicate that the generation of tertiary and benzylic radicals are more favored than the breakage of other C–C bonds (primary and secondary). Also, the homolysis of the carbon at β-position of an aromatic ring is more favored (301 kJ/mol) than that at the alpha position (389 kJ/mol).

As reported in the literature, the homolytic cleavage of carbon–carbon bonds (C–C) in aromatic compounds is very difficult because of the resonance stabilization energy, which contributes an extra 25 kJ/mol to C–C bonds in benzene [Gray 2015]. Polynuclear aromatic compounds have slightly lower stabilization energies than benzene. For example, naphthalene has average stabilization energy of 23 kJ/mol of C–C bonds. Resonance stabilization renders aromatic bonds unbreakable at normal process temperatures (< 600°C) unless the aromatic character is first eliminated by hydrogenation [Gray 2015].

The cracking of a complex hydrocarbon mixture such as heavy oil and bitumens is determined by the reactivity of its components. In general, a hierarchy of cracking reactions can be established from the most reactive to the least reactive [Gray 1994, Blanchard and Gray 1997]:

TABLE 4.2
Bond Dissociation Energy for the Homolytic Cleavage Reactions of Selected Compounds

Bond	Reactant[a]	Radicals Generated[b]	ΔH at 298K (kJ/Mol)[c]	Type[d]
C–H	$H_3C–H$	$CH_3{}^\bullet + H^\bullet$	435	Primary C.
	$H–CH_2CH_2CH_3$	$\bullet CH_2CH_2CH_3 + H^\bullet$	410	Primary C.
	$H–CH=CH_2$	$\bullet CH=CH_2 + H^\bullet$	427	Secondary C.
	$H–C_6H_5$	$\bullet C_6H_5 + H^\bullet$	431	Phenyl C.
	$H–CH_2C_6H_5$	$\bullet CH_2C_6H_5 + H^\bullet$	356	Benzylic C.
C–C	$H_3C–CH_3$	$H_3C^\bullet + {}^\bullet CH_3$	368	Primary C.
	$H_3C–C_6H_5$	$H_3C^\bullet + {}^\bullet C_6H_5$	389	Carbon in alpha position
	$H_3C–CH_2C_6H_5$	$H_3C^\bullet + {}^\bullet CH_2C_6H_5$	301	Carbon in beta position
	$(CH_3)_3C–C(C_6H_5)_3$	$(CH_3)_3C^\bullet + {}^\bullet C(C_6H_5)_3$	63	Tertiary and benzylic C.
C–S	$H_3C–SH$	$H_3C^\bullet + {}^\bullet SH$	305	Thiol
	$H_3C–SC_6H_5$	$H_3C^\bullet + {}^\bullet SC_6H_5$	285	Phenyl sulfide
	$H_3C–SCH_2C_6H_5$	$H_3C^\bullet + {}^\bullet SCH_2C_6H_5$	247	Benzylic sulfide
C–N	$H_3C–NH_2$	$H_3C^\bullet + {}^\bullet NH_2$	331	Alkyl amine
	$H_3C–NHC_6H_5$	$H_3C^\bullet + {}^\bullet NHC_6H_5$	285	Phenyl alkyl amine
	$H_3C–N(CH_3)C_6H_5$	$H_3C^\bullet + {}^\bullet N(CH_3)C_6H_5$	272	Phenyl dialkyl amine
	$(C_6H_5)H_2C–NHCH_3$	$(C_6H_5)H_2C^\bullet + {}^\bullet NHCH_3$	289	Benzyl alkyl amine
C–O	$H_3C–OCH_3$	$H_3C^\bullet + {}^\bullet OCH_3$	335	Alkyl ether
	$H_3C–OC_6H_5$	$H_3C^\bullet + {}^\bullet OC_6H_5$	381	Phenyl ether

Dean [1999]

[a] Where "–" denotes the bond broken during the homolytic cleavage.

[b] Radical species generated where "•" denotes the unpaired electron.

[c] Dissociation energy at 298K.

[d] Type of atom(s) or compound(s) involved in the homolytic cleavage reaction. Where C=Carbon atom.

$$\text{Paraffins} > \text{Linear Olefins} > \text{Naphthenes} > \text{Cyclic Olefins} > \text{Aromatics} \qquad (4.4)$$

Consistent with the values for C–C bonds, the formation of free radicals containing benzylic- and phenyl-sulfides, amines, and ethers have lower dissociation energies that the alkylates counter-parts (Table 4.2). Thus, during cracking reaction, the former species react faster than the latter.

In the Propagation step, the free radical generated by homolysis ($R_1\bullet$) can further react with stable molecules to form new free radicals ($R_3\bullet$) via H-transfer reaction (Equation 4.5). This reaction is very fast and leads to the transferring of a radical from one molecule to another.

$$R_1^\circ + R_3 - H \rightarrow R_1 - H + R_3^\circ \qquad (4.5)$$

Where R_1 and R_3 are hydrocarbons.

During the propagation step, one of the main reactions is the decomposition of the free radical into olefins and the generation of a new free radical of lower molecular weight. An example of this reaction can be seen in Equation 4.6.

$$R_4CH_2 - CH_2 - CH_2^\circ \rightarrow R_4CH_2^\circ + CH_2 = CH_2 \qquad (4.6)$$

This type of reaction is referred to as β-scission and leads to the production of α-olefins. The newly formed primary free radical ($R_4CH_2\bullet$) can continue to undergo β-scission to give ethylene and a smaller radical until, ultimately, a methyl radical is formed [Gates *et al.* 1979]. For alkyl aromatics, the reaction is swift and leads to the production of α-olefins to an aromatic ring (Equation 4.7).

$$C_6H_5CH_2CH_2CH_2^\circ \rightarrow CH_3^\circ + C_6H_5CH = CH_2 \qquad (4.7)$$

Depending on the type of free radical and configuration of the hydrocarbon molecules, other chemical reactions can occur such as radical addition and rearrangement [Gray 2002], isomerization, cyclization, and aromatic condensation [Blanchard and Gray 1997, Huc 2011]. Also, secondary reactions of the primary products are plausible leading to a chain reaction mechanism during the propagation step.

Termination is the last step of thermal cracking and involves the recombination of two radicals to form a bond and a non-radical molecule. Equation 4.8 depicts this type of reaction.

$$R_1^\circ + R_2^\circ \rightarrow R_1 - R_2 \qquad (4.8)$$

Where R_1 and R_2 are hydrocarbons.

Under thermal cracking and Visbreaking conditions, the recombination of two radicals is a very thermodynamically downhill event due to the low concentration of radical species and the small probability of two radicals colliding with one another. In other words, the Gibbs free energy barrier is very high for this reaction, mostly due to entropic rather than enthalpic considerations. Thus, termination reactions occur during cool down and storage in which initiation and propagation processes are vastly reduced and enough time has elapsed for the recombination of the radicals (Equation 4.8).

It is important to mention that the tendency to bond breaking reaction, also known as crackability, tends to increase with the molecular weight (or boiling range) of the starting material. Larger molecules have more bonds which can easily rupture which, in turn, increase the probability of breaking [Gray 1994, Blanchard and Gray 1997].

Similar cracking chemistry occurs during the thermal-only subsurface upgrading processes. For example, during steam injection or *in-situ* combustion recovery processes, the operating temperature at downhole condition leads to thermal cracking of the crude oil molecules and the concomitant generation of lighter products. The literature review on this subject is presented in Chapters 6, 7, and 9.

4.1.2.2 Visbreaking Process

This process was developed in the late 1930s to produce more desirable and valuable products and involves the viscosity reduction and hydrocarbon-bond breaking, hence, the name of Visbreaking. It is a relatively mild form of thermal cracking (temperature range of 450–480°C) operating at a relatively low pressure. The thermal reactions are not allowed to proceed to completion and are interrupted by quenching [Joshi et al. 2008, Huc 2011, Speight 2012]. During Visbreaking, most of the chemical reactions are thermally driven so there is no selectivity in bond cleavage.

There are two types of Visbreaking technologies, "Coil" and "Soaker" type [Noguchi 1991, Joshi et al. 2008]. The first process consists of a reaction furnace and a fractionation section. The feedstock passes through the reaction furnace where the cracking reactions take place at 480°C. The effluent is quenched and then separated into fractions such as gas, gasoline, gasoil, and visbroken residue. In the "Soaker" type, the furnace is operated under milder conditions (450°C), and a soaking drum is used to extend the residence time to similar conversion as the "Coil" counterpart [Noguchi 1991]. Due to the lower temperature, the "Soaker" type technology is reported to have 30% lower fuel consumption of the latter [Noguchi 1991].

The typical yield and properties of the visbroken products using an Iranian VR as feed can be seen in Table 4.3. The data are almost the same for both the "Coil" and "Soaker" type of Visbreaking processes [Noguchi 1991]. As shown, there is the generation of ~20% of distillable materials (gases, gasoline, and middle distillates) in the visbroken products. More importantly, there is a three-fold reduction in the viscosity of the residue in comparison with the feed (from 100,000 to 35,000). This feature is one of the most significant advantages of the Visbreaking process and the reason why there is continued interest in applying it downhole at subsurface conditions. This topic is covered in Chapter 6.

On the other hand, one of the main disadvantages of the Visbreaking process is that there is very little or no changes in the specific gravity, API gravity, or sulfur content between the Visbreaking feed and the visbroken products (Table 4.3).

As mentioned previously, thermal cracking is not a selective process because all molecules tend to react according to their chemical structures and compositions and can interact strongly with each

TABLE 4.3

Yield and Properties of the Visbroken Products Using an Iranian VR as Feed

	Yield (wt. %)[a]	Specific Gravity[b]	°API Gravity[c]	Viscosity at 50°C (cSt)[d]	% S (wt. %)[e]
Iranian VR deed	–	1.013	8.2	100,000	3.5
Gases (C$_4^-$)	2.25	–	–	–	–
Gasoline (C$_5$-165°C)[f]	4.75	N-A	N-A	N-A	N-A
Middle distillates (165–350°C)[f]	13.1	0.864	32.3	2.3	2.1
Residue (350°C+)[f]	79.9	1.033	5.5	35,000	3.8

Noguchi [1991]

[a] Typical yield weight percentage.
[b] Specific gravity measured at 15.5°C.
[c] API Gravity calculated as (141.5/Sp. Gravity) – 131.5.
[d] Viscosity measured at 50°C in cSt.
[e] Weight percentage of sulfur.
[f] Distillate range.

other via H-transfer reaction. In Figure 4.3, the typical product distribution for the Visbreaking (480°C) and Delayed Coking (507°C) processes is presented. As seen, as temperatures increase, the product slate favors gases and coke. This figure and Table 4.3 are examples of the Boduszynski's Continuum Model as described in Section 2.3.4 in which the properties of the thermally treated feed increases continuously with regard to aromaticity, hydrogen deficiency, and tendency to form coke [Boduszynski 2015]. More details on the product distributions and reaction kinetics can be found elsewhere [Joshi *et al.* 2008, Huc 2011, Speight 2012]:

Generally, in any upgrading unit, the economics improve as the conversion of the residue is enhanced. However, the severity of Visbreaking is limited by the stability of the final product as asphaltenes become destabilized due to an increase of light saturates in the cracked product and the aromaticity of the cracked asphaltenes. This topic is discussed in Section 4.3.

Approximatively 130 Visbreaking facilities have been built worldwide [Huc 2011]. The main variables that affect the economics of the process are the fuel and power costs due to the relatively high temperature of operation of the coil or soaker units. Despite the significant limitations outlined above, the Visbreaking remains an important, relatively inexpensive, bottom-of-the-barrel upgrading process in many areas of the world

4.1.3 COKING AND DELAYED COKING

Coking Processes were invented in the 1930s and are well-established technologies. They are the most widely used processes for the upgrading of petroleum resids with an installed capacity of more than 6 MMBD worldwide [Huc 2011]. As shown in Figure 4.4, coking can be visualized as a higher severity Visbreaking process. While the latter generates a small amount of gases and pitch, the former is operated to maximize the production of distillable materials (gases, gasoline, and middle distillates) with the cogeneration of solid (coke) or semi-solid (pitch) byproducts. Thus, coking is a relatively severe form of thermal cracking (temperature range of 490–510°C) also operating at relatively low pressure.

One of the main advantages of coking processes is that they are insensitive to feedstock characteristics such as heteroatoms content (sulfur, nitrogen, oxygen), metals concentrations, and the presence of asphaltenes. These processes can handle many types of feeds (e.g., atmospheric and vacuum residues, bitumen, solvent deasphalted tars, whole heavy crudes, fluid catalytic cracked bottoms, etc.), and the liquid yields depend on the hydrogen-to-carbon ratio of the starting material.

As Visbreaking, the chemistry of high-temperature cracking involves the generation, propagation, and termination of free radicals. In the next section, a description of the coking

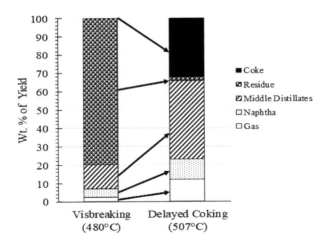

FIGURE 4.3 Product distribution for Visbreaking and coking processes and the average process temperature.

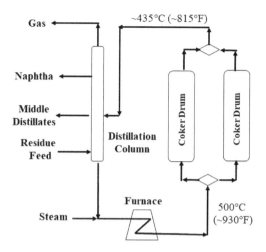

FIGURE 4.4 Flow diagram of the Delayed Coking.

mechanism is presented as an introduction to the subsurface upgrading processes that utilize this route.

4.1.3.1 Coking Mechanism

As described in Section 4.1.2, the coking mechanism proceeds through thermal cracking of C–H, C–C, C–S, C–N, and C–O bonds. The exact coking mechanism is complex which makes it somewhat difficult to provide a precise description of the changes and chemical pathways that are occurring inside the coker unit. At the severe coking temperature (490–510°C), several chemical reactions are taking place simultaneously such as dealkylation [Huc 2011], dehydrogenation [Speight 2007], cyclization [Reynierst *et al.* 1994], demethylation [Bozzanoa and Dente 2005], aromatization [Reynierst *et al.* 1994, Speight 2007], and aromatic condensation [Huc 2011]. Examples of the latter reactions are shown in Equations 4.9–4.14:

Dealkylation:

$$(4.9)$$

Dehydrogenation:

$$+ \quad H_2 \qquad (4.10)$$

Cyclization:

$$(4.11)$$

Demethylation:

$$+ \ \overset{\bullet}{C}H_3 \qquad\qquad (4.12)$$

Aromatization:

$$+ \ 2H_2 \qquad\qquad (4.13)$$

Aromatic condensation:

$$\longrightarrow \ \text{Coke} \qquad (4.14)$$

Reactions 4.9–4.13 lead to the growth of the carbon layer with the simultaneous generation of coke (Equation 4.14). It has been found [Reynierst *et al.* 1994, Bozzanoa and Dente 2005, Speight 2007, Huc 2011] that the primary reactions involve the dealkylation of the alkyl side chains attached to the aromatic nuclei to generate gases (C1–C10, CO_2, and olefins) and insoluble intermediates. Successive aromatization (Equation 4.13) and aromatic condensation (Equation 4.14) reactions of the latter lead to the formation of a dense-solid material known as coke. Most of the coking-forming reactions are endothermic and cause a drop in temperature of ~50°C in the coker drum.

Coking processes can also be visualized as a series of successive thermal reactions (Equation 4.15) going from the vacuum residue feed (VR) to coke, as evidenced by reduction of the hydrogen-to-carbon molar ratio. In Equation 4.15, the intermediate products are C7-insolubles or asphaltenes. This chemical pathway can be rationalized as another example of the Boduszynski's Continuum Model (Section 2.3.4) in which the hydrogen deficiency increases gradually and continuously as the petroleum fraction is thermally treated [Boduszynski 2015]. In fact, a phase separation kinetic model for coke formation has been reported by Wiehe, which considers the induction period before coke formation [Wiehe 1993]. If the thermal process continues, graphite is formed with an H/C molar ratio lower than 0.6 (Equation 4.15).

$$\text{VR} \xrightarrow{\Delta} \underset{\substack{\text{(Asphaltenes)}\\ 1.05\text{-}1\text{-}2}}{\text{C7-Insolubles}} \xrightarrow{\Delta} \underset{0.6\text{-}0.8}{\text{Coke}} \xrightarrow{\Delta} \underset{<0.6}{\text{Graphite}}$$

$$\text{H/C} = 1.4\text{-}1\text{-}5 \qquad\qquad\qquad (4.15)$$

Where Δ = thermal process.

The involvement of asphaltenes or C7-insolubles during coke and sediment formation has been reported during heavy oil upgrading. This reaction depends on the degree of polynuclear condensation of the feedstock, the average number of alkyl groups in the polynuclear aromatic system, and the H/C molar ratio of the asphaltenes [Storm *et al.* 1997, Speight 2007]. Additionally, polyaromatic containing heteroatoms (sulfur, nitrogen, and oxygen) are believed to be heavily involved in the coking mechanism and are known to increasing coking tendency and change the coke morphology [Siskin 2006].

Various types of commercial coking processes are Delayed Coking, Fluid Coking, and Flexicoking. The latter is a continuous fluidized bed technology which converts heavy residue to lighter more valuable product and substantially eliminates the coke production. The most common coking technology is Delay Coking which is described in the next section.

4.1.3.2 Delayed Coking

Delayed coking is a well-developed commercial process that operates on a semi-continuous basis at 480–515°C heating at 90 psi. Figure 4.4 shows the flow diagram of the Delayed Coking [Speight 2007]. The main facilities consist of a feed furnace, a fractionator, and two or more coke drums equipped with high-pressure water jets for docking [Noguchi 1991].

The AR or VR feeds are fed directly to the fractionator where they are flushed and combined with the recycle oil at the bottom. The combined feed is mixed with steam and pumped to the coker furnace where it is exposed to cracking temperature for a short time. The residence time is low enough, so no cracking takes place. Then, the heated feed/steam mixture is sent to the coke drums where most of the bond breaking and solid separation occur yielding the coke. The drums are switched over at preset time intervals (typically several hours) to remove the deposited coke by drilling a hole through the center and lateral cuts as well, with high-pressure water. The overhead vapor from the coke drums enters the fractionator to be separated into gas, naphtha, fuel oil, and recycle oil.

Table 4.4 reports the properties of the feed and yield of products for three different feeds using Delayed Coking [Noguchi 1991]. The yields and the qualities of the products are functions of the properties of the feedstock and operating conditions. The main feed properties that affect coke yield and quality are the Conradson Carbon, specific gravity, and level of contaminants. In general, as the pressure or recycle ratio increases, the gas and coke increase (see Figure 4.3) while the liquid yields decrease.

In 1983, Audeh and Yan carried out kinetic, tracer, and quenching studies and reported that the feed entering the coker at the incipient coking temperature undergoes a two-stage thermal decomposition to gas and liquid products and coke. In the first stage, thermal cracking takes place producing gas and liquid at a fast rate, with simultaneous aromatization reactions (e.g., Equation 4.13)

TABLE 4.4

Properties of the Feed and Product Yields for Delayed Coking

	Thermal Tar[a]	California Residue[b]	Midcontinent Residue[c]
Feedstock Properties			
Specific gravity	1.09	0.986	0.984
Wt.% sulfur	0.56	1.6	0.38
Conradson carbon (wt. %)	8.6	9.6	11.3
Product Yields (wt. %)			
Gas	18.1	12.0	6.5
Naphtha	0.9	15.7	16.0
Gasoil	21.1	50.7	56.5
Coke	59.9	21.6	21.0

Noguchi [1991].

[a] Outlet temperature = 507°C, drum pressure of 50 psi, and recycle ratio 1.08.

[b] Outlet temperature = 496°C, drum pressure of 60 psi, and recycle ratio 0.3 v/v.

[c] Outlet temperature = 487°C, drum pressure of 30 psi, and recycle ratio 0.1 v/v.

which result in the formation of non-volatile "semi-coke." The volatile compounds leave the coking zone rapidly. Steam injected with the feed (Figure 4.4) facilitates the removal of volatiles to minimize secondary cracking [Audeh and Yan 1983].

Hot feed continues to enter the drum and pushes the "semi-coke" upward with little or no mixing. Meanwhile, in the second stage, the "semi-coke" continues to undergo cracking/pyrolysis, leading to more gas, liquid, and coke, at a slower rate. The authors reported that, at 490°C, the reaction rate for the first stage is ~11-fold higher than that of the second stage [Audeh and Yan 1983].

Favorable economics has made coking the choice for primary upgrading at field sites [Ramirez-Corredores 2017]. Foster Wheeler, which offers Visbreaking, Solvent Deasphalting (SDA), and Delayed Coking technologies reported an economic comparison of these three processes under comparable conditions. The net present value was very similar for Visbreaking and SDA (US$427 and 391 MM, respectively) while coking almost duplicated this value (US$756 MM) [Elliott 2004]. The economics of Delayed Coking are driven by the differential between transportation fuels and high-sulfur residual fuel oil [Ramirez-Corredores 2017].

Finally, it is important to mention that all coke distillate product contains very high diolefins, olefins, nitrogen, and sulfur contents compared to those found in virgin or catalytically processed streams. Thus, severe hydrotreatment and/or hydrocracking of the coker fluids are mandatory before the latter can be used to make finished products [Huc 2011].

4.2 HYDROGEN DONATION

As examined in Figure 4.1, the other logical pathway to increase the hydrogen content and improve the properties of HO/B is hydrogen addition using thermal (no catalyst) or catalytic routes [Noguchi 1991, Huc 2011, Gray 1994, 2015, Solari and Baumeister 2007, Speight 2007, Ramirez-Corredores 2017]. A few examples of these pathways are hydrovisbreaking, hydrocracking, hydrotreating, hydroprocessing, and the use of slurry catalysts.

All the technologies mentioned above have potential applications in subsurface conditions (Chapter 7 and 8). Thus, the following sections cover the chemistry and mechanistic pathways involved in the HO upgrading processes via hydrogen addition. The process flow diagrams, catalyst compositions, and catalytic pathways are described as well. Comparisons between the different upgrading technologies in terms of conversion, quality of products, and cost are presented.

4.2.1 HYDROVISBREAKING

This process involves the use of hydrogen gas or hydrogen-donor compounds to enhance the properties of the upgraded products further. Figure 4.5 shows the block diagrams and a comparison between Visbreaking and Hydrovisbreaking processes [Noguchi 1991]. As seen, both technologies preheat the feed at temperatures as high as 480°C but operate at different pressures. Visbreaking uses ~350 psi whereas Hydrovisbreaking works at higher pressures (1,600 psi). For a Venezuelan crude oil, the presence of hydrogen or hydrogen donors leads to an increase of the residue conversion from 20% to 25% and viscosity reductions of the products from 2–3- to 3–3.5-fold [Noguchi 1991].

There are three Hydrovisbreaking processes germane to subsurface upgrading, and they differ from each other in the source of hydrogen and if they are purely thermal or catalytic. Those processes use hydrogen gas (thermal), hydrogen donor compounds (thermal), and water (catalytic), and they are described next.

4.2.1.1 Use of Hydrogen Gas

The Tervahl H process, offered by IFP/Axens/Technip, is one example of a commercially available Hydrovisbreaking technology [Peries *et al.* 1988, Noguchi 1991]. As shown in Figure 4.5, the feed and hydrogen-rich stream are heated and allowed to react by passing them through the soaker drum. The reaction products are mixed with recycled hydrogen and separated in the hot and cold

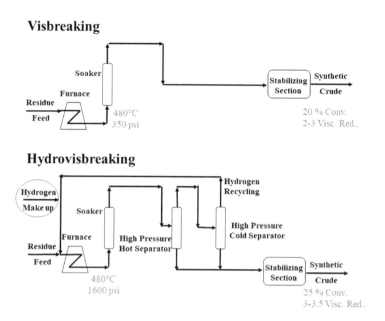

FIGURE 4.5 Block diagrams and comparison between Visbreaking and Hydrovisbreaking Processes.

separators. The liquids are sent to the stabilizer section where purge gas and synthetic crude are produced. As mentioned before, higher residue conversion and lower viscosity are obtained than when using the Visbreaking route. This process is proposed for those installations that need viscous oils to be transported or stored such as oil pipelines, oil terminals, and refineries, steam plants, or petrochemical plants. It is proposed that this type of process reduces the need for diluent, heating, tracing, and insulation. It is also suggested that the lower viscosity and pour point leads to lower pumping costs [Noguchi 1991].

Gray reports pilot plant data for Visbreaking and Hydrovisbreaking of Canadian Cold Lake bitumen [Gray 2015]. Table 4.5 shows that the first process reduces the viscosity from 55,000 cSt to 495 cSt at 20°C whereas the second provides a further reduction down to 340 cSt. The effects are even more pronounced in the percentages of sulfur removal and residue conversion. For Visbreaking, the S-removal and heavy fraction conversion with respect to the feed are 9% and 19%, respectively. Whereas for Hydrovisbreaking the values increase to 14% and 35%. It is important to point out the product stabilities were comparable, and the lengths of the furnace operating cycles were equivalent [Gray 2015].

Hydrovisbreaking has been considered several times for commercial applications, but its commercial uses have been limited for three reasons: First, the hydrogen gives an only limited improvement in comparison with Visbreaking. Second, the use of high hydrogen pressure increases the CAPEX and OPEX of the process, and third, the recovery of the hydrogen gas from the cracked gases to enable recycling of this valuable material adds additional complexity and cost to the process [Gray 2015]. The uses of H_2 or H_2-precursors for the subsurface upgrading of HO/B are discussed in Sections 7.1 (thermal) and 8.3 (catalytic).

4.2.1.2 Use of Hydrogen Donor Solvents

Hydrogen donor solvents for coal liquefaction were first used in Germany in the 1930s and 1940s. Since then, several processes have been developed such as Solvent Refined Coal (SRC), Exxon Donor Solvent (EDS), and the H-Coal [Alemán-Vázquez *et al.* 2016].

Hydrogen donors are chemical compounds that can easily transfer hydrogen to the HO/B. There is a variety of hydrogen donors reported in the literature which differs in the capacity for donating hydrogen, the rate of reaction, operating conditions, and cost [Alemán-Vázquez *et al.* 2016]. Two of

TABLE 4.5

Pilot Plant Data for Visbreaking and Hydrovisbreaking of Canadian Cold Lake Bitumen[a]

Property	Feed	Visbreaking	Hydrovisbreaking
°API gravity	10.3	12.2	14.7
Viscosity (cSt) at			
20°C	55,000	495	340
50°C	2,020	106	55
100°C	95	15	9.4
Wt. % sulfur[b]	4.19	3.80 (9%)	3.60 (14%)
Pour point (°C)	−3	−9	−15
Reaction yields			
< 375°C	26.2	34	41.5
375–500°C	19.1	21.5	23.1
> 500°C[c]	54.7	44.5 (19%)	35.4 (35%)

Gray [2015].

[a] Operating conditions were not the same for the two processes, but the product stabilities were comparable and then lengths of the furnace operating cycles were equivalent.

[b] Number in brackets are the percentages of sulfur removal vs. the feed.

[c] Number in brackets are the percentages of residue conversion.

the most commons hydrogen donors are 1,2,3,4-tetrahydronaphthalene, or tetralin and the decahydronaphthalene or decalin (cis or trans), and they react by the following pathway:

$$\text{(4.16)}$$

As seen for each mol of decalin that reacts, one mol of tetralin and two moles of hydrogen are available to be transferred to hydrogen receptors. Whereas in going from the tetralin to dihydronaphthalene, one additional mol of H_2 is generated. Finally, the production of the fully aromatic naphthalene leads to the generation of one mol of hydrogen. These reactions are very fast at Hydrovisbreaking conditions.

Bedell et al. [1993] evaluated the hydrogen transfer efficiency in coal liquefaction between three species of hydrogen-rich donors such as cyclic olefins, hydroaromatics (as tetralin), and cycloalkanes (naphthenes). They found out that for 30 min thermal reaction condition at 380°C, the amount of hydrogen release from an equivalent weight of donors was in the following order [Bedell 1993]:

$$\text{Cyclic olefins > hydroaromatics > cycloalkanes} \qquad (4.17)$$

Ancheyta et al. [Alemán-Vázquez et al. 2016] carried out a literature review in the use of hydrogen donors for the partial upgrading of heavy petroleum. The authors reported that several types of organic compounds can serve as useful hydrogen donors for upgrading HO/B. Partially hydrogenated polynuclear aromatics of pyrene, anthracene, phenanthrene, fluoranthene, and basic nitrogen compounds such as quinoline and benzoquinolines can function as effective hydrogen transfer agents and with better performance than conventional donors such as tetralin or decalin. For example, the hydrogen donation reaction of tetrahydroanthracene proceeds as follows:

$$+ \ 2H_2 \qquad (4.18)$$

The reason for the higher reaction rates for tetrahydroanthracene seems to be associated with the more favorable thermodynamics for high-number ring polynuclear aromatics (5–6) in comparison with lower number counterparts [Alemán-Vázquez *et al.* 2016]. Additionally, oxygen-containing compounds such as alcohols, cyclic ethers, and occasionally formic and ascorbic acids were found to be effective hydrogen donors. The choice of donor is based on the nature of the reaction, its availability, and solubility in the reaction medium.

Since the use of hydrogen donor for coal liquefaction, there have been several attempts to measure the hydrogen donor ability of either a specific compound or a given oil-derived solvent. Methods based on the reaction test with benzoyl peroxide followed by NMR spectroscopy [Le Roux *et al.* 1982], dehydrogenation of coal-derived solvents using sulfur as hydrogen acceptor [Aiura *et al.* 1984], ^{1}H-NMR spectroscopic data [Awadalla *et al.* 1985], and tetralin as a chemical probe [Li *et al.* 2016] have been reported in the literature. However, there is no universal tool to quantify the hydrogen donor ability of a given solvent. One plausible explanation could be the fact that hydrogen donation during thermal cracking reactions is strongly dependent on the process conditions, type of feed (coal, bitumen, heavy crude oil, etc.), and temperature and pressure. These operating conditions strongly affect the hydrogenation–dehydrogenation equilibria of the hydrogen-donating species as well as the ability of the receptor molecules to react [Alemán-Vázquez *et al.* 2016].

Table 4.6 shows the comparison between Visbreaking and Hydrovisbreaking using hydrogen gas and hydrogen donors in terms of reaction conditions (temperature and pressure), the percentage of residue conversion, viscosity reduction, and hydrogen consumption. Consistent with the results presented in Table 4.5, Hydrovisbreaking leads to higher residue conversion (25% vs. 20%) and viscosity reduction (0.3–0.5) than Visbreaking. These improvements are achieved at expense of hydrogen consumption, either in the gas phase or as donor solvent. However, in the presence of the latter, lower pressure can be used with the concomitant reductions in CAPEX and OPEX. Thus, hydrogen donors are a convenient way to transfer hydrogen without the need for a high-pressure system.

TABLE 4.6

Comparison between Visbreaking, Hydrovisbreaking Using Hydrogen Gas and Hydrogen Donors[a]

Process	Temp. (°C)[b]	Pressure (psi)[c]	% Residue Conv.[d]	Relative Viscosity[e]	Hydrogen Consumption[f]
Visbreaking	480	350	20%	1	–
Hydrovisbreaking using hydrogen gas	480	1,600	25%	0.3–0.5	Yes
Hydrovisbreaking using hydrogen donors	480	350	25%	0.4–0.5	Yes

[a] Data taken from the literature at approximatively the same reaction conditions [Noguchi 1991, Gray 2015].
[b] Temperature of the soaker drum.
[c] Pressure in the cracking section.
[d] Percentage of 540°F+ residue conversion vs. the feed.
[e] Relative viscosity of the visbroken product vs. Visbreaking Process.
[f] Hydrogen gas or donor solvent consumptions.

Several hydrogen donors derived from refinery streams containing multi-ring hydroaromatic components have been used as additives to improve the results obtained by Visbreaking and Hydrovisbreaking processes. Del Bianco and coworkers [1995] reported that the Visbreaking performance of Belaym VR could be substantially enhanced by using the 200–390°C fraction of fresh or hydrogenated fuel oils as hydrogen donor solvents. Improvements were measured regarding residue conversion, less olefinic and desulfurized naphtha fractions, and inhibition of coke formation.

Similarly, Chen et al. [2014] used the 350–420°C distillation cut of a coker gas oil as a hydrogen donor for the Visbreaking of Karamay VR. The authors found higher residue conversion, increase in the coke induction period (from 17 to 27 min), and a decrease in the liquid-coke particle size (from 0.307 to 0.186 μm) in the presence of the H-donor solvent in comparison with conventional Visbreaking. Chen et al. proposed that hydrogen shuttling between asphaltene and distillation cut is the main reason for the better performance of Visbreaking in the presence of hydrogen donors [Chen et al. 2014].

However, no measurements of the hydrogen donor ability nor the amount of hydrogen transferred from the refinery streams to the upgraded visbroken product were carried out. Also, no molecular characterizations before or after Hydrovisbreaking were done to understand the species responsible for the hydrogen transfer.

4.2.1.3 Mechanism of Hydrovisbreaking

From the fundamental point of view, the mechanistic pathways of Hydrovisbreaking are based on the generation of free radicals, as described for thermal cracking (Section 4.1.2.1). However, most of the hydrogen transfer mechanisms are poorly understood, and little is known about the chemistry of Hydrovisbreaking of HO/B [Gray 2015, Alemán-Vázquez et al. 2016].

In the case of hydrogen gas, the mechanism involves the following fundamental steps:

Initiation:

$$R_1 - R_2 \rightarrow R_1 + R_2^\circ \tag{4.19}$$

Propagation:

$$R_1^\circ + H_2 \rightarrow R_1 - H + H^\circ \tag{4.20}$$

$$H^\circ + R_3 - H \rightarrow R_3^\circ + H_2 \tag{4.21}$$

Termination:

$$R_1^\circ + R_2^\circ \rightarrow R_1^\circ - R_2 \tag{4.22}$$

Where R_1–R_2 is the parent hydrocarbon and $R_1^\bullet$, $R_2^\bullet$, and $R_3^\bullet$ are hydrocarbon free radicals.

Reaction 4.19 is the homolytic cleavage of the hydrocarbon R_1–R_2 as shown in Equation 4.1. In the presence of hydrogen gas, the propagation step involves the reactions 4.20 and 4.21 in which the free radical species ($R_1^\bullet$ and $R_2^\bullet$) react with H_2 to yield the saturated hydrocarbon (R_1–H or R_2–H). This mechanism spawned the idea of *capping of free radicals* [Curran et al. 1967]. The role of hydrogen sources (molecular hydrogen or donor solvents) is to stabilize the reactive radical species.

Furthermore, in the presence of hydrogen donor solvents, alternative free radical reactions can occur such as hydrogen transfer and disproportionation. The following reaction mechanisms of tetralin (Equations 4.23–4.26) can be proposed [Alemán-Vázquez et al. 2016].

Hydrogen Transfer:

$$R_1^\circ + Tetralin \rightarrow R_1 - H + Tetralin^\circ \tag{4.23}$$

Disproportionation:

$$Tetralin^\circ \rightarrow Dihydronaphthalene + H^\circ \tag{4.24}$$

Hydrogen Transfer:

$$H^\circ + \text{Tetralin} \rightarrow \text{Tetralin}^\circ + H_2 \tag{4.25}$$

Termination:

$$R_1^\circ + \text{Tetralin}^\circ \rightarrow R_1 - H + \text{Dihydronaphthalene} \tag{4.26}$$

Where $R_1{}^\bullet$ is a hydrocarbon free radical generated in Equation 4.19 and tetralin$^\bullet$ is a free radical in the benzylic position of tetralin (see Equation 4.16). Also, dihydronaphthalene can further react with the generation of naphthalene, as depicted in Equation 4.16. The mechanistic pathway shown in Equations 4.23–4.26 can also be extrapolated to other varieties of hydrogen donors, whose base structures are similar to the tetralin.

Hydrogen donors have been recognized to be extremely important to control coking reactions. By reacting very actively with olefins, hydrogen donors can eliminate the pathway for polymerization and coke formation, therefore suppressing coking reactions (Equation 4.27). Gray and McCaffrey [2002] reacted n-hexadecane and tetralin together and observed that the yield of olefins was highly suppressed. The rate of cracking of n-hexadecane in tetralin was half the rate seen for the pure compound. Both observations were consistent with the formation of benzylic type tetralin radicals by the abstraction of α-hydrogen from tetralin, thereby the yield of coke was suppressed [Blanchard and Gray 1997].

Olefin Removal:

$$R_4 - C = C + \text{Tetralin} \rightarrow R_4 - C - C + \text{Dihydronaphthalene} \tag{4.27}$$

where R_4 is a hydrocarbon chain. Hydrogens were omitted for simplicity.

The uses of thermal and catalytic hydrogen donors for the surface upgrading of heavy oils and bitumens are discussed in Sections 7.3 and 8.4, respectively.

4.2.1.4 Use of Water. Aquaconversion

Unlike traditional thermal cracking processes in which conversion is limited by the coke formation and products stability, Aquaconversion is a Hydrovisbreaking technology (Figure 4.8) that uses a metal-containing catalyst to transfer hydrogen from water to the feedstock [Pereira *et al.* 1998, Speight 2012]. This process was developed by Petroleos de Venezuela (PDVSA-Intevep) and is commercially licensed jointly by UOP, Intevep, and FosterWheeler.

Figure 4.6 shows the block diagram in which the feed is heated in the presence of steam and an oil-soluble catalyst system. Thermal cracking occurs both in the heater and the soaker drum (reactor) together with a hydrogen transfer to allow achievement of higher residue conversion (36%) and product stability (P-value = 1.2) than the conventional Visbreaking technology (28% and 1.15, respectively) [Pereira *et al.* 1998].

The reaction products (Figure 4.6) are separated in a fractionation column to yield naphtha, gas oil, and vacuum gasoil (VGO). The bottom of the fractionator is sent to a catalyst recovery unit (not shown) to partially recover the catalyst for reuse. An essential aspect of Aquaconversion technology is that it does not produce any solid by-product, such as coke, nor require any hydrogen source or high-pressure equipment. The use of water as a hydrogen source is an essential improvement vs. hydrogen donor solvent.

The Aquaconversion process can also upgrade HO/B to produce lighter synthetic crude oils for upstream applications. Thus, the need for an external diluent (see Section 3.2.3) and its transport over vast distances is eliminated. As can be seen in Table 4.7, the Aquaconversion process led to an increase of ~7°API, viscosity reduction of three orders of magnitude, 20% of hydrodesulfurization, TAN reduction of 98%, reduction of ~11% of Conradson carbon, and a slight increase in liquid yields. Thus, the Aquaconversion process converts an HO/B to synthetic crude without diluents

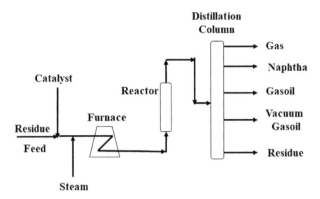

FIGURE 4.6 Block diagram of the Aquaconversion Process.

TABLE 4.7
Properties of a Virgin and Synthetic Crude Oils by Using the Aquaconversion Process[a]

Property	Virgin Heavy Oil	Synthetic Crude Oil
°API gravity	8.2	15.0
Viscosity at 100°F (cSt)	140,000	82
Wt. % of sulfur[b]	3.69	2.95 (20%)
TAN (mgKOH/g)[c]	3.3	< 0.2 (98%)
Conradson carbon (wt. %)[d]	14.5	13.0 (11.5%)
Yield (v/v %)	100.0	102.3

[a] Data taken from UOP brochure No. UOP 2699D-79.
[b] Weight percentage of sulfur by elemental analysis. Number in bracket represents the percentage of dihydrodesulfurization.
[c] Total acid number. Number in bracket represents the percentage of reduction of TAN with respect to the virgin heavy oil.
[d] Weight percentage of Conradson carbon. Numbers in brackets represents the percentage of reduction of Conradson C. with respect to the virgin heavy oil.

---

or capital-intensive bottom-of-the-barrel upgrading technology. The Aquaconversion process was commercially demonstrated by modifying an existing 38,000- BBL/D soaker Visbreaker at the ISLA refinery in Curacao. However, no commercial development has been disclosed to date.

As reported by Carrazza and coworkers [1997], the catalyst is composed of two metals supported on a low surface area material such as silica, aluminosilicates, aluminas, cokes, or carbon-based solids. The first metallic component is selected from non-noble Group VIII metals such as iron, cobalt, nickel. The second metal is selected from the group consisting of potassium or sodium [Carrazza *et al.* 1997].

The mechanism of the Aquaconversion process is believed to proceed through hydrogen capping of the free radicals formed concurrently with the thermal cracking [Pereira *et al.* 1998]. The catalyst facilitates the dissociation of water into two hydrogens and one oxygen free radicals (Equation 4.28). As shown in Equation 4.29, the highly reactive hydrogen free radicals (H•) is transferred to the hydrocarbon radicals (R$_1$•) generating a saturated hydrocarbon compound R$_1$–H.

Water dissociation

$$H_2O \xrightarrow{\text{Catalyst}} 2H^\circ + O^\circ \qquad (4.28)$$

Hydrogen Addition

$$R_1^\circ + H^\circ \xrightarrow{\quad Catalyst \quad} R_1 - H \tag{4.29}$$

where $R_1{}^\bullet$ is a hydrocarbon free radical generated in Equation 4.19. $H^\bullet$ and $O^\bullet$ are hydrogen and oxygen free radicals, respectively [Pereira *et al.* 1998].

The mild hydrogen transfer reaction allows higher conversion to be achieved without precipitation of asphaltenes as in conventional Visbreaking technology. As in typical steam reforming, the oxygen free radical [$O^\bullet$] leads to the formation of carbon dioxide as a by-product [Pereira *et al.* 1998]. Because water is always present at reservoir conditions, its use as hydrogen source has been extensively studied since the 1970s. Thermal and catalytic concepts and processes are discussed in Sections 7.2 and 8.2, respectively.

4.2.1.5   Use of Methane as Hydrogen Source

Fundamental studies using deuterium labeling and ^{1}H- and ^{2}D-NMR experiments have shown that methane can be thermally activated and become involved in heavy oil upgrading reactions [Ovalles *et al.* 1995, Ovalles 2003]. As Hydrovisbreaking, the mechanism proceeds via a free radical pathway as shown in Equations 4.30 and 4.31. Cracked fragments of hydrocarbon molecules ($R^\bullet$) abstract hydrogen from methane to produce methyl radicals ($CH_3{}^\bullet$). In turn, the latter species react with additional hydrocarbon molecules ($R–R'$) producing methylated species ($R–CH_3$) and further free radicals ($R^\bullet$) (4.32). Termination reaction can further generate methylated species as shown in Equation 4.33 [Ovalles *et al.* 1995, Ovalles 2003].

Initiation:

$$R - R' \rightarrow R^\circ + R'^\circ \tag{4.30}$$

Propagation:

$$R^\circ + CH_4 \rightarrow R - H + CH_3^\circ \tag{4.31}$$

$$CH_3^\circ + R - R' \rightarrow R - CH_3^\circ \left(R'CH_3^\circ \right) + R^\circ \left(R'^\circ \right) \tag{4.32}$$

Termination:

$$R^\circ + CH_3^\circ \rightarrow R - CH_3 \tag{4.33}$$

As with water, methane (gas or dissolved) is always present in heavy oil and bitumen reservoirs. Thus, its use as a convenient and inexpensive hydrogen source has been suggested [Ovalles *et al.* 2003]. This topic is discussed in Sections 7.3.3 and 8.4.1.

4.2.2   CATALYTIC CRACKING AND HYDROCRACKING

During the upgrading of HO/B either at the surface or subsurface, one of the fundamental steps is the C–C bond breaking with the concomitant generation of lighter products. In refining units, catalytic-based processes are the most widespread technologies used to improve the conversion, product quality, and economic prospects of the downstream operations. Catalytic cracking is the thermal decomposition of petroleum-derived hydrocarbons in the presence of an acidic catalyst such as silica aluminates, zeolites, or mixtures of both [Speight 2007], whereas catalytic hydrocracking is the combination of the cracking reactions with hydrogenation. Thus, hydrocracking catalysts have a dual function in which one part is responsible for the C–C bond breaking (acid site) and the other for hydrogenation (metal site).

The objectives of using a catalyst are to lower the energy barrier for a reaction pathway and increase the rate of conversion. In all catalytic reactions, such as catalytic acid cracking and hydro-cracking, the fundamental steps are the following [Gray 2015]:

1. Diffusion of reactants from the bulk to the catalyst active site (for catalytic cracking is an acid site and for hydrocracking is acid- and metal-containing sites)
2. Coordination of the reactants with the catalyst active site(s)
3. Reaction at the site (such as C–C bond breaking or hydrogenation)
4. Diffusion of the products away from the site(s) to the bulk

Based on the steps mentioned above, the two primary chemical reactions that occur in catalytic cracking and hydrocracking are acid catalysis and hydrogenation. Next, both topics are discussed in detail.

4.2.2.1 Acid Catalysis

As mentioned, thermal cracking reactions (Section 4.1.2) follow a free radical mechanism whereas catalytic cracking is an ionic process involving carbonium ions. The latter species are hydrocarbon ions having a positive charge on a carbon atom. The formation of carbonium ions during catalytic cracking can occur by:

1. Addition of a proton from an acid site of the catalyst to an olefin
2. Abstraction of a hydride ion (H^+) from a hydrocarbon by the acid catalyst or by another carbonium ion

A critical reaction of carbonium ions is the rearrangement by hydrogen-atom and carbon shifts [Gates *et al.* 1979]. The former leads to a double-bond isomerization (Equation 4.34) while the latter to skeletal rearrangement by methyl group shift (Equation 4.35).

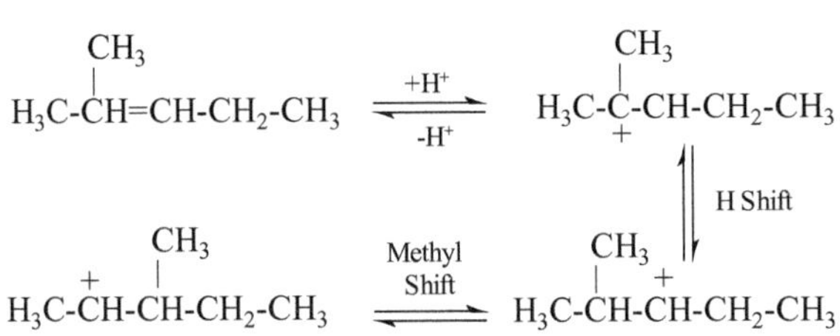

$$H_2C=CH\text{-}CH_2\text{-}CH_2\text{-}CH_3 \qquad H_3C\text{-}CH=CH_2\text{-}CH_2\text{-}CH_3$$

$$\begin{array}{c} +H^+ \\ -H^+ \end{array} \qquad \begin{array}{c} +H^+ \quad -H^+ \end{array}$$

$$H_3C\text{-}CH\text{-}CH_2\text{-}CH_2\text{-}CH_3$$
$$\overset{+}{}$$

(4.34)

Carbonium
Ion

$$\underset{\overset{\displaystyle |}{\underset{}{}}}{CH_3} \qquad \qquad \underset{\overset{\displaystyle |}{\underset{}{}}}{CH_3}$$

$$H_3C\text{-}\overset{\displaystyle CH_3}{\underset{\displaystyle |}{C}}H=CH\text{-}CH_2\text{-}CH_3 \quad \underset{-H^+}{\overset{+H^+}{\rightleftharpoons}} \quad H_3C\text{-}\overset{\displaystyle CH_3}{\underset{\displaystyle |}{\overset{|}{C}}}\text{-}CH\text{-}CH_2\text{-}CH_3$$

H Shift

Methyl
Shift

$$H_3C\text{-}\overset{\displaystyle CH_3}{\underset{\displaystyle |}{\overset{+}{C}}}H\text{-}CH\text{-}CH_2\text{-}CH_3 \quad \rightleftharpoons \quad H_3C\text{-}\overset{\displaystyle CH_3}{\underset{\displaystyle |}{C}}H\text{-}\overset{+}{C}H\text{-}CH_2\text{-}CH_3$$

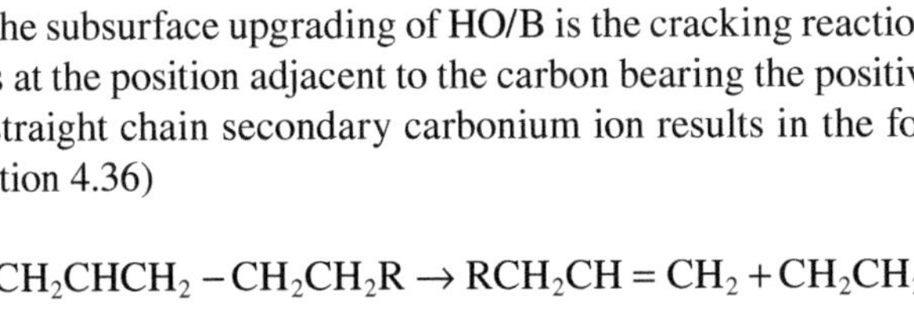

(4.35)

More germane to the subsurface upgrading of HO/B is the cracking reaction. Generally, the C–C bond breaking occurs at the position adjacent to the carbon bearing the positive charge (β-position). Thus, cracking of a straight chain secondary carbonium ion results in the formation of a primary carbonium ion (Equation 4.36)

$$RCH_2CHCH_2 - CH_2CH_2R \rightarrow RCH_2CH = CH_2 + CH_2CH_2R \qquad (4.36)$$

The primary ion can undergo a rapid hydrogen shift to give a more stable secondary carbonium ion (Equation 4.37).

$$\overset{+}{C}H_2CH_2R \xrightarrow{\text{H Shift}} CH_3\overset{+}{C}H_2R \tag{4.37}$$

The continued pattern of cracking of straight chain at the β-position leads to the formation of propylene in high yields. Ethylene is not formed by this mechanism. Thus, high yields of ethylene are indicative of free radical (thermal) cracking whereas high yields of propylene are indicative of catalytic cracking [Gates *et al.* 1979]. Figure 4.7 shows the detailed reaction scheme that explains the product distribution for the catalytic cracking of paraffins [Gates *et al.* 1979]. As seen, the primary products of the carbonium ion chemistry are gasoline hydrocarbons, butene, n-butane, and propylene. The secondary products, due to thermal reactions, are methane, ethane, ethylene, propane, and isobutane.

An important by-product of the catalytic cracking is coke (Figure 4.7). This thermodynamically favored material is formed by the condensation of an aromatic carbonium ion with other aromatic compounds present on the catalyst surface (e.g., reactions 4.13 and 4.14). Due to the high stability of the polynuclear aromatic carbonium ion, it can continue to grow for a relatively long time before a termination reaction occurs through the back donation of a proton [Gates *et al.* 1979]. It has been reported that the coke formation increases with acid strength and density. Thus, careful control of the acid-site properties is necessary.

A commercial cracking catalyst is composed of 3 to 25 wt. % of zeolites embedded in a silica–alumina matrix. Zeolites are a type of molecular sieve that forms a framework with cavities and channels (i.e., microporous) inside where cations, water, small molecules, and hydrocarbons may be adsorbed and reacted. Zeolites are not used alone because of their relatively higher cost in comparison with the silica–alumina. There are interactions between the acid site of the zeolite and the matrix that lead to a different acidity distribution than that found in starting materials. These differences are responsible for the higher catalytic activity for gas-oil cracking and better selectivity for gasoline production than non-zeolite-containing catalysts.

To illustrate an HO/B catalytic cracking process, the feed properties and the product yields for the Asphalt Residual Treating Process (ART process) are shown in Table 4.8 [Noguchi 1991]. This process uses a fluidized catalyst bed to achieve cracking without hydrogen addition. The coke generated is burned in a combustor vessel and the catalyst recycled to the reactor. As seen (Table 4.8), using a blend of Arab light and heavy VRs and Maya whole crude, the gasoline and fuel oil yields (vol. %) are 56% and 34%, respectively, with the concurrent generation of small amounts of gases [Noguchi 1991].

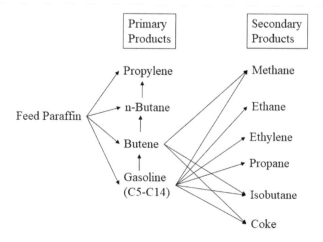

FIGURE 4.7   Detailed reaction scheme for catalytic cracking of paraffins.

TABLE 4.8

Properties of the Feed and Product Yields for the Asphalt Residual Treating Process[a]

Feedstock		Product Yields (Vol. %)[b]	
API gravity	14.9	Regular gasoline	42.5
Sulfur (wt. %)	4.10	Premium gasoline	14.1
Nitrogen (wt. %)	0.32	Fuel oil no. 2	32.4
Conradson C. (wt. %)	15.8	Fuel oil no. 6	2.1
C5 insolubles (wt. %)[c]	12.4	Propane	2.6
Vanadium (ppm)	264	Fuel gas	6.3

Noguchi [1991]

[a] Typical yield for a blend of Arab light and heavy VRs and Maya whole crude.

[b] Values normalized to 100 vol. %.

[c] Amount of the feed insoluble in pentane.

4.2.2.2 Hydrogenation

Formally, hydrogenation involves the addition of hydrogen to hydrocarbon samples in the presence of a metal-containing catalyst and under relatively high hydrogen pressure (1,000–3,000 psi) at elevated temperatures (> 350°C). The hydrogenation reaction must be distinguished from thermal hydrogen donation which, as described in Section 4.2, does not require the use of a catalyst. As shown in Figure 4.1, hydrogen addition to HO/B leads to an increase of the hydrogen-to-carbon molar ratio from 1.5–1.6 to greater than 1.6. Thus, hydrogenation of HO/B and their fraction is one of the most critical upgrading reactions at subsurface and surface conditions. This topic is discussed in Chapter 8.

During the catalytic addition of hydrogen, other chemical reactions are occurring at the same time such as hydrocracking, isomerization, cyclization, condensation, and sulfur, nitrogen, and metal removals. The removal of S, N, O, and metals is described in Section 4.2.3 whereas hydrocarbon isomerization (Equations 4.34–4.35), cyclization (Equation 4.11), and condensation (Equation 4.14) are mostly the same as discussed earlier. This section concentrates on hydrogenation aspects that are occurring during hydrocracking of hydrocarbon samples.

During hydrocracking, one of the leading chemical pathways is the hydrogenation of olefins by the metal-containing catalyst (Figure 4.8). The mechanism of olefin hydrogenation in homogeneous and heterogeneous catalysts has been extensively studied in the literature [Parshall 1980, Bond 2005, Zaera 1996]. In general, it involves the coordination of the olefin ($RCH_2=CH_2$) and hydrogen to the metal site or sites, followed by hydrogen transfer to generate metal–alkyl ($M–CH_3$) and metal hydride ($M–H$) intermediates. Addition of the second hydrogen atom is followed by the generation of the saturated hydrocarbon ($RCH_2–CH_3$).

The effect of hydrogen pressure on naphthenic hydrocarbons leads to ring opening followed by immediate saturation of each end of the fragment by the hydrogenation function of the hydrocracking catalyst [Speight 2007]. For example, methyl cyclopentane is converted to 2-methyl pentane, 3-methyl pentane, and n-hexane (Equation 4.38):

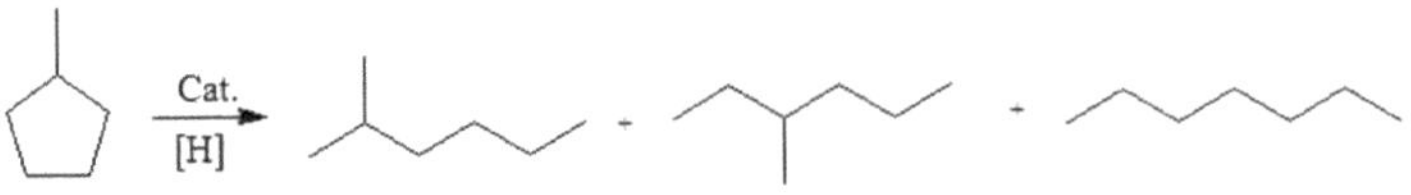

$$(4.38)$$

Aromatic and especially polyaromatic hydrocarbons are resistant to hydrogenation under mid conditions. During hydrocracking, aromatic compounds are converted to naphthenic rings which, in turn, suffer scissions of the alkyl side chains and further crack to paraffin (Equation 4.38)

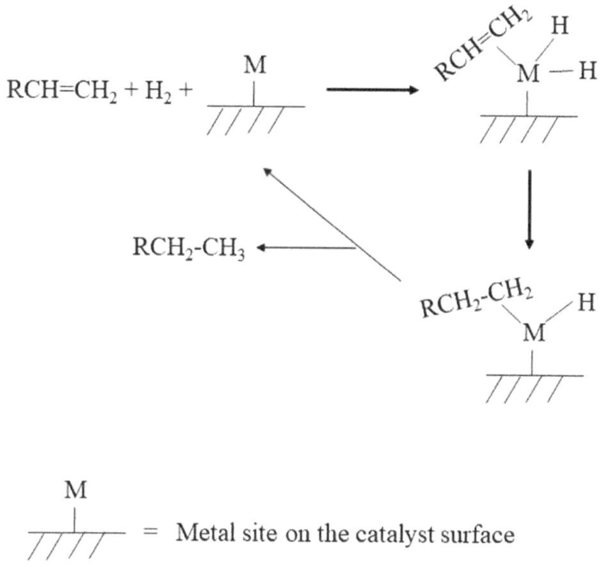

$$RCH=CH_2 + H_2 + \underset{/\!/\!/\!/}{\overset{M}{|}} \longrightarrow \underset{/\!/\!/\!/}{\overset{RCH=CH_2 \quad H}{\overset{\diagup \quad |}{M-H}}}$$

$$RCH_2\text{-}CH_3 \longleftarrow \qquad \underset{/\!/\!/\!/}{\overset{RCH_2\text{-}CH_2 \quad H}{\overset{\diagdown \diagup}{M}}}$$

$$\underset{/\!/\!/\!/}{\overset{M}{|}} = \text{Metal site on the catalyst surface}$$

FIGURE 4.8 General mechanism for olefin hydrogenation on a metal site on the catalyst surface.

There are several hydrocracking technologies available in the market. They use different reactor designs such as fixed bed, ebullated-bed, moving-bed, or slurry-phase reactors to upgrade HO/B and their residues [Sahu *et al.* 2015, Ramirez-Corredores 2017]. The catalyst comprises active metals (Co, Mo, Ni) and noble metals (Pt, Pd) supported on alumina, mixed silica-alumina and micro- and mesoporous zeolites [Robinson 2007]. Table 4.9 shows the typical properties of the feed and product yields for hydrocracking process [Noguchi 1991]. As seen, an Arabian Light VGO (boiling point range 360–530°C) is converted to middle distillates (124–295°C) and light VGO (295–375°C) with more than 93% vol. yield. Also, the API gravity of the products is very high (> 38°API), and the level of contaminants is very low (S < 5 ppm and N < 1 ppm).

Metal-catalyzed hydrogenation reactions have been suggested as subsurface upgrading processes [Pereira-Almao 2012]. This topic is discussed in detail in Chapter 8.

TABLE 4.9
Typical Properties of the Feed and Product Yields and Properties for Hydrocracking Process[a]

Feedstock		Product Yields and Properties				
API gravity	22.6	Fraction	(Vol. %)[b]	°API Gravity	Sulfur (ppm)	Nitrogen (ppm)
Sulfur (wt. %)	2.24	Naphtha (115–124°C)	2.2	70.5	< 2	< 1
Nitrogen (ppm)	620	Middle. dist. (124–295°C)	51.6	57.5	< 2	< 1
Conradson C. (wt. %)	0.34	VGO (295–375°C)	42.3	44.0	< 5	< 1
Asphaltenes (ppm)	< 100	Residue (375°C+)	3.8	38.0	< 5	< 1

Noguchi [1991]

[a] Typical yield for an Arab light VGO (360–530°C).
[b] Values normalized to 100 vol. %.

4.2.2.3   Comparison between Catalytic Cracking and Hydrocracking

Table 4.10 summarizes the differences between catalytic cracking and hydrocracking. As seen, whereas the pressure is low in the first process, high values are needed for the second. The contrary applies to the temperature, i.e., catalytic cracking operates in the 480–538°C range whereas hydrocracking between 350–425°C range.

During hydrocracking, all the *primary* reactions of catalytic cracking occur (Figure 4.7), but some of the secondary ones are inhibited or stopped due to the presence of hydrogen. For example, the yields of olefins and the further reaction of these compounds are substantially diminished. Thus, the high partial pressure of hydrogen acts to favor the slightly heavier molecules by hydrogenating the olefins that easily crack into smaller ones. This a general explanation of why hydrocrackers lead to the production of heavier middle distillate and VGO fractions while catalytic cracking units produce naphtha or gasoline hydrocarbons.

As shown in Table 4.10, both processes have a substantial increase in the product volume. Due to the hydrogenation reaction, hydrocracking produces up to 140 vol. % uplift with respect to the feed. Also, hydrocracked products have much lower sulfur and nitrogen contents than those generated in catalytic crackers. On the other hand, due to the need for hydrogen production and high H_2-pressure, capital cost favors the latter in comparison with hydrocracking.

4.2.3   HYDROTREATMENT AND HYDROPROCESSING

These two processes are synonymous. They are based on the catalytic addition of hydrogen (see Figure 4.1) to the heavy oils and bitumens and their fractions but without simultaneous catalytic cracking. Most of the C–C bond breakage occurs due to thermal reactions. The role of the catalyst is to stop or delay these reactions (i.e., "damage control"). These processes are aimed at increasing the hydrogen-to-carbon molar ratio of the products and hence enhancing their quality and value [Speight 2007]. The primary purposes of Hydroprocessing of HO/B are the following [Gray 2015]:

1. Hydrogenation of olefins and converting aromatics to naphthene compounds (as described in Section 4.2.2.2) and decreasing coke formation
2. Removal of sulfur (Hydrodesulfurization, Section 4.2.3.1) and nitrogen (Hydrodenitrogenation, Section 4.2.3.2) in the presence of catalyst and hydrogen
3. Removal of metals (mostly vanadium and nickel) from the vacuum residue and asphaltene fractions (Hydrodemetallization, Section 4.2.3.3)
4. Removal of oxygen (Hydrodeoxygenation, Section 4.2.3.4)

TABLE 4.10

Comparison between Catalytic Cracking and Hydrocracking

Conditions	Catalytic Cracking	Catalytic Hydrocracking
Pressure range	Low (< 100 psi)	High (1,500–2,100 psi)
Temperature range	High (480–538°C)	Moderate (350–415°C)
Olefin production	High	Very low
Vol. increase of products	112–118 vol. %	115–140 vol. %
Sulfur content of products	Moderate to high	Very low
Nitrogen content of products	Moderate to high	Very low
Capital costs	Moderate	High

Robinson and Dolver [2007].

A wide variety of metals are catalytically active for the hydroprocessing. Platinum, palladium, ruthenium, rhodium, iridium, osmium, rhenium, cobalt, and iron are the most commonly used [Marafi and Furimsky 2017]. However, those catalysts are easily poisoned by sulfur-containing compounds, so more resistant metal sulfides are frequently preferred. Pure or mixed metal catalysts based on tungsten, cobalt, chromium, and molybdenum sulfides are the most commercially utilized.

For the upgrading of heavy oil and bitumens and their fractions, hydroprocessing is carried out at pressures and temperatures in the range of 2,000–3,000 psi (13.8–20.7 MPa) and 380–450°C, respectively. Table 4.11 shows the typical performance reported in the literature for the hydroprocessing of HO/B and their fractions [Speight 2007, Solari and Baumeister 2007, Verstraete *et al.* 2011, Gray 2015, Ramirez-Corredores 2017]. As seen, high levels of sulfur and metals removals and residue conversion (> 90%) are obtained by hydroprocessing technologies. Also, significant values of nitrogen and oxygen removals (> 50%) are reported as well. To achieve these high levels of performance, hydrogen consumption and space velocities are employed in the range of 135–760 Nm^3/m^3 (800–4,500 SCF/bbl) and 0.1–0.6 h^{-1}, respectively.

The main difference of commercially available hydroprocess technologies is the type of reactor used to achieve the levels of performance mentioned above. Those types are:

- Fixed bed processes: It consists of several multi-bed adiabatic reactors operating in series in which hydrogen gas and liquid feed are flowing co-currently from top to bottom.
- Moving-bed processes: It is focused on increasing the cycle length of the Hydroconversion by continuously adding and renewing the catalyst.
- Ebullated-bed process: It has the same goal as the moving-bed reactor only that the catalyst bed is held in a fluidized state by using a liquid recycle.
- Slurry-bed process: It is aimed at the full conversion of the HO/B into light*er* and higher-quality fractions by using dispersed or unsupported metal catalysts. Due to the potential applications of this type of catalytic system to the subsurface upgrading of HO/B, Section 4.2.4 discusses this topic in detail.

More germane subsurface upgrading of HO/B is the chemistry involved in HDS, HDN, HDM, and HDO.

4.2.3.1 Hydrodesulfurization

As shown in Table 2.2, heavy oils and bitumens contain between 2.9–4.4 wt.% of sulfur [Meyer *et al.* 2007]. Thus, the removal of sulfur from HO/B is one of the most important upgrading reactions at subsurface and surface conditions. There are two types of sulfur-containing compounds present in petroleum. These are aliphatic and aromatic sulfur. The first family is composed of mercaptans (R-S-H, where R is an aliphatic carbon chain), sulfides (R_2S), and disulfides (RS-SR).

TABLE 4.11

Typical Performance Reported in the Literature for the Hydroprocessing of HO/B and Their Fractions

Process	Acronym (Section)	Range Reported
Hydrodesulfurization	HDS (Section 4.2.3.1)	80–98%
Hydrodenitrogenation	HDN (Section 4.2.3.2)	20–60%
Hydrodemetallization	HDM (Section 4.2.3.3)	80–99.5%
Hydrodeoxygenation	HDO (Section 4.2.3.4)	20–60%
Percentage residue conversion	% residue conv. (Section 4.4)	35–99%

Speight [2007], Solari [2007], Verstraete [2011], Gray [2015], Ramirez-Corredores [2017].

The removal of sulfur from these aliphatic molecules is relatively easy and normally is carried out under thermal conditions. Equations 4.39–4.41 show the HDS reactions of these sulfur-containing compounds:

$$R-S-H+H_2 \rightarrow R-H+H_2S \tag{4.39}$$

$$R_2S+2H_2 \rightarrow 2R-H+H_2S \tag{4.40}$$

$$R-S-S-R+3H_2 \rightarrow 2R-H+2H_2S \tag{4.41}$$

In HO/B, most of the sulfur is forming S-containing aromatic rings (benzothiophenes and dibenzothiophenes), and the presence of a catalyst is necessary to achieve sulfur removal. HDS has been widely studied, and two general pathways have been observed. In the first one, the direct removal of the sulfur atom in one step is proposed. In the second mechanism, hydrogenation of the attached aromatic rings takes place first followed by generation of hydrogen sulfide.

Figure 4.9 shows the catalytic cycle for the HDS of dibenzothiophene (DBT) with a molybdenum-nickel sulfide catalyst (**I**). Firstly, hydrogen reacts with the Ni/Mo sulfide surface to generate two hydrogen atoms bonded to the sulfur ligands (**II**). Next, H$_2$S is produced by opening a coordination site for the DBT bonding to the Ni/Mo sulfide (**III**). Further reaction with H$_2$ leads to biphenyl (**IV**) and regenerates the molybdenum-nickel sulfide site (I). Thus, the sulfur on the Ni/Mo sulfide surface is an integral part of the activity of the catalyst and accounts for the high selectivity of these catalyst systems for HDS of HO/B and their fractions [Gray 2015].

4.2.3.2 Hydrodenitrogenation

As shown in Table 2.2, heavy oils and bitumens contain between 0.4–0.6 wt. % of nitrogen [Meyer *et al.* 2007]. In general, nitrogen is one of the hardest heteroatoms to remove from petroleum and its fractions [Gray 2015, Ovalles *et al.* 2013]. Figure 4.10 shows some examples of basic and non-basic nitrogen-containing compounds that can be found in heavy crude oils and bitumens [Furimski and Massoth 2005, Grange and Vanhaeren 1997, Prado 2017]. As shown in Figure 4.10, the nitrogen atom can be predominantly found in five- (pyrrole, indole, and carbazoles) and six-membered (pyridine, quinoline, and acridines) aromatic rings. The available information on the composition of various feeds clearly indicates a predominance of highly alkylated N-compounds [Furimski and

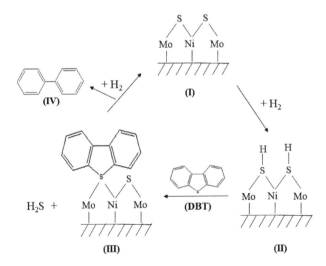

FIGURE 4.9 Catalytic cycle for the HDS of dibenzothiophene (DBT) with a molybdenum-nickel sulfide catalyst.

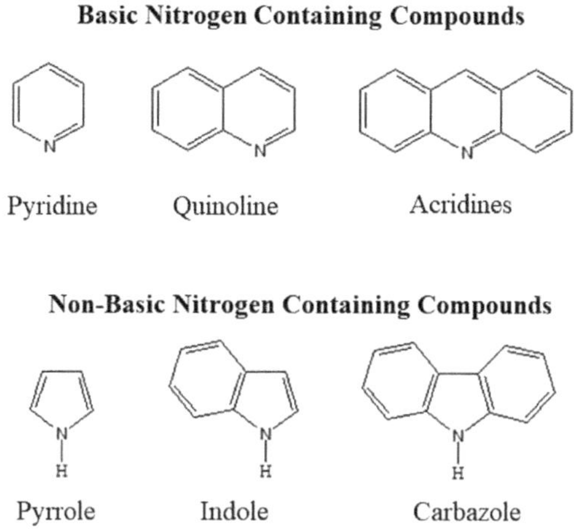

Basic Nitrogen Containing Compounds

Pyridine   Quinoline   Acridines

Non-Basic Nitrogen Containing Compounds

Pyrrole   Indole   Carbazole

FIGURE 4.10   Examples of basic and non-basic nitrogen-containing compounds.

Massoth 2005]. As examples, the HDN reactions of pyrrole and pyridine can be seen in Equations 4.42 and 4.43.

$$\text{Pyrrole: } C_4H_4NH + 4H_2 \rightarrow C_4H_{10} + NH_3 \tag{4.42}$$

$$\text{Pyridine: } C_5H_5N + 5H_2 \rightarrow C_5H_{12} + NH_3 \tag{4.43}$$

Under hydroprocessing conditions, the non-basic nitrogen compounds are rapidly hydrogenated to basic N-compounds. It was reported that the ratio of non-basic/basic N-rings depends on the feed origin [Shin *et al.* 2000]. During cracking of an atmospheric residue, basic nitrogen compounds concentrated in the heavier fractions and residues leading to lower reactivity and catalyst deactivation [Furimski and Massoth 2005].

During HDN, the C–N bonds of the nitrogen-containing rings cannot be cleaved directly. Denitrogenation over hydrotreating catalysts (i.e., sulfide NiMo/Al$_2$O$_3$) requires prior hydrogenation followed by heterocyclic ring opening. Figure 4.11 shows the reaction network for the HDN of quinoline, **I**, [Furimski and Massoth 2005]. As seen, several equilibrium reactions are occurring

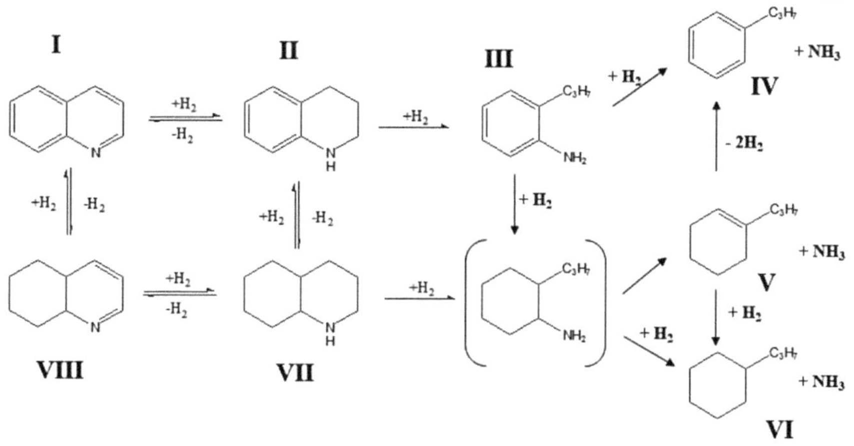

FIGURE 4.11   Reaction network for the HDN of Quinoline.

simultaneously, and depending on the temperature, and hydrogen pressure, several intermediates compounds (**II, III, VII, and VIII**) and HDN products (**IV, V, and VI**) have been proposed.

4.2.3.3 Hydrodemetallization

As shown in Table 2.2, heavy oils and bitumens contain between 180–340 ppm of vanadium and 60–90 ppm of nickel [Meyer *et al.* 2007]. These metal compounds may cause significant detrimental impact during refining processes, leading to the deactivation of catalysts used for hydrocracking and hydrotreatment of HO/B [Ali and Abbas 2006, Gray 2015, Zhao *et al.* 2016, Ramirez-Corredores 2017]. Thus, it is highly recommended to remove vanadium and nickel from petroleum fractions before upgrading.

Generally, vanadium and nickel complexes found in petroleum have been classified into porphyrins and non-porphyrins. Porphyrins are compounds that have a structure formed by four pyrrole rings linked by four methylene groups, and in the case of petroporphyrins, they contain a metal in the center of the molecule. Figure 4.12 shows two of the possible structures of vanadyl (V=O) porphyrins found in petroleum [McKenna *et al.* 2009]. These compounds correspond to Etioporphyrins (ETIO) and deoxophylloerythroetio porphyrins (DPEP). Although it was proposed that non-porphyrins found in HO/B contain atypical porphyrin or pseudo aromatic tetradentate systems, no non-porphyrin molecules have been identified in crude oil until the present time [Zhao *et al.* 2016].

The reactions of the removal of vanadium and nickel in hydroprocessing conditions (HDM) are more complex than those discussed for HDS and HDN. It most probably involves the initial destruction of the metal-trapping porphyrin ring by hydrogenation of one of the pyrrole groups followed by the loss of the vanadium and nickel as metal sulfides on the catalyst surface [Gray 2015]. It was proposed that the resulting metal sulfides also have some catalytic hydrogenation activity. However, the surface areas of such deposited sulfides are much lower than the originally metal-containing phases, so the loss of catalyst deactivation is expected.

4.2.3.4 Hydrodeoxygenation

Hydrodeoxygenation (HDO) reactions are of importance in subsurface upgrading processes of HO/B that involve Aquathermolysis (Chapters 7 and 8) and *in-situ* combustion (Section 3.1.5 and

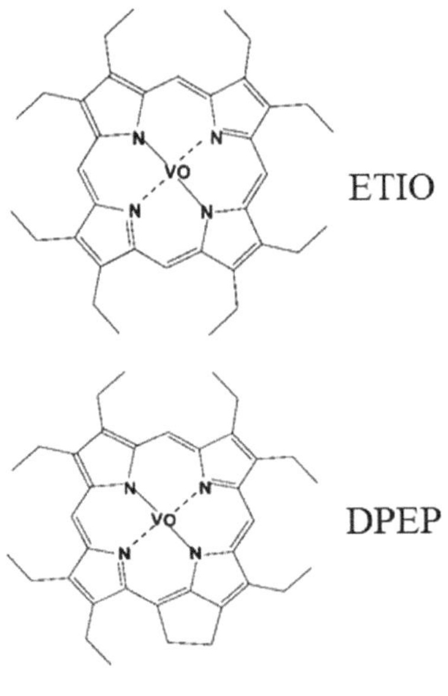

FIGURE 4.12   Possible structures of vanadyl porphyrins found in petroleum.

Chapter 9) processes. In general, the incomplete combustion of hydrocarbons leads to the generation of oxygen-containing compounds which need to be deoxygenated during processing and refining.

Additionally, as shown in Table 2.2, heavy oils and bitumens contain between 1.6–2.5 wt. % of indigenous oxygen [Meyer *et al.* 2007]. Most of the oxygen compounds are concentrated in the most polar fractions where they can reach a concentration as high as 10 wt. %. As shown in Figure 4.13, there are several oxygen-containing groups present in the HO/B such as carboxylic acids, phenols, alcohols, esters, ketones, and ethers [Strausz and Lown 2003, Gray 2015]. Among those compounds, naphthenic acids (Figure 4.13) and their esters are the ones that have received considerable attention [Ramirez-Corredores 2017]. These materials are the culprits of many operational issues such as desalter upsets, corrosion in unexpected areas, fouling, catalyst poisoning, off-specs and degradation of finished products, and environmentally harmful discharges [Ramirez-Corredores 2017].

Under thermal conditions (200–400°C), carboxylic acid groups can be decomposed to generate carbon dioxide and the alkyl chain (Equation 4.44) [Ramirez-Corredores 2017]. In the presence of nitrogen-containing compounds, the reaction leads to olefin production and an alkyl substituted amine (Equation 4.45) [Ovalles *et al.* 1998b]:

$$R-COOH \rightarrow R-H + CO_2 \qquad (4.44)$$

$$\overset{\displaystyle L}{\underset{\displaystyle R-CH-CH_2-COOH}{|}} \xrightarrow{\Delta} RCH_2=CH_2 + HL + CO_2 \qquad (4.45)$$

Where $L=-NR_3$, $-NHR_2$, $-NH_2R$, and R = alkyl chain

During hydrotreating (i.e., hydrogen pressure and metal supported catalysts), most of the oxygen-containing compounds are quantitatively removed with the generation of carbon dioxide as the

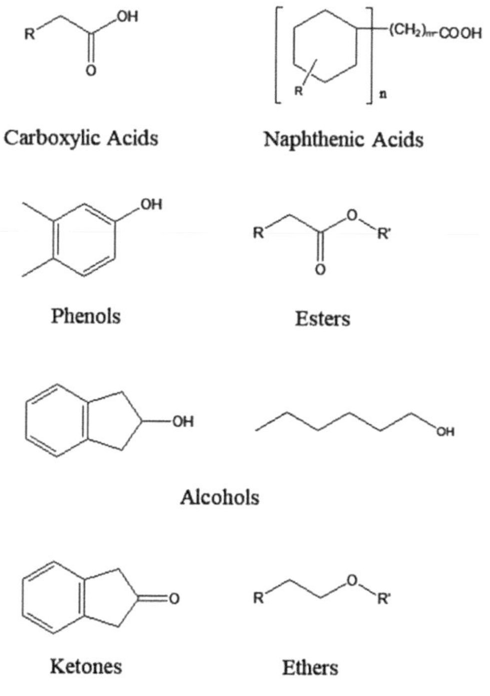

FIGURE 4.13 Representative oxygen-containing groups present in the HO/B.

main product [Mortensen *et al.* 2011, Ramirez-Corredores 2017]. However, this chemistry has been considerably less studied than for the sulfur and nitrogen counterparts [Gray 2015].

4.2.3.5   Catalytic Use of Methane

CH_4 is the hydrocarbon with the highest H/C molar ratio. Thus, its use as hydrogen source has always been of interest to the petroleum industry [Ovalles 2003]. However, methane is rather inert with a C–H bond energy of 435 kJ/mol (see Table 4.2). For this reason, the use of catalytic systems is recommended to accelerate the activation and conversion of methane under hydrogen addition processes. Using X-Ray Photoelectron Spectroscopy (XPS), Nuclear Magnetic Resonance (^{1}H-NMR), Mass Spectrometry (MS), and Secondary Ion Mass Spectrometry (SIMS) experiments, Ovalles *et al.* studied the mechanism of methane activation using MoS_2 catalyst and CH_4 at 420°C and 0.3 MPa [Ovalles *et al.* 1998a]. As shown in Equation 4.46, they found that the methane decomposes to form CH_x (where x = 1, 2, or 3) and H_{4-x} species on the surface of the MoS_2 catalyst. The adsorbed CH_x on the catalyst surface can be incorporated into the hydrocarbon molecules to form methylated products (R–CH_3), and subsequently, the H_{4-x} species are available to hydrogenate cracked fragments and remove sulfur in the form of H_2S [Ovalles *et al.* 1998a].

$$CH_4 \ + \ \boxed{Catalyst} \ \longrightarrow \ \begin{array}{cc} CH_x & H_{4-x} \\ | & | \\ \multicolumn{2}{c}{\boxed{Catalyst}} \end{array} \qquad (4.46)$$

Where x = 1, 2 or 3, and catalyst = MoS_2.

Egiebor and Gray found methyl and dimethyl products by GC analysis of the donor solvent (tetralin) which was attributed to direct alkylation by reaction with methane in their iron-catalyzed coal liquefaction experiments [Egiebor and Gray 1990]. The incorporation of methyl groups, coming from methane, into the heavy crude oil molecules was confirmed by isotopic carbon distribution measurements (^{13}C/^{12}C) using ^{13}CH$_4$ as a source of hydrogen [Ovalles *et al.* 1998a]. However, the addition of methane during upgrading reaction is minimal (estimated value 0.01% w/w), and the role of the thermal processes (i.e., free radicals as depicted in Equations 4.30–4.33) in the amount of CH_4 incorporated by this pathway cannot be neglected [Ovalles *et al.* 1998a]. The catalytic use of methane as a source of hydrogen for the subsurface upgrading of HO/B is discussed in Section 8.4.

4.2.4   Use of Slurry Catalysts and Processes

The use of slurry metal catalysts represents an alternative with excellent potentiality for the subsurface and surface upgrading of heavy crude oils and bitumens mainly due to the possibility of obtaining a high dispersion of the active metal species with the concurrent high activity toward hydroprocessing. Generally, these catalytic systems are not recovered after the upgrading reaction and have been successfully applied to the liquefaction of coal [Curtis and Pellegrino 1989], the hydrocracking [Robinson 2007, Sahu *et al.* 2015] and hydroconversion [Gray 2015, Ramirez-Corredores 2017] of heavy oil residues, and heavy and extra-heavy crude oil upgrading [Del Bianco *et al.* 1994, Ovalles *et al.* 1998a].

The preparation of slurry catalysts is as follows. A power [Sahu *et al.* 2015] or water- or oil-soluble metal catalyst precursor [Ramirez-Corredores 2017] is typically introduced to the feed, and it decomposes under thermal treatment and gives rise to a slurry of fine solid particles. The sulfided catalyst is subsequently produced by the *in-situ* reaction of the metal precursor with sources of sulfur in the feed. The oil-soluble precursors are typically dissolved in a lighter oil fraction to ensure a high dispersion into the more dense and viscous feed. The water-soluble precursors are used in the form of water-in-oil emulsion [Ramirez-Corredores 2017].

Slurry-phase hydroprocessing converts residue in the presence of hydrogen under severe process conditions: 430–450°C and 2,000 to 3,000 psig (13,891 to 20,786 kPa). In slurry-phase processes, the catalyst is homogeneously distributed in the reactor which allows achievement of residue conversion levels higher than 90% [Robinson and Dolver 2007, Ramirez-Corredores 2017]. Figure 4.14 shows the typical residue conversion for several residue conversion processes [Srinivasan 2016]. As seen, the slurry-catalyst based technology LC-Slurry can generate a higher level of liquid yields than other typical hydrotreating technologies such as Coking, LC-Fining, LC-Fining-Coking, and LC-Max [Srinivasan 2016]. The applications of slurry metal catalysts and processes to the subsurface upgrading of HO/B are described in Chapter 8.

4.3 STABILITY OF UPGRADED PRODUCTS

In general, the term "stability" is referred to the formation of undesirable gums, sediments, or deposits when a petroleum-derived material is produced, stored, transported, or converted. Gum formation is the generation of *soluble* organic material, whereas sediment or deposit is the *insoluble* organic material. Depending on the source, the latter may contain coke and inorganic components trapped during precipitation and solid formation.

Storage stability is a term often used to describe the ability of the liquid to remain in storage over extended periods of time without appreciable deterioration, as measured by gum and/or sediment generation. Also, thermal stability is defined as the ability of the liquid to withstand relatively high temperatures for short periods without the formation of sediments (i.e., carbonaceous deposits or coke) [Speight 2007].

Therefore, it is not surprising that one of the main concerns of subsurface and surface upgrading processes is the stability of the converted HO/B and their fractions. If these soluble or insoluble materials are allowed to form, fouling can take place in valves, pumps, furnaces, heat exchangers, distillations columns, and other upstream and downstream hardware with the concomitant increase in downtime and decrease in reliability and revenues. To illustrate this point, Figure 4.15 shows a transversal cut of a tube from a heat exchanger of a refinery crude unit. As seen, a solid was found deposited on the walls with a reduction of 20–30% of the active area [Rogel *et al.* 2015b].

4.3.1 CAUSES OF INSTABILITY AND COMPATIBILITY OF BLENDS

Several reasons have been proposed for the formation of sediments and coke during thermal and catalytic upgrading of petroleum-derived materials. First, at high severity conditions maltenes and asphaltenes undergo drastic chemical changes. Oils and resins are cracked and hydrogenated, so the

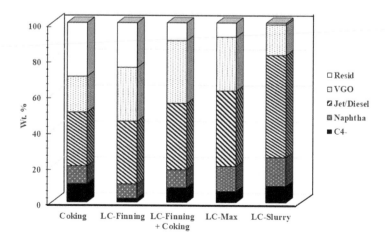

FIGURE 4.14 Typical residue conversion yields for several residue conversion processes. Data taken from Srinivasan [2016].

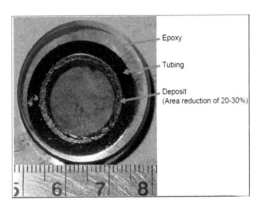

FIGURE 4.15 Photograph of a heat exchanger tube after asphaltene deposition with an area reduction of 20–30% [Rogel 2015b].

aromaticity and solvent power of the resins/oils decreases. At the same time, the breaking of the side chains and naphthenic rings occurs in the asphaltene molecules leaving the growing aromatic cores mostly unaffected and incompatible with the oil. These changes make the oil product more paraffinic and the unconverted asphaltene cores more aromatic and condensed than those in the feed. In consequence, a disturbance in the asphaltene–resin interactions occurs leading to the precipitation of asphaltene as sediments [Kunnas and Smith 2011, Ovalles *et al.* 2015b]. Additionally, as temperature increases, the rates of thermal cracking reactions increase more rapidly than those of hydrogen addition. Thus, hydrogen transfer limitations occur which can lead to the growth of aromatic structures in the asphaltenes making them more prone to precipitate once these compounds leave the reactor zone. Finally, as sediment irreversibly deposits on catalysts, reactors, and other vessel walls, coke forming reactions start occurring which invariably lead to fouling of the downstream equipment [Kunnas and Smith 2011, Ovalles *et al.* 2015b].

Another cause of instability of upgrading products is the presence of olefins and diolefins [Carbognani *et al.* 2015]. Downhole production of olefinic compounds can occur as a consequence of production mechanisms, such as subsurface thermal processes or *in-situ* combustion (Section 3.1.5). In downstream operation and during thermal cracking (Section 4.1.2), one of the primary reactions is the decomposition of free radicals into olefins (Equation 4.6). These unwanted components readily polymerize giving origin to gums and solid deposits [Carbognani *et al.* 2015]. The methods to measure olefins in petroleum products are described in Section 4.3.3.

Instability of crude oils and upgrading products also occurs during dilution or blending. Under certain process conditions, i.e., pressure, temperature, flow regime, composition, etc., the mixing of petroleum-containing materials leads to asphaltene precipitation and plugging of upstream and downstream process equipment [Rodríguez *et al.* 2016]. Because of blending, the solvent power of the maltenes might not be enough to keep the less soluble asphaltenes in solution; therefore, they precipitate out. To avoid these problems, it is first necessary to evaluate the compatibility of crude oil blends through a study of asphaltene stability. This type of testing can also be used to study those virgin oils known as "self-incompatible." These raw materials present insoluble asphaltenes, which, in combination with other inorganic solids as well as waxes, can form sediments leading to fouling in transport and process equipment [Rodríguez *et al.* 2016].

4.3.2 How to Measure Asphaltene Stability of Petroleum Products

Different tests have been used to predict asphaltene stability in virgin and upgraded crude oils. Recently, this topic was reviewed by Ancheyta and coworkers [Guzman *et al.* 2017]. In general, there are two types of methods, i.e., using compositional measurements or titration techniques.

In the first, the SARA compositions are determined, and the stability is related to the ratio of the components. Two of these methods are the asphaltene–resins ratio and the Colloidal Instability Index or CII as shown in Equation 4.47 [Asomaning and Watkinson 2000, Guzman *et al.* 2017].

$$CII = \frac{(wt.\% \, Saturates + wt.\% \, Asphaltenes)}{(wt.\% \, Aromatics + wt.\% \, Resins)} \tag{4.47}$$

Both compositional methodologies assume that crude oils are colloidal systems in which some components act as asphaltene stabilizers (resin and aromatic contents) whereas others contribute to their destabilization (saturate and asphaltene contents). Asphaltene/resin ratios are widely used because both fractions are heavy, non-volatile, and can be quantified precisely. Resins are considered as "natural peptizers" of asphaltenes by keeping them into solution. The asphaltene/resin ratio should be larger than 0.35 for the system to be considered stable [Asomaning and Watkinson 2000]. In the case of the CII, values larger than 0.9 indicate that the system is prone to asphaltene precipitation while values lower than 0.7 indicate a sample with stable asphaltenes. The stability of crude samples with values between 0.7 and 0.9 is uncertain [Asomaning and Watkinson 2000]. These methods, although widely used, tend to predict the stability of virgin and upgraded crude oils incorrectly. The main weakness is that they only consider sample composition and *not* the chemical characteristics of the components. Its fundamental assumption is that oil components derived from feeds from different origin and composition behave similarly, which is not necessarily true.

The second group of methods uses flocculation onset titration techniques and measures the amount of a precipitant solvent needed to destabilize asphaltenes contained in hydrocarbon mixtures [Guzman *et al.* 2017]. The higher the amount of precipitant added the more stable is the sample. This onset of precipitation is related to the solvency of the oil and the asphaltenes. Several standard test methods have been developed based on this principle [ASTM 2001, 2005a, 2005b, 2005c]. The main differences between these methods are the detection technique, the solvent used to dissolve the sample (toluene, methylnaphthalene, etc.), and the titrant agent or precipitant solvent used to precipitate the asphaltenes (heptane, n-cetane, etc.).

Despite these differences, all the asphaltene stability methods use similar equations to determine the parameters. Heithaus compatibility parameters are probably the most widely used to describe stability and compatibility of virgin and upgraded crude oils and their fractions. Three parameters are calculated based on the flocculation ratio of the sample at different concentrations in an aromatic solvent. These parameters are the peptizability of the asphaltenes (Pa), the solvent power of the maltenes (Po), and the overall compatibility of the system or P-value (P). The formula used is shown in Equation 4.48:

$$P = \frac{Po}{1 - Pa} \tag{4.48}$$

In these titration methods, the greater the values of the parameters are, the more stable the sample is. In general, a petroleum sample with a P-value closer or equal to one is considered unstable. Our experience indicates that flocculation onset titration procedure was proven to be a simple and reliable tool for measurement of asphaltene stability during surface and subsurface upgrading of HO/B.

Other methodologies have been developed based on flocculation measurements. One of the most successful is the one reported by Wiehe and Kennedy [2000]. In this case, two parameters are used: the insolubility number (IN), which measures the degree of insolubility of the asphaltenes, and the solubility blending number (SBN), which measures the solvency of the oil for asphaltenes. The ratio SBN/IN is a measurement of the compatibility. When this ratio is one or less than one, incompatibility is predicted. If it is higher than one but close to it, the oil is nearly incompatible. A number much larger than one is very compatible [Wiehe and Kennedy 2000].

Recently two High-Performance Liquid Chromatography (HPLC) methods were developed that are faster and more economical than conventional compositional and titration methodologies

[Rogel *et al.* 2010, 2015]. They are based on on-column precipitation [Rogel *et al.* 2010] or in-line filtration [Rogel *et al.* 2015]. They have been shown to effectively correlate with conventional standard methods [ASTM 2001] and represent a step forward in a new and improved generation of asphaltene stability methods.

4.3.3   HOW TO MEASURE OLEFINS IN PETROLEUM PRODUCTS

Several analytical techniques have been used for the determination of olefins and diolefins in petroleum-containing compounds. This topic has been the subject of several reviews since 1992 [Badoni *et al.* 1992, Kaminski *et al.* 2005, deAndrade *et al.* 2010, Carbongnani 2015]. Two methods have gained wide acceptance. Those are the Bromine number determination and olefin in crude oils by proton-nuclear magnetic resonance (H-NMR). The first method was initially applied to distillates < 315°C and has been modified for fractions boiling up to 550°C [Lubeck and Cook 1992]. Bromine number analysis is significantly influenced by structural effect and by the presence of heteroatoms from small molecular weight compounds [Ceballo *et al.* 1998]. This method can provide olefins contents in wt. % and/or vol. %.

For complex materials like whole HO/B for which structural and heteroatomic effects cannot be determined, a ^{1}H-NMR method is practiced by Canadian petroleum producers [CCQTA 2005]. This method requires two spectra per sample, the first one for the neat material and the second for the 1-decene spiked material. The methodology reports mass % olefins as 1-decene equivalent. Recently, Carbognani and coworkers [2015] reported a universal calibration for mono-olefin analysis by the H-NMR method. These authors found several parameters to affect H-NMR determination. Thus, the proposed universal calibration was only used as a reasonable estimation. For olefin-enriched samples (> 10 wt. % olefins), the relative errors were lower than 10%.

Conjugated diolefin analysis was also found feasible via H-NMR, however, overlapping aromatic signals led to a manual integration protocol and the use of an internal standard to achieve a workable alternative [Carbongnani 2015].

4.4   RELATIONSHIPS BETWEEN RESIDUE CONVERSION AND ASPHALTENE STABILITY

As mentioned in Section 1.3.1, HO/B upgrading processes are classified as low, medium, and high according to the percentage of residue conversion (Figure 1.7). This parameter measures the upgrading of heavy crude oil or bitumen based on the conversion of the fraction with a boiling point greater than 538°C$^+$ (1,000°F$^+$) and is calculated with the following Equation 4.49:

$$\% \, Conv. \, 1000°F^+ = \frac{wt.\% \, Distillables \, 1000°F^+ \, in \, Product - wt.\% \, Distillables \, 1000°F^+ \, in \, Feed}{wt.\% \, Distillables \, 1000°F^+ \, in \, Feed} \times 100 \qquad (4.49)$$

Generally, in subsurface and surface upgrading processes the stability of the product decreases as the residue conversion increases. Several causes have been proposed for the formation of solids, sediment, asphaltenes, and coke during thermal and catalytic upgrading of petroleum-derived materials. First, at under thermal conditions, maltenes and asphaltenes undergo drastic chemical changes. Oils and resins are cracked and hydrogenated; therefore, the aromaticity and solvent power of the resins/oils decrease. At the same time, the breaking of the side chains and naphthenic rings occurs in the asphaltene molecules, leaving the growing aromatic cores mostly unaffected and incompatible with the oil. These changes make the oil product more paraffinic and the unconverted asphaltene cores more aromatic and condensed than those in the feed. In consequence, a disturbance in the asphaltene–resin interactions occurs, leading to the precipitation of asphaltene as solid and sediments [Bartholdy *et al.* 2001, Ovalles *et al.* 2015b, Stanislaus *et al.* 2005, Wandas 2007].

In the presence of hydrogen, as the temperature increases, the rates of thermal cracking reactions increase more rapidly than those of hydrogen addition. Thus, hydrogen transfer limitations occur, which can lead to the growth of aromatic structures in the asphaltenes, making them more prone to precipitate once these compounds leave the conversion zone. Finally, as solid irreversibly deposits on catalysts, reactors, and other vessel walls, coke forming reactions start occurring, which invariably lead to fouling of the equipment (temperature range of 200–350°C). Next, the effect of residue conversion on asphaltene stability is presented.

4.4.1 Thermal Processes

As mentioned, products coming from processes such as Visbreaking, catalytic cracking, coking, and Delayed Coking among others are prone to asphaltene instability as the residue conversion increases. Figure 4.16 shows the Visbreaking conversion of Athabasca vacuum residue vs. stability value (i.e., P-value) [Carbognani *et al.* 2008]. As seen the limiting stability P-value of 1.1 was reached when conversion approached ~30 wt. %. These results were attributed to changes in the resin-to-asphaltene ratio as increases in later fraction and simultaneous decreases in the first were observed which closely matched the Visbreaking process severity [Carbognani *et al.* 2008]. Additionally, it was found that product stabilities depend on asphaltene contents. Higher asphaltene contents are observed to correlate with smaller P values, i.e., more unstable products. The same results have been reported by other authors [diCarlo and Janis 1992, Omole *et al.* 1999, Singh *et al.* 2004].

4.4.2 Hydrogen Donation Processes

Bartholdy and Andersen studied the changes in asphaltene stability during hydrotreating by a flocculation onset titration procedure [Bartholdy and Andersen 2000, Bartholdy *et al.* 2001]. As the reaction temperature was increased, the H/C ratio of the asphaltenes was reduced. This change was also reflected in the critical solubility parameter for the onset of asphaltene precipitation. At low severity hydroprocessing, the stability of the product increased, but when the severity was raised, the asphaltene stability was reduced. [Bartholdy and Andersen 2000]. The authors found that a correlation between H/C and the critical solubility parameter which indicates that although only a fraction of the asphaltenes precipitates at flocculation onset, the chemical changes take place throughout the whole molecular distribution [Bartholdy *et al.* 2001].

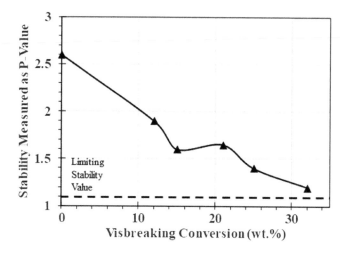

FIGURE 4.16   Visbreaking conversion of Athabasca vacuum residue vs. stability value (P-value). Data taken from Carbognani *et al.* [2008]. Line drawn to show tendency.

REFERENCES

Aiura, M., Masunaga, T., Moriya, K., Kageyama, Y., 1984, "Chemistry of Solvents in Coal Liquefaction: Quantification of Transferable Hydrogen in Coal-Derived Solvents", *Fuel*, 63(8), 1138–1142.

Alboudwarej, H., Beck, J., Svrcek, W. Y., Yarranton, H. W., Akbarzadeh, K., 2002, "Sensitivity of Asphaltene Properties to Separation Techniques", *Energy Fuels*, 16(2), 462–469.

Alemán-Vázquez, L. O., Torres-Mancera, P., Ancheyta, J., Ramírez-Salgado, J., 2016, "Use of Hydrogen Donors for Partial Upgrading of Heavy Petroleum", *Energy Fuels*, 30(11), 9050–9060.

Ali, M. F., Abbas, S., 2006, "A Review of Methods for the Demetallization of Residual Fuel Oils", *Fuel Proc. Technol.*, 87(7), 573–584.

Andersen, S. I., 1995, "Effect of Precipitation Temperature on the Composition of n-Heptane Asphaltenes, Part 2", *Fuel Sci. Technol. Int.*, 13(5), 579 and references therein.

Asomaning, S., Watkinson, A. P., 2000, "Petroleum Stability and Heteroatom Species Effects in Fouling of Heat Exchangers by Asphaltenes", *Heat Transf. Eng.*, 21(3), 10–16.

ASTM International, 2001, *ASTM D6703, Standard Test Method for Automated Heithaus Titrimetry*, ASTM International, West Conshohocken, PA.

ASTM International, 2005a, *ASTM D7060, Standard Test Method for Determination of the Maximum Flocculation Ratio and Peptizing Power of the Maximum Flocculation Ratio and Peptizing Power in Residual and Heavy Fuel Oils (Optical Detection Method)*, ASTM International, West Conshohocken, PA.

ASTM International, 2005b, *ASTM D7112 Standard Test Method for Determining Stability and Compatibility of Heavy Fuel Oils and Crude Oils by Heavy Fuel Oil Stability Analyzer (Optical Detection)*, ASTM International, West Conshohocken, PA.

ASTM International, 2005c, *ASTM D7157 Standard Test Method for Determination of Intrinsic Stability of Asphaltene-Containing Residues, Heavy Fuel Oils, and Crude Oils (n-Heptane Phase Separation; Optical Detection)*, ASTM International, West Conshohocken, PA.

Audeh, C. A., Yan, T. Y., 1983, "Process of Coke Formation in Delayed Coking", In: *ACS Symposium Series*, L.F. Albright, R. T. K. Baker, Ed., American Chemical Society, Washington, DC, p 295–308.

Awadalla, A., Cookson, D. J., Smith, B. E., 1985, "Coal Hydrogenation: Quality of Start-Up Solvents and Partially Process-Derived Recycle Solvents", *Fuel*, 64(8), 1097–1107.

Badoni, R. P., Bhagat, S. D., Joshi, G. C., 1992, "Analysis of Olefinic Hydrocarbons in Cracked Petroleum Stocks: A Review", *Fuel*, 71(3), 483–491.

Bartholdy, J., Andersen, S. I., 2000, "Changes in Asphaltene Stability during Hydrotreating", *Energy Fuels*, 14(1), 52–55.

Bartholdy, J., Lauridsen, R., Mejlholm, M., Andersen, S. I., 2001, "Effect of Hydrotreatment on Product Sludge Stability", *Energy Fuels*, 15(5), 1059–1062.

Bedell, M. W., Curtis, C. W., Hool, J. H., 1993, "Comparison of Hydrogen Transfer from Cyclic Olefins and Hydroaromatic Donors under Thermal and Catalytic Liquefaction Conditions", *Energy Fuels*, 7(2), 200–207.

Blanchard, C. M., Gray, M. R., 1997, "Free Radical Chain Reactions of Bitumen Residue", Div. Fuel Chem., Preprint of Papers Presented at 213th ACS Nat. Meeting, San Francisco, CA, 42, 137–138.

Boduszynski, M. M., 2015, "Petroleum Molecular Composition Continuity Model". In: *Analytical Methods in Petroleum Upstream Applications*, C. Ovalles, C. Rechsteiner, Ed., CRC Press, Boca Raton, Chapter 1, 1–29 and references therein.

Bond, G. C., 2005, *Metal-Catalyzed Reactions of Hydrocarbons*, Springer, New York, p 291–292.

Bozzanoa, G., Dente, M., 2005, "A Mechanistic Approach to Delayed Coking Modelling", In: Europa Symposium on Computer Aided Process Engineering", L. Puigjaner and A. Espuña, Ed., Elsevier, p 15.

Carbognani, L., González, M. F., Lopez-Linares, F., Sosa-Stull, C., Pereira-Almao, P., 2008, "Selective Adsorption of Thermal Cracked Heavy Molecules", *Energy Fuels*, 22(3), 1739–1746.

Carbognani, L., Lopez-Linares, F., Wu, Q., Trujillo, M., Carbognani, J., Pereira-Almao, P., 2015, "Analysis of Olefins in Heavy Oil, Bitumen and Their Upgraded Products", In: *Analytical Methods in Petroleum Upstream Applications*, C. Ovalles, C. E. Rechsteiner Jr., Ed., CRC Press, Boca Raton, FL Chapter 5, pp 81–110 and references therein.

Carrazza, J., Pereira, P., Martinez, N., 1997, "Process and Catalyst for Upgrading Heavy Hydrocarbon", U.S. Patent No. 5,688,741, November 18.

CCQTA, Canadian Association of Petroleum Producers, 2005 "Olefins in Crude Oil by Proton NMR Method". Also identified as MAXXAM: CAPP Olefin by NMR, version 1.04.

Ceballo, C. D., D'Ambrosio, F., Torres, N., 1998, "Estimation of Olefins to Aromatics Ratio (O/A) in Cracked Naphthas by Bromine Number Assay", *Pet. Sci. Technol.*, 16(1&2), 179–189.

Chen, Q. Y., Gao, Y., Wang, Z. X., Guo, A. J., 2014, "Application of Coker Gas Oil Used as Industrial Hydrogen Donors in Visbreaking", *Pet. Sci. Technol.*, 32(20), 2506–2511.

Colyar, J., 2009, "Has the Time for Partial Upgrading of Heavy Oil and Bitumen Arrived?", *Pet. Technol. Q.*, 4th Q., 43–56.

Curran, G. P., Struck, R. T., Gorin, E., 1967, "Mechanism of the Hydrogen-Transfer Process to Coal and Coal Extract", *Ind. Eng. Chem. Proc. Des. Dev.*, 6(2), 166–173.

Curtis, C. W., Pellegrino, J. L., 1989, "Activity and Selectivity of Three Molybdenum Catalysts for Coal Liquefaction Reactions", *Energy Fuels*, 3(2), 160–168 and references therein.

Dean, J., 1999, *"Lange's Handbook of Chemistry"*, 15th Ed., McGraw Hill, Chapter 4, pp 4.43–4.46.

deAndrade, D. F., Fernandez, D. R., Miranda, J. L., 2010, "Methods for the Determination of Conjugated Dienes in Petroleum Products: A Review", *Fuel*, 89(8), 1796–1805.

Del Bianco, A., Garuti, G., Pirovano, C., Russo, R., 1995, "Thermal Cracking of Petroleum Residues: 3. Technical and Economic Aspects of Hydrogen Donor Visbreaking", *Fuel*, 74(5), 756–760.

Del Bianco, A., Panariti, N., Di Carlo, S., Beltrame, P. L., Carniti, P., 1994, "New Developments in Deep Hydroconversion of Heavy Oil Residues with Dispersed Catalysts. 2. Kinetic Aspects of Reaction", *Energy Fuels*, 8(3), 593–597.

diCarlo, S., Janis, B., 1992, "Composition and Visbreakability of Petroleum Residues", *Chem. Engineer. Sci.*, 47(9–11), 2695–2700.

Egiebor, N. O., Gray, M. R., 1990, "Evidence for Methane Reactivity during Coal Pyrolysis and Liquefaction", *Fuel*, 69(10), 1276–1282.

Elliott, J. D., Stewart, M. D., 2004, "Cost Effective Residue Upgrading: Delayed Coking, Visbreaking & Solvent Deasphalting", Proc. 4th Russian Refining Technol. Conf. Moscow, Russia, 23–24 September, p 26.

Furimski, E., Massoth, F. E., 2005, "Hydrodenitrogenation of Petroleum", *Catal. Rev.*, 47(3), 297–489.

Gates, B. C., Katzer, J. R., Schuit, G. C. A., 1979, *Chemistry of Catalytic Processes*, 1st Ed., McGraw-Hill, New York, Chapter 1, pp 8–10.

Grange, P., Vanhaeren, X., 1997, "Hydrotreating Catalysts, an Old Story with New Challenges", *Catal. Today*, 36(4), 375–391.

Gray, M. R., 1994, *Upgrading Petroleum Residues and Heavy Oils"*, Marcel Dekker, New York, April.

Gray, M. R., 2015, *Upgrading Oilsands Bitumen and Heavy Oil*, The University of Alberta Press, Edmonton, and references therein.

Gray, M. R., McCaffrey, W. C., 2002, "Role of Chain Reactions and Olefin Formation in Cracking, Hydroconversion, and Coking of Petroleum and Bitumen Fractions", *Energy Fuels*, 16(3), 756–766.

Guzmán, R., Ancheyta, J., Trejo, F., Rodríguez, S., 2017, "Methods for Determining Asphaltene Stability in Crude Oils", *Fuel*, 188, 530–543, and references therein.

Huc, A.-Y., Ed., 2011, *Heavy Crude Oils. From Geology to Upgrading. An Overview*, Technip, Paris, France, and references therein.

Joshi, J. B., Pandit, A. B., Kataria, K. L., Kulkarni, R. P., Sawarkar, A. N., Tandon, D., Ram, Y., Kumar, M. M., 2008, "Petroleum Residue Upgradation via Visbreaking: A Review", *Ind. Eng. Chem. Res.*, 47(23), 8960–8988.

Kaminski, M., Kartanowicz, R., Gilgenast, E., Namiesnik, J., 2005, "High-Performance Liquid Chromatography in Group-Type Separation and Technical or Process Analytics of Petroleum Products", *Crit. Rev. Anal. Chem.*, 35(3), 193–216.

Kunnas, J., Smith, L., 2011, "Improving Residue Hydrocracking Performance", *Digit. Refining, PTQ*, Q3.

Le Roux, M., Nicole, D., Delpuech, J., 1982, "Performance Indices for Coal Liquefaction Solvents", *Fuel*, 61(8), 755–760.

Li, Q., Liu, D., Song, L., Wu, P., Ya, Z., Li, M., 2016, "Investigation of Solvent Effect on the Hydro-Liquefaction of Sawdust: An Innovative Reference Approach Using Tetralin as Chemical Probe", *Fuel*, 164, 94–98.

Lubeck, A., Cook, R. D., 1992, "Bromine Number Should Replace FIA in Gasoline Olefins Testing", *Oil Gas J.*, Dec. 28, 86, 89.

Marafi, M., Furimsky, E., 2017, "Hydroprocessing Catalysts Containing Noble Metals: Deactivation, Regeneration, Metals Reclamation, and Environment and Safety", *Energy Fuels*, 31(6), 5711–5750.

McKenna, A. M., Purcell, J. M., Rodgers, R. P., Marshall, A. G., 2009, "Identification of Vanadyl Porphyrins in a Heavy Crude Oil and Raw Asphaltene by Atmospheric Pressure Photoionization Fourier Transform Ion Cyclotron Resonance (FT-ICR) Mass Spectrometry", *Energy Fuels*, 23(4), 2122–2128.

Meyer, R. F., Attanasi, E. D., Freeman, P. A., 2007, "Heavy Oil and Natural Bitumen Resources in Geological Basins of the World", U.S. Geological Survey Open-File Report 2007-1084, available online at http://pubs.usgs.gov/of/2007/1084/.

Mortensen, P. M., Grunwaldt, J.-D., Jensen, P. A., Knudsen, K. D., Jensen, A. D., 2011, "A Review of Catalytic Upgrading of Bio-Oil to Engine Fuels", *App. Catal. A Gen.*, 407(1–2), 1–19.

Noguchi, T., 1991, *Heavy Oil Processing Handbook*, Research Association for Residual Oil Processing, Tokyo, Japan.

Omole, O., Olieh, M. N., Osinovo, T., 1999, "Thermal Visbreaking of Heavy Oil from the Nigerian Tar Sand", *Fuel*, 78(12), 1489.

Ovalles, C., Filgueiras, E., Morales, A., Rojas, I., de Jesus, J. C., Berrios, I., 1998a, "Use of a Dispersed Molybdenum Catalyst and Mechanistic Studies for Upgrading Extra-Heavy Crude Oil Using Methane as Source of Hydrogen", *Energy Fuels*, 12(2), 379–385.

Ovalles, C., Garcia, M. del C., Lujano, E., Aular, W., Bermudez, R., Cotte, E., 1998b, "Structure/Interfacial Activity Relationships and Thermal Stability Studies of Cerro Negro Crude Oil and Its Acid, Basic and Neutral Fractions", *Fuel*, 77(3), 121–126.

Ovalles, C., Hamana, A., Rojas, I., Bolívar, R. A., 1995, "Upgrading of Extra-Heavy Crude Oil by Direct Use of Methane in the Presence of Water. Deuterium Labeled Experiments and Mechanistic Consideration", *Fuel*, 74(8), 1162.

Ovalles, C., Rogel, E., Lopez, J., Pradhan, A., Moir, M. E., 2013, "Predicting Reactivity of Feedstocks to Resid Hydroprocessing by Using Asphaltene Characteristics", *Energy Fuel*, 27(11), 6552.

Ovalles, C., Rogel, E., Moir, M. E., Morazan, H., 2015a, "Effect of Temperature on the Analysis of Asphaltenes by the On-Column Filtration/Redissolution Method", *Fuel*, 146, 20–27 and references therein.

Ovalles, C., Rogel, E., Morazan, H., Moir, M. E., Dickakian, G., 2015b, "Method for Determining the Effectiveness of Asphaltene Antifoulants at High Temperature: Application to Residue Hydroprocessing and Comparison to the Thermal Fouling Test", *Energy Fuels*, 29(8), 4956–4965 and references therein.

Ovalles, C., Vallejos, C., Vasquez, T., Rojas, I., Ehrman, U., Benitez, J. L., Martinez, R., 2003, "Downhole Upgrading of Extra-Heavy Crude Oil Using Hydrogen Donors and Methane Under Steam Injection Conditions", *Pet. Sci. Tech.*, 21(1–2), 255–274.

Parshall, G. W., 1980, *Homogeneous Catalysis*, Wiley, New York, p 36.

Pereira, P., Marzin, R., Zacarías, L., López-Trosell, I., Hernández, F., Córdova, J., Szeoke, J., Flores, C., Duque, J., Solari, R. B., 1998, "Aquaconversion™: A New Option for Residue Conversion and Heavy Oil Upgrading", *Vision Tecnologica*, 6 (1), 5–14.

Pereira-Almao, P., 2012, "In Situ Upgrading of Bitumen and Heavy Oils via Nanocatalysis", *Can. J. Chem. Eng.*, 90(2), 320–329.

Peries, J. P., Quignard, A., Farjon, C., Laborde, M., 1988, "Thermal and Catalytic Asvahl Processes under Hydrogen Pressure for Converting Heavy Crudes and Conventional Residues", *Oil & Gas Sci. Technol. Rev. IFP*, 43(6), 847–853.

Prado, G. H. C., Rao, Y., de Klerk, Arno, 2017, "Nitrogen Removal from Oil: A Review", *Energy Fuel*, 31(1), 14–36.

Ramirez-Corredores, M. M., 2017, *The Science and Technology of Unconventional Oils: Finding Refining Opportunities*, 1st Ed., Elsevier, London, Chapter 5, p 387 and references therein.

Reynierst, G. C., Fromenti, G. F., Kopinke, F.-D., Zimmermann, G., 1994, "Coke Formation in the Thermal Cracking of Hydrocarbons. 4. Modeling of Coke Formation in Naphtha Cracking", *Ind. Eng. Chem. Res.*, 33(11), 2584–2590.

Robinson, P. R., Dolver, G. E., 2007, "Commercial Hydrotreating and Hydrocracking", In: *Hydroprocessing of Heavy Oils and Residua*, J. Ancheyta, J. G. Speight, Ed., CRC Press, Boca Raton, Chapter 10, pp 281–312.

Rodríguez, S., Ancheyta, J., Guzmán, R., Trejo, F., 2016, "Experimental Setups for Studying the Compatibility of Crude Oil Blends under Dynamic Conditions", *Energy Fuels*, 30(10), 8216–8225 and references therein.

Rogel, E., Ovalles, C., Moir, M. E., 2010, "Asphaltene Stability in Crude Oil and Petroleum Materials by Solubility Profile Analysis", *Energy Fuel*, 24(8), 4369–4374.

Rogel, E., Ovalles, C., Moir, M. E., 2015b, "On-Column Filtration Asphaltene Characterization Methods for the Analysis of Produced Crude Oils and Deposits from Upstream Operations", In: *Analytical Methods in Petroleum Upstream Applications*, C. Ovalles, C. Rechsteiner, Ed., CRC Press, Boca Raton, FL, Chapter 9, pp 161–176.

Rogel, E., Ovalles, C., Vien, J., Moir, M. E., 2015a, "Asphaltene Solubility Properties by the In-Line Filtration Method", *Energy Fuels*, 29(10), 6363–6369.

Sahu, R., Song, B. J., Im, J. S., Jeon, Y.-P., Lee, C. W., 2015, "A Review of Recent Advances in Catalytic Hydrocracking of Heavy Residues", *J. Ind. Eng. Chem.*, 27, 12–24 and references therein.

Shin, S., Sakanishi, K., Mochida, I., Grudoski, D. A., Shinn, J. H., 2000, "Identification and Reactivity of Nitrogen Molecular Species in Gas Oils", *Energy Fuels*, 14(3), 539–544.

Singh, J., Kumar, M. M., Saxena, A. K., Kumar, S., 2004, "Studies on Thermal Cracking Behavior of Residual Feedstocks in a Batch Teactor", *Chem. Engineer. Sci.*, 59(21), 4505–4515.

Siskin, M., Kelemen, S. R., Eppig, C. P., Brown, L. D., Afeworki, M., 2006, "Asphaltene Molecular Structure and Chemical Influences on the Morphology of Coke Produced in Delayed Coking", *Energy Fuels*, 20(3), 1227–1234.

Solari, R. B., Baumeister, A. J., 2007, "Review of Heavy Oil Upgrading", 2nd AIChE/SPE Workshop, Houston, TX, 22–27 April.

Speight, J. G., 2007, *The Chemistry and Technology of Petroleum*, 4th Ed., CRC Press, Boca Raton, FL, and references therein.

Speight, J. G., 2012, "Visbreaking: A Technology of the Past and the Future", *Sci. Iran. C*, 19(3), 569–573.

Speight, J. G., 2017, *Handbook for Petroleum Refinery*, 1st Ed., CRC Press, Boca Raton, FL, and references therein.

Srinivasan, B., 2016, "Europe Largest Slurry Hydrocracker", *Pet. Technol. Q.*, 4th Q., 145–146.

Stanislaus, A., Hauser, A., Marafi, M., 2005, "Investigation of the Mechanism of Sediment Formation in Residual Oil Hydrocracking Process through Characterization of Sediment Deposits", *Catal. Today*, 109(1–4), 167–177.

Storm, D. A., Decanio, S. J., Edwards, J. C., Sheu, E. Y., 1997, "Sediment Formation During Heavy Oil Upgrading", *Pet. Sci. Technol.*, 15(1&2), 77–102.

Strausz, O. P., Lown, E. M., 2003, *The Chemistry of Alberta Oil Sands, Bitumens and Heavy Oils*, Alberta Energy Research Inst., Calgary, Alberta, Canada.

Verstraete, J., Guillaume, D., Auberger, M. R., 2011, "Catalytic Hydrotreatment and Hydroconversion: Fixed Bed, Moving Bed, Ebullated Bed, and Entrained Bed", In A.-Y. Huc, Ed., *Heavy Crude Oils: From Geology to Upgrading: An Overview*, Technip, Paris, France.

Wandas, R., 2007, "Structural Characterization of Asphaltenes from Raw and Desulfurized Vacuum Residue and Correlation between Asphaltene Content and the Tendency of Sediment Formation in H-Oil Heavy Products", *Pet. Sci. Technol.*, 25(1–2), 153–168.

Wiehe, I. A., 1993, "A Phase Separation Kinetic Model for Coke Formation", *Ind. Eng. Chem. Res.*, 32(11), 2447–2454.

Wiehe, I. A., Kennedy, R. J., 2000, "Application of the Oil Compatibility Model to Refinery Streams", *Energy Fuels*, 14(1), 60–63.

Zaera, F., 1996, "On the Mechanism for the Hydrogenation of Olefins on Transition-Metal Surfaces: The Chemistry of Ethylene on Pt(111)", *Langmuir*, 12(1), 88–94 and references therein.

Zhao, X., Xu, C., Shi, Q., 2016, "Porphyrins in Heavy Petroleums: A Review", *Struct. Bond*, 168, 39–70.

5 Physical Separation

As shown in Figure 1.9, there are several routes reported in the literature for subsurface upgrading (SSU) of heavy oils and bitumens (HO/B). One of the most straightforward concepts is the physical separation of the HO/B into lighter and heavy fractions to increase rates and percentages of recovery, and at the same time, improve the properties of the produced oil. Two physical separation routes have been reported in the literature. Those are steam distillation and solvent deasphalting. In the first concept, the SSU of heavy oils is coupled to steamflooding processes whereas, in the second, the action of a hydrocarbon solvent (also known as anti-solvent) leads to asphaltene precipitation downhole via solvent deasphalting (SDA). Except for the steam distillation, the main advantage of SDA processes is that they do not require external energy sources because they are operated at reservoir temperature [Pourabdollah and Mokhtari 2013] or in the presence of warm solvent [Nenninger and Nenninger 2005]. In the next sections, these two routes are described in detail.

Other topics are covered in this chapter such as the physical and numerical simulations of downhole solvent deasphalting and asphaltene precipitation, and several field tests carried out such as VAPEX, Cyclic Propane Injection, and N-Solv. This chapter finishes discussing the use of asphaltene precipitants to further improve the economic prospect of the SDA routes.

5.1 STEAM DISTILLATION

The technique known as steam distillation has been known since the last century and is commonly used for the separation and purification of high-boiling organic compounds [Vogel 1974, Wu 1977]. According to Dalton's partial pressure law, when two or more gases are mixed at a constant temperature, each one acts as if it were alone and the sum of their partial pressure is equal to the total pressure of the system. In the case of two immiscible liquids in which of one them is water (P_w) and the other is a volatile oleic phase (P_{oil}), the total pressure of the system (P_T) is equal to the following Equation 5.1.

$$P_T = P_w + P_{oil} \qquad (5.1)$$

When those two liquids are distilled, the boiling point of the mixture is the temperature at which the sum of the partial pressures is equal to the environment. This boiling temperature is lower than that of the more volatile component. Because one of them is water, steam distillation leads to the co-evaporation of fractions present in the oleic phase at **much lower** temperatures than those obtained if the oil were alone. Thus, an increment of the production of distillable materials is found along with steam [Vogel 1974, Wu 1977].

Based on extensive lab experiments, the oil production mechanism by steam distillation was proposed [Blevins *et al.* 1984, Hong 1994, Ovalles *et al.* 2002], and it is shown in Figure 5.1 [Duerksen and Hsueh 1983, Wu and Elder 1983]. First, it is important to point out that the scheme shown in this figure is one-dimensional whereas the field process is a phenomenon occurring in the three dimensions of space. As seen, after steam heating, a fraction of the crude oil is vaporized and transported along with the steam front until it condenses to form a hot water/condensed crude zone. The vaporization, transport, and condensation of crude fractions is a dynamic process which leads to a constant displacement of the light components from the injector toward the producing wells. In turn, this may lead to an increase in the API gravity and reduction of the viscosity of the produced oil (upgrading) due to a disproportionate content of light fractions. This mechanism is valid for light, medium, and heavy oils and depends strongly on the amount of distillable materials (> 30% wt.) present in the original crude oil [Blevins *et al.* 1984, Hong 1994]. As mentioned in Section 3.1.2.2

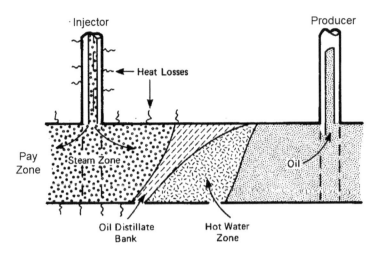

FIGURE 5.1 One-dimensional mechanism for oil production by steam distillation [Blevins *et al.* 1984]. Reprinted with permission from the Society of Petroleum Engineering.

and shown in Figure 3.6, steam distillation is one of the oil recovery mechanisms for steamflooding of heavy oils [Hong 1994].

Sharpe and coworkers carried out a compositional numerical simulation of steamflooding of a typical California heavy oil [Sharpe *et al.* 1995]. Results indicate that the nonlinear steam distillation/*in-situ* solvent generation process is capable of significantly reducing oil saturations upstream of the advancing condensation zone. This hydrocarbon condensate (or solvent) dissolves the original oil and leads to subsurface upgrading by viscosity reduction, density reduction (i.e., API increase), and oil swelling via thermal expansion [Sharpe *et al.* 1995].

For the heavy oil studied, an initial waterflood residual saturation of 20% was reduced to 6.6% by steamflooding [Sharpe *et al.* 1995]. The authors concluded that 1-D and 2-D mechanistic simulation models using fine grids were required to accurately resolve the relevant physics. An appropriate pseudo-component equation-of-state which is tuned to laboratory data was also essential [Sharpe *et al.* 1995].

Liu and coworkers carried out three steam distillation experiments employing saturated steam, steam superheated by 10°C, and steam superheated by 40°C using heavy oil samples derived from the Jin-1 Block oilfield in China [Liu *et al.* 2018]. The steam distillation rates obtained with steam superheated by 40°C, steam superheated by 10°C, and saturated steam were 11.2%, 10.9%, and 10.4%, respectively. These results indicated that the oil rate increased with increasing degree of superheating. The authors also observed that the density and viscosity of residual oil increased, the contents of saturated hydrocarbons and resin decreased, and the contents of aromatic hydrocarbons and asphaltenes increased. Conversely, analysis of the distilled materials indicated that lightweight components with smaller carbon numbers were removed first by steam distillation, which in turn led to lower density (i.e., higher API) of the upgraded oil [Liu *et al.* 2018].

Even though during steam injection processes (cyclic or continuous), the generation of a lighter fraction via steam distillation has been observed, the results have not been optimized to the production of upgraded crude oils. Thus, this SSU route has not been thoroughly studied in the literature.

5.2 PHYSICAL SIMULATIONS OF DOWNHOLE SOLVENT DEASPHALTING

As presented in Section 4.1.1, refinery-based solvent deasphalting (SDA) is a carbon rejection route (Figure 4.1) and well-established technology. Several commercially proven processes are available in the market and have enjoyed several degrees of success throughout the years. The subsurface upgrading of heavy oils and bitumens via solvent deasphalting borrows concepts developed and commercialized in refining and manufacturing (i.e., Downstream, see Figure 1.8) and applies them to upstream operations. In line with this reasoning, the addition of hydrocarbon solvents or anti-solvents to induce the subsurface asphaltene precipitation has been studied since the 1990s [Mokrys and Butler 1993, Das 1998, Das 2005]. In this way, a deasphalted crude oil (DAO) is generated *in-situ*, which in turn increases the oil rates and the percentages of oil recovery, leaving behind the asphaltene fraction.

Since 1980 [Redford and McKay 1980], the vast majority of articles found in the literature are concentrated on evaluating the benefit of solvent and solvent deasphalting on production rates and oil recovery. The fundamentals of steam and solvent co-injection were discussed in Section 3.1.3. However, except for viscosity, *no other* physical and chemical properties, or compositional changes of the produced oils in comparison with the original crude oils, are seldom mentioned or not reported [Al-Murayri *et al.* 2016, Bayestehparvin *et al.* 2016, Jamaloei *et al.* 2012, Gupta *et al.* 2012, Jha *et al.* 2012, Hosseininejad-Mohebati *et al.* 2012]. Thus, the effect of subsurface upgrading (SSU) cannot be precisely determined. In this section, PVT and one- and two-dimensional SDA experiments (also known as physical simulations) are described for HO/B from several parts of the world. Also, hot solvent experiments are presented as well (see also Section 3.1.4). These experiments are aimed at determining the degree of upgrading of the crude oil not only in terms of viscosity reductions but increases in the API gravity and decreases of asphaltene, heteroatom, and metal contents.

5.2.1 PVT AND ONE-DIMENSIONAL EXPERIMENTS

Vallejos *et al.* carried out the addition of liquid propane to a heavy crude oil from the Orinoco basin (8.9°API) at several solvent-to-crude ratios using a pressure, volume, and temperature (PVT) cell at 1,100 psi and 139°F (60°C) [Vallejos *et al.* 2002]. After separation of the top fraction and flashing the propane (Figure 5.2), the API gravity of the upgraded product (DAO) did not change until a 1:1 v/v solvent/crude ratio (SvOR) was used (11°API). A maximum value of 21°API was reported using a relatively high SvOR of 8:1 v/v. In the same experiments, reductions of the asphaltene content of the upgraded products were also observed (see Figure 5.3). As before, a 1:1 volumetric ratio was necessary to detect a significant decrease in the asphaltene fraction (from 16.5 to 14.4 wt. %), and a minimum value of 0.01 wt. % was obtained at 8:1 solvent/crude ratio. Consistent with these results, reductions in the vanadium contents (up to 96 wt. %) and ten-fold reduction of the viscosity of the upgraded products were also disclosed [Vallejos *et al.* 2002]. Similar results were reported by Mokrys and Butler for Canada's Cold Lake and Lloydminster

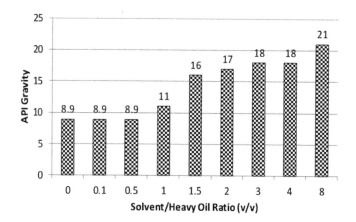

FIGURE 5.2 API gravities of the upgraded products (DAO) as a function of the solvent (C3)/Venezuelan crude ratio. Data were taken from Vallejo [2002].

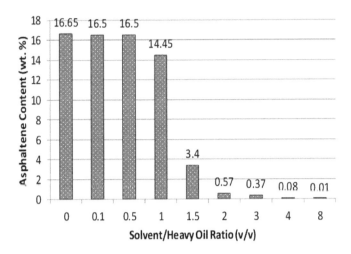

FIGURE 5.3 Asphaltene contents (wt. %) of the upgraded products (DAO) as a function of the solvent (C3)/ Venezuelan crude ratio. Data were taken from Vallejo [2002].

heavy oils in which significant reductions in viscosity (~2–3 orders of magnitude) and metal contents were found in the deasphalted oils when the propane/dead oil mass ratios reached 0.5 w/w [Mokrys and Butler 1993].

Ovalles *et al.* carried out studies of the onset of flocculation of asphaltene precipitation with propane at typical Orinoco reservoir conditions (50°C or 122°F and 1,087 psi). The results are presented in Figure 5.4 [Ovalles *et al.* 2014]. The experiments were carried out with live crude oil containing 72 SCF of methane per STB of crude oil in the absence of sand. Increasing amounts of propane in increments of 5 vol. % and up to 75 vol. % were added to the live crude oil. In a typical determination, both the heavy crude oil and propane were mixed and pumped (at 300 cc/h) through the optical probe, and images were recorded until achieving a stationary stage around 10 hrs. As seen in Figure 5.4, asphaltene flocs were observed at propane concentration as low as 10 vol. %. Also, the amount of asphaltene particles increased with the propane concentration from 10% to 30 vol. %. Between 30% and 45%, asphaltene precipitation took place, and significant amounts of

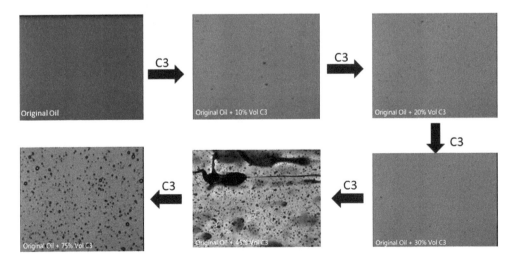

FIGURE 5.4 Micrographs of the titration of live Venezuelan heavy crude oil with propane at reservoir conditions (50°C or 122°F and 1,087 psi) in the absence of sand.

flocs can be clearly seen. At 75%, the asphaltenes were totally precipitated and were found deposited throughout the instrument.

During the study of the onset of flocculation of asphaltene precipitation with propane, asphaltene particle size distributions were determined as shown in Figure 5.5. Consistent with Figure 5.4, the number of asphaltene particles increased with the volume percentage of propane. In general, the particles sizes were in the 4 to 14 μm range [Ovalles et al. 2014].

Ovalles et al. also carried out phase behavior studies using a PVT cell as shown in Figure 5.6. Three volumetric propane/Orinoco crude oil ratios were studied (2.5, 4, and 9). A 4:1 C3/crude oil blend was stirred in the cell, and after approximately five days, a clear separation between the propane-rich phase (C3-solubles or DAO) at the top and the propane-poor phase (C3-insolubles) at the bottom could be seen (see right side of Figure 5.6). Both samples were collected and flashed to remove propane. As reported by Vallejos et al. [2002], the results showed an increase of the API from 8.9°API of the original crude oil to 18–20°API for the DAO. Similarly, the viscosity of C3-insolubles was in the order of ~10^6 cP whereas the original oil was ~68,000 cP at 104°F (40°C).

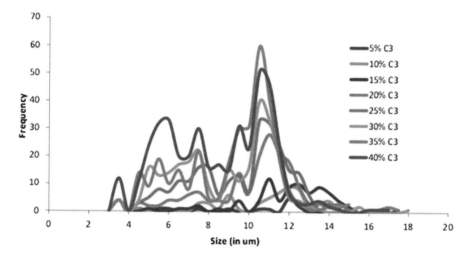

FIGURE 5.5 Asphaltene particle size distributions with different propane concentrations at reservoir conditions (50°C or 122°F and 1,087 psi).

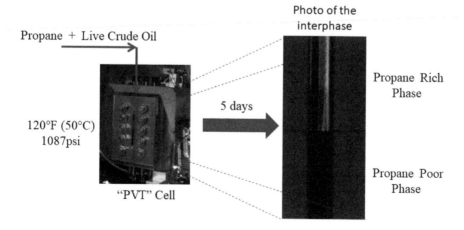

FIGURE 5.6 Phase behavior studies using a PVT cell of asphaltene precipitation of live Venezuelan heavy crude oil and propane (C3/crude = 4:1 v/v) at 122°F and 1,087 psi.

For the 2.5 C3/crude oil blend, no clear separation was observed in the PVT cell even after several days of static conditions. [Ovalles *et al.* 2014, 2017]. These results agree with those reported by Mitchel and Speight [Mitchell and Speight 1973] as discussed in Section 2.4.1 and Figure 2.9. The characterization of these two phases was carried out, and the results were used to feed an asphaltene numerical simulation model (see Section 5.4).

Consistent with previous findings, Nourozieh *et al.* [2012] studied the effect of the propane/ Alberta bitumen ratio on equilibrium compositions at different temperatures and pressures (50– 100°C and 2–8 MPa). The results showed that the propane/oil mixtures partition into solvent- (at the top) and asphaltene-enriched (at the bottom) phases at a given condition. The simulated distillation (SimDis, see Section 2.3) data showed that the light components were extracted by propane into the solvent-enriched phase, whereas the heavy fractions remained in the asphaltene phase. Similarly, the density of the top layer was lower than that measured for the bottom liquid whereas the viscosity had the opposite trend, i.e., top phase had a lower value than the bottom [Nourozieh *et al.* 2012]. Unfortunately, no more details were published.

More recently, Oliveira and coworkers [2017] characterized by physical and chemical methods the top and bottom fractions separated from a Brazilian heavy crude (12.7°API gravity) at different propane/crude oil ratios and high T (< 100°C) and P (< 15 MPa) conditions for 24 h. Interestingly, the authors published a photograph with the aspect of the liquid extracts (DAO) after removal of C3 at different crude oil/propane ratios (see Figure 5.7). As seen, at low propane ratio, the color of the DAO is dark and became lighter as the propane/heavy oil ratio increased.

As seen in Figure 5.8, Oliveira *et al.* reported that the API gravity increased from 12.7°API of the original crude to ~18–20°API for the top layer (red trace). On the other hand, the API gravity bottom layer (C3-insolubles) decreased (black line) as the propane/heavy crude oil ratio increased from two to ten [Oliveira *et al.* 2017]. These results are consistent with those reported by Vallejos *et al.* [2002] and Ovalles and coworkers [2014, 2017].

Additionally, Oliveira and coworkers [2017] reported that the H/C molar ratio of the DAO increased from 1.65 to 1.68 when the C3/heavy oil increased from 1.9 to 10. Also, the aromaticity as measured by ^{13}C-NMR of the bottom layer is higher than that for the top phase (DAO).

Moreno-Arciniegas and Babadagli [2014] studied the effect of asphaltene flocculation on the produced fluid and its deposition on the rock surface using a PVT cell during solvent injection (propane, n-hexane, and n-decane) of two Canadian heavy crude oils up to 120°C. The authors found that the percentages of asphaltenes decreased with temperature and the chain length of the three paraffinic solvents [Moreno-Arciniegas and Babadagli 2014], as discussed in Section 2.4.1 and Figure 2.9. These results agree with those published by Anderson [Anderson 1995], Ovalles *et al.* [2015], and Mitchel and Speight [Mitchell and Speight 1973].

Furthermore, as shown in Table 5.1, the API gravity of the deasphalted oils (i.e., top layer after solvent removal) increased from 8.7°API and 10.2°API of the original Canadian crude oils to 14.5°API

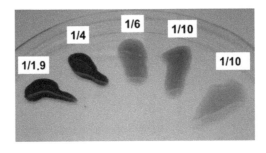

FIGURE 5.7 Aspect of the liquid extract (DAO) after removal of C3 at different Brazilian heavy crude oil/ propane ratios. Reprinted with permission from Oliveira *et al.* [2017]. Copyright (2017) American Chemical Society.

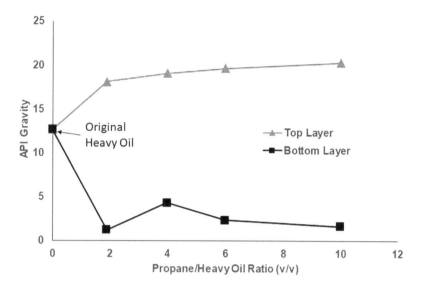

FIGURE 5.8 API gravity of the original crude oil and top and bottom layers in function of the propane to Brazilian heavy oil ratio (v/v). Lines were drawn to show the tendency. Data were taken from Oliveira *et al.* [2017].

and 15.9°API using propane [Moreno-Arciniegas and Babadagli 2014]. As seen in Table 5.1, smaller increases were observed using n-hexane and n-decane. Similar results were reported for the DAO viscosity (results not shown in Table 5.1) [Moreno-Arciniegas and Babadagli 2014].

Ovalles *et al.* carried out one-dimensional propane flood experiments in a vertically oriented displacement cell using a live Venezuelan crude oil [Ovalles *et al.* 2017]. The percentages of oil recovery and masses of oil produced are shown in Figure 5.9 (lines were drawn to show a tendency). As seen, the oil rates (square marker) are higher at the beginning of the flood reaching 83–85% of oil recovery (diamond marker) at the end of the run. Propane breakthrough was observed after ~6 h. After 150 h, the amount of oil produced was minimal. The experiment finished after the injection of 5 PV of propane.

In Figure 5.10, changes in H/C molar ratio and API gravity are shown as a function of the eluted time. H/C ratio and API gravity initially increased as the deasphalting took place until they reached a maximum at around 80 h and then started to decrease until the end of the run. It is important to point out that the decrease in H/C ratio and API gravity coincided with a change in the color of effluents from dark brown to amber. This phenomenon is an indication that an important change in the composition of the effluent took place. Also, once this change happened (> 80 h), the percentage of recovery effluent was very low (0.2 to 1.1% recovery as shown in Figure 5.9) pointing out that

TABLE 5.1

Effect of the Type of Solvent on the API Gravity of the Deasphalted Oils (DAO) during PVT Experiments

		Solvent[a]		
	Original Crude[b]	Propane[b]	n-Hexane[b]	n-Decane[b]
Canadian crude oil A	8.67	14.5	12.53	10.17
Canadian crude oil B	10.22	15.91	15.26	11.24

Data taken from Moreno-Arciniegas and Babadagli [2014].

[a] Solvent used during the deasphalting process.

[b] API gravity of the deasphalted crude oil (DAO).

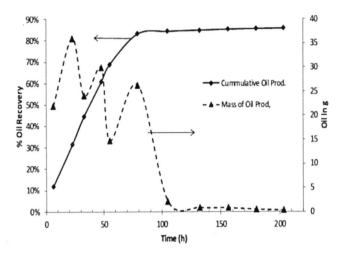

FIGURE 5.9 Percentage of oil recovery and mass of oil produced during the propane flood experiment of live Venezuelan heavy crude oil at 122°F and 1,100 psi (lines were drawn to show tendency).

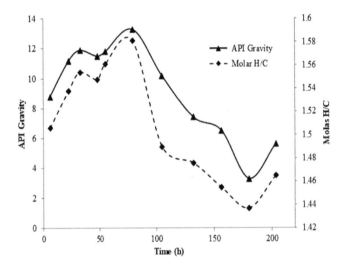

FIGURE 5.10 Molar H/C ratio and API gravity of the produced oil from the propane flood as a function of time (lines were drawn to show tendency).

these products correspond probably to minor material that moved slower than the main components through the core.

On the other hand, molecular weights and C7-asphaltene contents of the samples initially decreased until they reach a minimum and then, increased again (see Figure 5.11). The decrease in asphaltene content was expected as part of the deasphalting process, and therefore a reduction in the molecular weight was observed. However, as in the previous case, there was a radical change in behavior for the last five effluents. For these liquids, the asphaltene contents and molecular weight increased as a function of the time. This finding is consistent with the elution of heavy components that have more significant amounts of dissolved asphaltenes and therefore, high molecular weights. Figure 5.12 shows the correlation between asphaltene content and molecular weight of the C3 flood effluents.

After the propane flood, the C3-asphaltenes were isolated from the inlet, middle, and outlet of the sand pack by solvent extraction with 10% methanol in dichloromethane at 120°C for 3 h [Ovalles *et al.* 2012]. As can be seen in Table 5.2, the percentage of asphaltenes at the inlet was

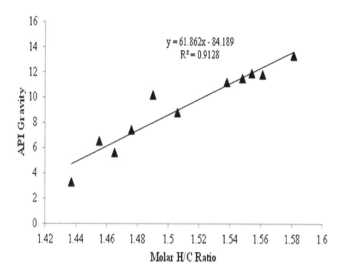

FIGURE 5.11   API gravity vs. molar H/C ratio of the produced oil from the propane flood.

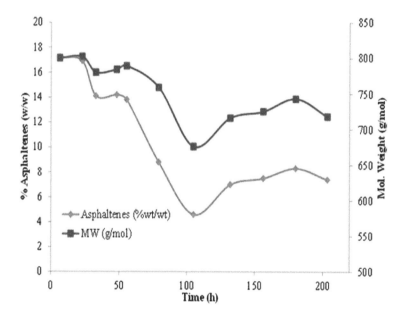

FIGURE 5.12   Percentage of asphaltenes and molecular weight of the produced oil from the propane flood as a function of time (lines were drawn to show tendency).

two-fold higher than at the outlet of the cell (3.1 wt. % vs. 1.5 wt. %). The higher asphaltene content found at the inlet was attributed to the relatively higher concentration of propane in comparison with the outlet [Ovalles *et al.* 2017]. However, the asphaltene characteristics were quite similar as revealed by Table 5.2. In fact, H/C molar ratios (~1.2), percentages of sulfur (5.1–5.2 wt. %), metal contents (V ≈ 1,300 ppm and Ni ≈ 300 ppm), molecular weights (1,000–1,060 g/mol), and densities (~1.14/cc) were within the order of magnitude of the errors of the techniques [Ovalles *et al.* 2017].

Mamora *et al.* studied the steam-propane injection for the Hamaca (Venezuelan, 8°API) and Duri (Indonesian, 19°API) heavy crude oils using a vertical cell containing a mixture of sand, oil, and water [Mamora *et al.* 2003]. The physical simulations were run under superheated steam at 170°C and 50 psig for Hamaca crude and at 260°C and 500 psig for Duri and propane/steam mass

TABLE 5.2

Characterization of the C3 Asphaltenes Extracted from the *Post-Mortem* Sands after the Propane Flood Experiment[a]

Section	% Asphalt Extracted (w/w)[b]	Molar H/C[c]	S (% wt.)[d]	V (ppm)[e]	Ni (ppm)[f]	Mw (g/mol)[g]	Density (g/mL)[h]
Inlet	3.1%	1.253	5.21	1,340	317	1,164	1.130
Middle	2.7%	1.240	5.14	1,283	296	1,108	1.149
Outlet	1.5%	1.196	5.23	1,275	305	1,015	1.136

[a] Average of two determinations. Mass balances in the 98–101% range.
[b] Percentage of asphaltenes (wt./wt.) after solvent extraction of 10% methanol in dichloromethane at 120°C for 3 h.
[c] Hydrogen-to-carbon molar ratio.
[d] Percentage of sulfur by elemental analysis.
[e] Vanadium content by elemental analysis.
[f] Nickel content by elemental analysis.
[g] Molecular weight by Size Exclusion Chromatography.
[h] Density at 60°F.

ratios ranging from 0:100 to 10:100. Results showed that the start of oil production was accelerated by 23% (Hamaca) and 30% (Duri) and that, in the presence of the solvent, steam injectivity was up to three times higher [Mamora *et al.* 2003].

Figure 5.13 shows the API gravity of the produced oil as a function of time for the steam only and the propane/steam runs for the Duri crude oil [Mamora *et al.* 2003]. As seen, the API gravity increased from 19°API to 21–23°API. Consistent with the results by Ovalles *et al.*, Hamaca crude oil showed similar increases of API gravity from 8°API to 14–16°API. The authors proposed a mechanism that involves the propane traveling in the gas phase with steam and condensing in the cooler part of the cell. By this way, solvent deasphalting was taking place with the concomitant reduction of viscosity and an increase in the API gravity of the produced oil [Mamora *et al.* 2003].

Moreno-Arciniegas and Babadagli studied the type of solvent and operating conditions for the physical simulation of the downhole SDA process [Moreno-Arciniegas and Babadagli 2013]. These authors injected propane, n-hexane, n-decane, and a petroleum distillate into glass-bead pack

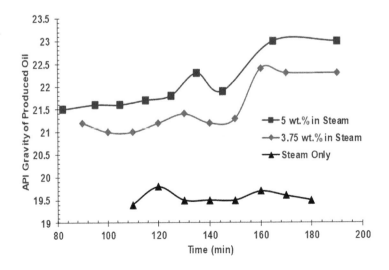

FIGURE 5.13   API gravity of the produced oil as a function of time for the Duri crude oil (lines were drawn to show tendency). Data were taken from Mamora *et al.* [2003].

systems saturated with two different Canadian heavy oils of 8.7°API and 10.3°API. For both crude oils, the recovery factor sharply increased with the temperature in the 50–120°C range. The maximum recovery (96%) was found using the distillate solvent at 50°C. This finding was attributed to the lower asphaltene precipitation caused by the presence of aromatic compounds. In the n-alkanes experiments, the lower recovery was explained in term of higher asphaltene precipitation with the concomitant decrease in the accumulated oil recovery. No mention of acceleration of the oil rate was reported [Moreno-Arciniegas and Babadagli 2013]. Based on these observations, these authors recommended using different solvents throughout the recovery processes, i.e., lighter ones initially and switching to heavier ones later [Moreno-Arciniegas and Babadagli 2014].

As shown in Table 5.3 [Moreno-Arciniegas and Babadagli 2013], the SARA analysis of the solvent-produced oils showed lower resin and asphaltene contents than those found in the original crude (8.7°API). As seen, the percentages of reduction of these fractions in comparison with the original crude oil were in the 50–60% and 20–40% ranges, respectively. In the petroleum distillates, the percentage of deasphalting is negligible (4%) and within the error of the analytical technique (±10%). As mentioned, this finding was attributed to the presence of the aromatic components in the later solvent in comparison with the other alkanes [Moreno-Arciniegas and Babadagli 2013].

Leyva-Gomez and Babadagli, carried out physical simulation experiments of solvent injection at elevated temperatures to recover heavy oil/bitumen from fractured carbonates [Leyva-Gomez and Babadagli 2016]. Three different solvents (propane, heptane, and distillate oil -naphtha) were injected at different temperatures representing a wide range of carbon number. Indiana limestone (outcrop) and vuggy naturally fractured carbonate samples (outcrop core samples from a producing formation in Mexico) were selected as core samples. The results revealed that heavy oil recovery increases with the solvent carbon number used. Naturally fractured carbonates had a lower recovery than Indiana limestone with naphtha, due to the heterogeneity and poor connectivity of fissures and vugs. Some parts of the rock were observed to be extremely tight (matrix), and no oil saturation was even possible. Solvent recoveries were in the 60–90% range. No changes in oil composition or accelerated production due to SSU were reported in comparison with the case without solvent [Leyva-Gomez and Babadagli 2016].

Hascakir et al. studied the effects of clays on steam deasphalting of Peace River bitumen from Alberta, Canada (8.8°API gravity, 54,000 cP viscosity at room temperature, and 34.3 wt. % n-pentane insoluble asphaltenes) in the presence of clays at 75 psi, from 30°C to 170°C [Coelho et al. 2016, 2017]. The sand–clay mixture mimicked the clay-rich layer in the Peace River formation which consisted of 15 wt. % clay (2.3 µm average particle size) and 85 wt. % sand [Bayliss and Levinson 1976]. The clay used in this study contained 90 wt. % kaolinite and 10 wt. % illite [Unal et al. 2015].

TABLE 5.3

SARA Analysis of the Original Canadian Crude Oil (8.7°API) and the Produced Oils after Solvent Injection into a Glass-Bead Pack with Different Types of N-Alkanes and a Petroleum Distillate[a]

Solvent Type[b]	Original Oil	Propane[c]	n-Hexane[c]	n-Decane[c]	Petroleum Distillated[c]
Saturates (wt. %)	17.24	50.19	34.25	34.98	37.12
Aromatics (wt. %)	38.60	29.48	40.72	40.10	36.19
Resins (wt. %)	32.66	13.34 (59%)	16.53 (49%)	15.72 (52%)	15.69 (52%)
Asphaltenes (wt. %)	11.50	7.00 (39%)	8.50 (26%)	9.20 (20%)	11.00 (4%)

Data taken from Moreno-Arciniegas and Babadagli [2013].

[a] Solvent flood experiments carried out in absence of steam in the 50–120°C range. SARA based on ASTM D 2007.

[b] Solvent used for the flooding.

[c] Numbers in bracket indicate percentages of reduction vs. the original crude oil.

Six displacement experiments (E1–E6) were carried as described in Figure 5.14. The cumulative oil recovery versus time for all the E1–E6 runs are shown in which clay and water contents are excluded [Coelho *et al.* 2016, 2017]. As seen, the cumulative oil recovery increased with the addition of solvent (E6: 9:1 C3-Steam—No clay) in comparison with the run with steam (E2: Steam—No clay). However, the presence of clays in reservoir rock affected oil production adversely in the absence (E1 vs. E2) and presence of propane (E6 vs. E5).

The impact of clay on oil recovery performance is attributed to its interaction with asphaltene and water [Neasham 1977, Redford and McKay 1980, Leontaritis *et al.* 1994, Coelho *et al.* 2016]. Additionally, kaolinite and illite are known to cause pore-filling, pore-lining, and pore-bridging, creating micro-porosity and tortuous flow paths [Neasham 1977]. Illite–water interactions lead to cementation which may reduce the permeability significantly [Kar *et al.* 2015]. The adverse effects of clays on oil recovery performance were also recognized by Smith *et al.* [2009] on portions of the Peace River reservoirs with much higher clay content (80–90%). However, Hascakir *et al.* showed that even at lower clay contents (15 wt. %), the recovery performance using steam flooding and solvent–steam co-injection was drastically reduced [Coelho *et al.* 2016, 2017].

The quality of the produced oil was initially analyzed by viscosity measurements. All generated oil samples exhibited higher viscosity than the initial bitumen due to the formation of water-in-oil emulsions, which could be further stabilized by the presence of clays [Kar *et al.* 2015, Coelho *et al.* 2016]. So, the oil quality was also assessed in terms of asphaltene contents. Figure 5.15 shows the normalized produced oil composition (wt. %) for the C5-asphaltenes and deasphalted oils (DAO) for the displacement experiments E1–E6 [Coelho *et al.* 2016, 2017]. As seen, lower asphaltenes contents indicated higher asphaltenes precipitation on the reservoir rock. As expected, the results showed that propane-steam-flooding produced higher quality oil (higher DAO content) than steam-flooding by decreasing the asphaltene contents (compare E3, E4, and E5 vs. E1 and E6 vs. E2) [Coelho *et al.* 2016, 2017].

Another critical finding extracted from Figure 5.15 is the positive impact of the presence of clays in reservoir rock on the oil quality. When clay-containing materials are present, the asphaltenes

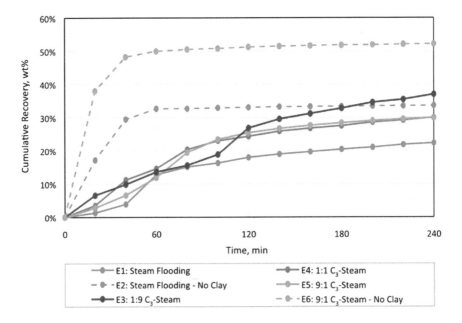

FIGURE 5.14 Cumulative oil recovery vs. time for displacement experiments E1–E6 using a Canadian bitumen. Clay and water contents are excluded [Coelho *et al.* 2016]. Reprinted with permission from Society of Petroleum Engineering.

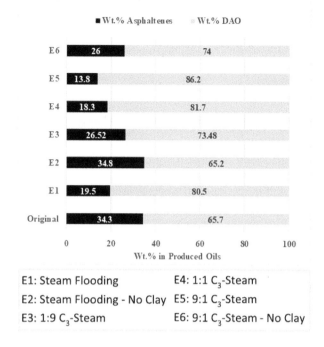

FIGURE 5.15 Normalized produced oil composition (wt. %) for C5-asphaltenes and the deasphalted oil (DAO) for displacement experiments E1–E6. Clay and water contents are excluded. Data were taken from Coelho *et al.* [2016].

content and the formation of water-in-oil emulsions [Coelho *et al.* 2016, 2017] were lower than the sand-only cases (compare E1 vs. E2 and E5 vs. E6). As previously reported [Kar *et al.* 2015, Mukhametshina *et al.* 2016], asphaltenes can be associated with clays and precipitate on the rock. Thus, a lower amount of asphaltenes remained available in the oil phase to stabilize water-in-oil emulsions. However, clays could also migrate to the oil phase, causing surface separation to be challenging and costly (compare E1, E3 E4, and E5 to E2 and E6) [Coelho *et al.* 2016, 2017].

Benson and Ovalles patented a process for producing upgraded heavy oil during SAGD operations by flowing a liquid phase additive into a near wellbore region of the steam chamber to control asphaltenes mobility [Benson and Ovalles 2017]. The additive was formulated to mobilize the asphaltenes within this region preventing blockage and formation damage. In this order of ideas, Yakubov *et al.* performed several experiments using C3, C4, C5, and C6 hydrocarbons with bitumen to evaluate the efficiency of several asphaltene inhibitors during solvent injection processes. The authors reported that the addition of asphaltene inhibitors is a way to prevent asphaltene precipitation and deposition in a reservoir [Yakubov *et al.* 2014].

Telmadarreie and Trivedi used a micromodel to imitate a fractured porous media to study asphaltene precipitation of a heavy crude oil (30,000 cP at 22°C) in the presence of three hydrocarbon solvents (nC5, nC7, and nC12) [Telmadarreie and Trivedi 2017]. A high-quality camera and Scanning Electron Microscopy (SEM) were utilized to capture images. Besides decreasing the amount of asphaltene precipitation, the deposited asphaltene distributed more widely in the studied porous medium as the carbon number of solvent increased (C5–C12). Furthermore, the authors observed that asphaltenes could be deposited perpendicular or parallel to the flow direction. The perpendicular deposits were in small pore throats and could block the diffusion path. However, the parallel deposits did not significantly restrict the flow pathways [Telmadarreie and Trivedi 2017].

From the results presented in this section, it can be concluded that the subsurface upgrading of HO/B by using solvent deasphalting has been sufficiently demonstrated in one-dimensional and

PVT physical simulation experiments for Venezuela, Canadian, Brazilian, and Indonesian crude oils. The effects of pressure, temperature, solvent type, solvent-to-steam ratios, type of reservoir, and presence of clays have been determined. However, more data are needed on the characterization of the upgraded oils (e.g., SimDis, elemental analysis, heteroatom content, asphaltene composition, and stability, etc.) at different operating conditions to fully understand the advantages and disadvantages of SSU-SDA technologies.

5.2.2 Two-Dimensional Experiments: VAPEX

The VAPEX process (Figure 5.16) utilizes two horizontal wells and is closely related to the SAGD process (Figure 3.7), but the steam chamber is replaced by a zone containing hydrocarbon vapor near its dew point. The solvent (also known as "dry" VAPEX) or solvent/steam mixture (also known as "wet" VAPEX) is injected in the top well whereas production is carried out using the bottom well. The main aim of the VAPEX process is to remove the asphaltenes *in-situ*, which in turn, reduces the viscosity of the produced oil thereby increasing the quality of the oil. This process has been the subject of several review articles since 1998 [Das 1998, 2005, Butler and Yee 2002, Upreti *et al.* 2007, Vargas-Vasquez and Romero-Zerón 2007, Pourabdollah and Mokhtari 2013].

In 1993, Mokrys and Butler [1993] described physical simulation experiments on the VAPEX process using Lloydminster heavy oil reservoirs from Canada (13°API oil gravity with 16 wt. % of asphaltene content). The physical simulations were conducted in a 2-D cell containing a pair of horizontal wells like that for an SAGD process. A large number of thermocouples were placed to measure the temperature distribution in the cell. The authors ran the process in two different ways, i.e., "dry" and "wet" VAPEX. First, propane was merely cycled through the cell. The second test involved the injection of propane with steam at pressures varying from 708–984 kPa (88–130 psig) and at room temperature.

Figure 5.17 shows the cumulative oil production and percentage of recovery versus time for "dry" VAPEX [Mokrys and Butler 1993]. As seen, the recovery reached ~50% after 9 h. According to the authors, this value represented five years of field time. Similarly, the viscosity of produced oil was reduced from 10,000 mPa s to 1,000 mPa s at 25% production and kept decreasing until the experiment was stopped at 50% production. The authors reported that once the cell was opened, it

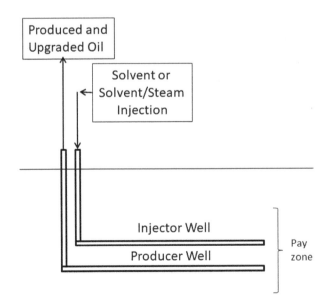

FIGURE 5.16 Schematic diagram of the VAPEX process.

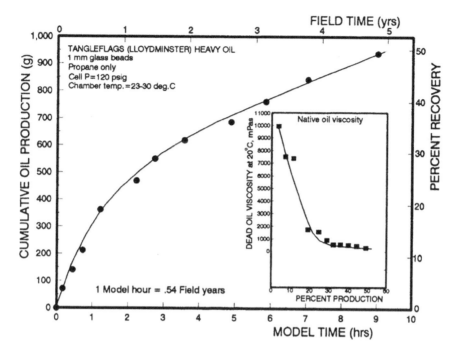

FIGURE 5.17   Cumulative oil production and percentage of recovery versus time for "dry" VAPEX (propane only) [Mokrys 1993]. Reprinted with permission from Society of Petroleum Engineering.

showed a large consolidated area of beads where the asphaltenes from the oil were deposited. The produced area of the cell was about 75% of the total while the amount of oil produced and weighed is only 50%. The difference was attributed to asphaltenes that remained in the 2-D cell [Mokrys and Butler 1993].

For wet VAPEX, the scaled model results showed that after three years of field time the steam–propane production of oil is comparable to that obtained by the SAGD method (steam only). The steam consumption for the steam–propane process was 69% lower than that for SAGD in the two-dimensional laboratory model. These results indicated that the dry and wet VAPEX processes led to recoveries that are roughly comparable with each other and that the steam–propane process was inherently more energy efficient than SAGD. The oil recovered by the steam–propane injection process was also reported to have a lower viscosity than the original oil by a factor of as much as 50-fold [Pourabdollah and Mokhtari 2013].

The mechanism for the wet VAPEX process was proposed to be in the following terms [Pourabdollah and Mokhtari 2013]: The steam formed a restricted hot zone near to the injector/producer wells, unable to travel further due to its relatively high dewpoint. Propane, by contrast, could go out beyond this zone, into the colder areas of the cell. Temperature measurements indicated that steam occupied a relatively small hot region local to the injector/producer wells. As the produced oil reached the hot zone, propane gas was removed from it and propagated back out toward the colder region in a kind of "internal recycling" process. As a result, the oil at the oil–propane interface benefited from the latent heat of solution of propane and spread progressively throughout the cell more efficiently than if steam alone had been used. Also, asphaltene deposition shifted the matrix wettability to be oil-wet and leading to decrease its permeability [Pourabdollah and Mokhtari 2013].

Upreti *et al.* [2007] reported that, during vapor extraction, the diffusion of solvent gas in HO/B was the first molecular phenomenon responsible for gas absorption and mixing with heavy oil and bitumen resulting in blends with lower viscosities. Thus, diffusion plays a significant role in VAPEX.

Das and Butler studied the effect of deposited asphaltenes in the permeability reduction in VAPEX process using propane and several Canadian heavy crude oils in a Hele-Shaw Cell [Das and Butler 1994]. The results showed that asphaltene deposition did not prevent the flow of oil through the reservoir for the proposed production scheme. On the contrary, flow rates were enhanced due to the viscosity reduction by SDA. It was observed that deasphalting takes place if the injected propane pressure was close to or higher than the vapor pressure of propane at the same temperature [Das and Butler 1994, Das 1998].

Haghighat and Maini [2010] studied the VAPEX process in a cylindrical model packed with 140–200 mesh sand with low permeability (~3 Darcy), using Mackay River Oil, propane and butane as solvents at 105 psi, and ambient temperature. The quality of the produced oil samples was evaluated through the viscosity measurements and SARA analysis. Results indicated lower oil production rates than expected due to asphaltene precipitation inside the porous media. On the other hand, five-fold reduction of the viscosity of the produced oil and 60% lower asphaltene content vs. the original crude oil were observed [Haghighat and Maini 2010]. The VAPEX process was field tested in Canada in 2005–2008. The results of the field tests are presented in Section 5.6.1.

5.2.3   HOT SOLVENT INJECTION

As discussed in Section 3.1.4, in solvent-only processes (also known as solvent flooding), the injection of hydrocarbon solvents is carried out cyclically or continuously into the reservoir to dissolve the oil without the need for thermal energy [Guo *et al.* 2016, Bayestehparvin *et al.* 2016]. In this case, warm VAPEX incorporates heat into the process by injecting superheated solvent [Haghighat *et al.* 2013]. This superheated vapor increases the temperature of the solvent–oil interface and causes an additional mixing due to condensation of the hydrocarbon at the interface [Rezaei *et al.* 2010].

Rezaei *et al.* carried out warm VAPEX experiments by injecting superheated pentane into rectangular packed physical models of 220 and 830 Darcy using Cold Lake bitumen (40,500 cP at 35°C) or Lloydminster heavy oil (5,400 cP at 35°C) at three levels of solvent vapor temperature (36, 43, and 50°C) [Rezaei *et al.* 2010]. Comparisons with conventional VAPEX were carried out by setting the temperature slightly higher than the dew point temperature of pentane at 37°C (±0.5°C). Figure 5.18 shows the effect of solvent temperature on the live oil production rates for the Cold Lake bitumen at high (830 Darcy, filled and hollow circles) and low (220 Darcy, filled and hollow squares) permeabilities. As seen, there was a significant increase in the oil production rate from the conventional VAPEX (hollow square and circle) to the warm VAPEX experiments at 36°C.

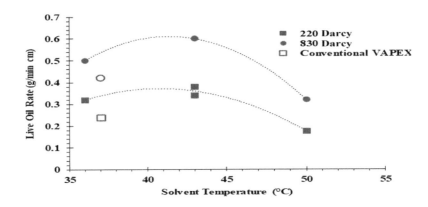

FIGURE 5.18 Live oil production rates (in grams per minute, per centimeter width of the model) for the warm and conventional VAPEX of Cold Lake bitumen at high (830 Darcy, filled and hollow circles) and low (220 Darcy, filled and hollow squares) permeabilities as a function of the solvent temperature. Data were taken from Rezaei [2010].

These increases in the live oil production rate at both permeability levels were attributed to solvent condensation upon the heat loss as well as convective mass transfer behavior of liquid condensate flowing on the interface of bitumen [Rezaei *et al.* 2010].

In Figure 5.18, the live oil production rate reached a maximum at the midlevel (46°C) while further increase in temperature (50°C) led to a decrease in the rate of production and solvent-to-oil ratio (results not shown). This reduction in the live oil production rate was explained by a lower amount of solvent condensing upon heat loss as well as by the lower solubility value of pentane in the bitumen at elevated temperature. Similar results were found for Lloydminster heavy oil (results not shown) [Rezaei *et al.* 2010].

Figure 5.19 shows the effect of solvent temperature on the percentages of C5-asphaltenes (A) and residual oil saturation inside the packed cell (B, in % of Pore Volume) for the warm VAPEX experiments of Cold Lake bitumen at high (830 Darcy) and low (220 Darcy) permeabilities [Rezaei *et al.* 2010]. As seen, at both permeabilities, the increase in the asphaltene content of the produced dead oil (Figure 5.19a) can be translated to a decrease in the extent of upgrading upon asphaltene precipitation. This reduction was expected because the increase in temperature causes an increase in the solubility of asphaltene fractions of the heavy oil. These results agree with those published by Anderson [Anderson 1995], Ovalles *et al.* [2015], and by Mitchel and Speight [Mitchell and Speight 1973], as discussed in Section 2.4.1 and Figure 2.9.

Similarly, the residual oil saturation increases with solvent temperature (Figure 5.19b), and for the case of the warm VAPEX experiments, a minimum value was determined at the solvent's

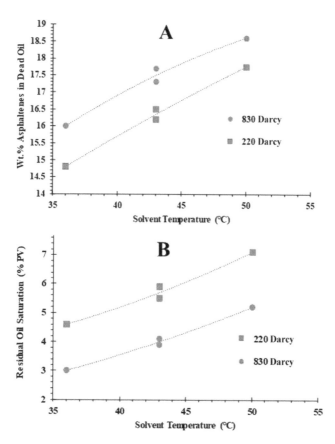

FIGURE 5.19 Percentage of C5-asphaltenes (A) and residual oil saturation inside the packed cell (B, in % of Pore Volume) for the warm VAPEX experiments of Cold Lake bitumen at high (830 Darcy) and low (220 Darcy) permeabilities as a function of the solvent temperature. Data were taken from Rezaei [2010].

bubble point temperature. In general, the asphaltene content analysis by Rezaei *et al.* revealed that asphaltene precipitation was more pronounced in warm VAPEX due to solvent condensation at the oil–vapor interface. Similar results were found for Lloydminster heavy oil (results not shown) [Rezaei *et al.* 2010].

Furthermore, researchers from the University of Calgary [Nourozieh *et al.* 2012] and Alberta [Pathak *et al.* 2012] have worked toward the fundamental understanding of hot solvent injection processes (no steam present). In general, it was found that, for low permeability media, asphaltene precipitation plays a significant role in oil recovery. The solvent type did not affect the recovery significantly, but the nature of the asphaltene precipitate could create a substantial impact on the dynamic of the process [Pathak *et al.* 2012].

Haghighat and Maini carried out warm VAPEX experiments in a large high-pressure physical model, packed with 250-Darcy sand, using propane as the solvent at 40, 50, and 60°C [Haghighat *et al.* 2013]. At the same injection pressure (817 kPa or 118 psi), there was an acceleration of oil rates at the early stages although the increasing temperature caused less dissolution of propane in bitumen. Increasing the temperature to 50°C without increasing the injection pressure improved production rate by 70%, and heating to 60°C was required to achieve the highest rate increase of 200%. When the pressure was increased to 1,550 kPa (225 psi), higher oil rates were observed due to the additional viscosity reduction caused by more solvent dissolution in bitumen. The results showed that the total rate of oil production was controlled by two mechanisms: (1) by solvent dissolution and oil mobilization at the boundaries of the vapor chamber, and (2) by gravity drainage beyond the vapor chamber toward the production well [Haghighat *et al.* 2013].

During the VAPEX process, the solvent stimulation of the near wellbore area is limited to a hot region created near the production well. This hot region is required to recover and recycle the solvent *in-situ*, to reduce the net solvent requirement and increase the upgraded oil production rate [Mokrys and Butler 1993, Frauenfeld *et al.* 2005, Rezaei *et al.* 2010, Ivory *et al.* 2010]. However, in the VAPEX process, it was observed that this stimulation procedure was not as effective in the presence of non-condensable gas (NCG). The accumulation of the NCG in the solvent chamber was found to inhibit the process of VAPEX during the *in-situ* solvent recycling process.

With these ideas in mind, the Nsolv process was developed to minimize or eliminate the poisoning effects of the NCG in VAPEX [Nenninger and Nenninger 2005]. In Nsolv, the solvent was allowed to condense at the bitumen interface by operating at the solvent's bubble point condition rather than the solvent dew point conditions, as it was conventionally implemented in the VAPEX process [Nenninger 2009]. With the bubble point conditions attained at the solvent–bitumen interface, the less volatile components of the oil also condensed along with the solvent; therefore, the vulnerability of the VAPEX process to the presence of less volatile components would be eliminated.

Additionally, the quality of produced oil is enhanced due to the *in-situ* upgrading through a solvent deasphalted mechanism. The Nsolv process was operated at higher pressures (within 10% of the reservoir's hydraulic fracturing pressure) and temperatures for better performance [Nenninger and Nenninger 2005, Nenninger and Dunn 2008, Nenninger 2009]. Based on an empirical correlation for up-scaling the VAPEX process results, it was estimated that Nsolv could produce considerably more oil than the SAGD process, with significantly lower energy consumption. The Nsolv process has been field tested since 2014, and the results are presented in Section 5.6.3.

5.3 SIMULATION OF ASPHALTENE PRECIPITATION

It is well known that asphaltene deposition can cause permeability reduction within the reservoir rock, sharp productivity decline of wells, and plugging of tubing, pumps, valves, and pipelines during production, transportation, and storage. Therefore, a model that accurately predicts asphaltene deposition is highly desirable [Cimino *et al.* 1995, Kelland 2009, Huc 2011, Al-Qasim and Bubshait 2017, Ramirez-Corredores 2017]. However, asphaltene precipitation does not necessary leads to the formation of asphaltene deposits. The simulation of asphaltene deposition in porous media at reservoir conditions is a challenging subject. Decades of experimental and theoretical research related to asphaltene behavior have only led to a partial understanding of asphaltene precipitation and deposition phenomena, with little scientific insight on how asphaltenes interact with the reservoir formation. As mentioned earlier, asphaltene deposition shifts the matrix wettability to be oil-wet, leading to decrease its permeability [Pourabdollah and Mokhtari 2013].

Currently, the simulation of asphaltene behavior and its effects on reservoir rocks is based on phenomenological modeling. Basically, it consists of the determination of model parameters by matching of core test data and using them to predict the effects of asphaltene precipitation and deposition on the behavior of a well(s) or a full field reservoir. As shown in Figure 5.20, the modeling supported by most commercial software consists in the mathematical description of several steps of asphaltene behavior in which precipitation and deposition are occurring simultaneously [Rogel 2012, Ovalles *et al.* 2017]. These steps are Asphaltene Precipitation Model, Asphaltene Flocculation Model, Asphaltene Deposition Model, Permeability Reduction, and Viscosity Model. These topics are discussed in the following sections [Rogel 2012, Ovalles *et al.* 2017].

5.3.1 PRECIPITATION MODELS

In the first step of Figure 5.20, the precipitated amount of asphaltenes is calculated based on the pressure, temperature, and composition of gases and liquids in the reservoir. Many thermodynamic models have been published in the open literature to perform such calculations. These models can be separated into two main groups: Those that consider the asphaltenes as colloidal entities dispersed in the crude oil with the help of the resins [Huc 2011, Leontaritis and Mansoori 1987, Victorov and Firoozabadi 1996, Wu *et al.* 1998, Pan and Firoozabadi 1998] and those that assume that asphaltenes are part of the crude oil forming true solutions. The latter models simulate the precipitation process as a liquid–liquid or liquid–solid equilibrium. The main problem with these approaches is that the more successful thermodynamic models require a large number of adjustable parameters (between three to five) which is impractical. Additionally, these thermodynamic models need extensive compositional characterization of the live crude oil and gases present in the reservoir [Eskin *et al.* 2016]. Commercial software packages (e.g., GEMS by Computer Modelling Group, Ltd or Eclipse by Schlumberger) can handle the thermodynamic modeling of asphaltene precipitation using an equation of states (EoS) and can work in the fully compositional model [Sammon 2003, Kohse and Nghiem 2004, Figuera *et al.* 2010, Foo *et al.* 2011].

An alternative to the use of a thermodynamic model is a table containing thermodynamic information obtained from PVT tests or equilibrium constants (K_j). It is also possible to evaluate

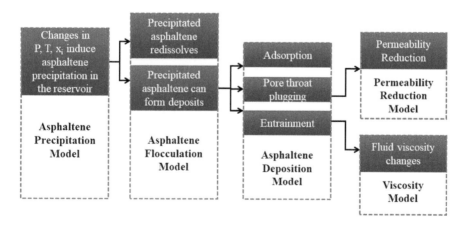

FIGURE 5.20 Steps supported by commercial software for the modeling of asphaltene precipitation and deposition.

asphaltene precipitation using a non-compositional model (black oil) just by using the experimental data as a direct input.

5.3.2 FLOCCULATION MODELS

The second step in Figure 5.20 is the flocculation modeling. It has been shown that asphaltene precipitation is a fairly reversible process [Hammami *et al.* 2000, Peramanu *et al.* 2001]. However, the redissolution process can take a considerably longer time that the precipitation step. To take into account this phenomenon, the simulation of asphaltene precipitation in reservoirs usually includes a flocculation model that takes into account the relatively long time required for the precipitated asphaltene to go back into solution. The reversibility of the process is modeled by considering that the solid, S1, is to be transformed using a chemical reaction into another solid, S2 [Kohse and Nghiem 2004]:

$$S1 \rightarrow S2 \tag{5.2}$$

The reaction rate for the formation of S2 can be calculated by:

$$r = k_{12}C_{11} - k_{21}C_{21} \tag{5.3}$$

Where k_{12} and k_{21} are the kinetic constants for the formation of S2 and its reverse reaction to S1, respectively. C_{11} and C_{21} are the concentrations of both species. Another important aspect of this mechanism is that solid S1 cannot form deposits, while solid S2 can. Values of $k_{12} = 0.01$ and $k_{21} = 0.0001$ have been used successfully to describe the CO_2 effect on asphaltene precipitation [Choiri 2010], $K_{12} = 0.1$ and $k_{21} = 0.08$ to study the asphaltene deposition in fracture reservoirs [Mirzabozorg *et al.* 2009], $k_{12} = 0.001$ and $k_{21} = 0.0001$ to evaluate asphaltene precipitation during the exploitation of the Marrat Reservoir in Kuwait [Yi 2009], and $k_{12} = 0.0001$ and $k_{21} = 0.00001$ to investigate the effects of diverse parameters on reservoir simulation performance [Al-Qasim 2011].

5.3.3 DEPOSITION MODELS

After the formation of asphaltene solid particles, the next step is to determine how these particles interact with the rock and identify the primary physical deposition processes (Figure 5.20). There are two different ways in which the asphaltene deposition process has been modeled. The simplest way is to consider that the deposition follows the Langmuir Adsorption Isotherm equation:

$$W_a = W_{a,max} \left(a\, C_2 / (1 + a\, C_2) \right) \tag{5.4}$$

Where W_a is the amount of adsorbed asphaltene, $W_{a,max}$ is the maximum amount of asphaltene that can be adsorbed, α is a constant, and C_2 is the asphaltene concentration in the fluid. This approach has been used in several research papers [Ali and Islam 1998, Kariznovi *et al.* 2008]. However, experimental evidence suggested that the dominant mechanism of asphaltene deposition is a physical blockage [Leontaritis 1998]. In fact, a careful study of asphaltene deposition during oil flow in core samples indicates that the formation of filtration cake (preceded by blockage of a significant part of the pore throats) is the deposition mechanism that better fits the data [Papadimitriou 2007, 2001]. Another important finding is the fact that the studied systems were in a dynamic state where deposited asphaltenes were drifted by the flow and either deposited again on clear areas of the core or left the core through its outlet [Papadimitriou 2001]. These findings seem to support the use of the second deposition model known as the Deep Bed Filtration Model (DBF) [Wan 2001, Boek *et al.* 2011]:

$$\frac{\delta V_2}{\delta t} = \alpha\, C_2\varphi - \beta\, V_2 \left(v_o - v_{cr} \right) + \gamma\, \mu_o C_2 \tag{5.5}$$

Where V_2 is the volume of deposited asphaltene, C_2 is the concentration of flowing flocculated asphaltene per volume of oil, v_o is the oil phase interstitial velocity, v_{cr} is the critical oil phase interstitial velocity, μ_o is the oil phase Darcy velocity, φ is the porosity, α is the surface deposition rate coefficient, β is the entrainment rate coefficient, and γ is the pore throat plugging rate coefficient defined as:

$$\gamma = \gamma_I \left(1 + \gamma_d V_2\right)$$ (5.6)

Where γ_I is the instantaneous pore throat plugging rate coefficient and γ_d is the snowball effect deposition constant.

This model takes into consideration that the asphaltene flocs can adsorb in the rock, a physical process by which the asphaltene precipitates adhere to the pore surface (first term of Equation 5.5) and are entrained and returned to the oil because of high, localized shear velocity (second term of Equation 5.5). This phenomenon happens when the interstitial speed is greater than some critical value. The pore space available for oil flow becomes smaller as asphaltene deposits accumulate. The interstitial velocity increases as the flow passages are narrowed down. When the interstitial velocity attains a specific critical value, some asphaltene deposits are swept away and moved with the flow. The third and last term represents the contribution because of the trapping of the flocs within the porous media. The pore throat plugging rate is directly proportional to the velocity u_o and the concentration in the liquid phase C_2.

DBF is available in commercial software packages. Its main drawback is that it requires at least four parameters. However, Boek *et al.* in a recent work showed that experimental data at short times could be reproduced by using only α (Equation 5.5), the surface deposition rate coefficient [Boek *et al.* 2011].

5.3.4 PERMEABILITY REDUCTIONS

Asphaltene precipitation decreases porosity and permeability and alters wettability of the reservoir rock as a consequence of the deposition (Figure 5.20). Also, it changes the crude oil characteristics. At least, two models have been proposed to evaluate changes in the reservoir rock: Resistance Factor Model [Nghiem 1998] and Power Law model [Kohse and Nghiem 2004].

In the Resistance Factor Model (RFM), the permeability changes in the rock related to asphaltene deposition are determined based on a permeability reduction factor. This factor (R_f) is calculated based on a maximum reduction factor ($R_{f,max}$) and the ratio between the actual amount of asphaltene deposit (W_a) and the maximum amount ($W_{a,max}$) [Nghiem 1998]:

$$R_f = 1.0 + \left(R_{f,max} - 1.0\right) W_a / W_{a,max}$$ (5.7)

Thus, the relative permeability (k_r) is calculated as:

$$k_r = k_{r0} / R_f$$ (5.8)

Where k_{ro} is the original relative permeability.

In the Power Law model, the permeability is related to the porosity by a power law:

$$k = a \left(\phi\right)^b$$ (5.9)

Where a and b are empirical coefficients that change depending on the type of rock. The porosity is calculated as a function of the initial porosity (ϕ_o) minus the amount of deposited asphaltenes V_2:

$$\phi = \left(\phi_o - V_2\right)$$ (5.10)

Porosity is calculated as the initial porosity minus the amount of deposited asphaltene. A resistance factor is calculated [Nghiem 1998].

The Power Law model [Kohse and Nghiem 2004] relates permeability (k) with porosity (φ) in the following way:

$$\frac{k_o}{k} = \left(\phi_o / \phi\right)^b \tag{5.11}$$

where k_o is the initial permeability and φ_o is the original porosity.

The resistance factor (R_f) is calculated as:

$$R_f = k_o / k = \left(\phi_o / \phi\right)^b \tag{5.12}$$

5.3.5 VISCOSITY MODEL

The last step in Figure 5.20 is the fluid viscosity changes. The effect of asphaltenes on the viscosity of HO/B was discussed in detail in Section 2.4. In this section, the most commonly used models are discussed to evaluate the effect of asphaltene depletion on the viscosity of heavy crude oils. In general, most of them require the knowledge of the viscosity of fluid μ_o in the absence of the asphaltenes [Al-Qasim 2011].

In the Generalized Einstein Model [Banchio and Nägele 1999], V_a is the volumetric concentration of asphaltenes in the fluid:

$$\mu / \mu_o = 1 + aV_a \tag{5.13}$$

Where a is a constant, 2.5 for spheres, and μ_o viscosity of the fluid when $V_a = 0$.

The Krieger and Dogherty Model [Wildemuth and Williams 1984] requires two parameters as shown in Equation 5.14. In this equation, the value of 2.5 represents ideal solution behavior and V_a is the volumetric concentration of asphaltenes. For spheres this is 0.65:

$$\mu/\mu_o = \left(1 - V_a/V_o\right) - 2.5V_o \tag{5.14}$$

In the Blending Model [Muñoz et al. 2016], extensive knowledge of the viscosity of different components present in the fluid is required, even though it can be reduced to only two components: Fluid without asphaltenes and asphaltenes.

$$\mu = \Sigma \, X_i\mu_I \tag{5.15}$$

where X_i and μ_I are the molar fraction and viscosity of component i in the blend.

5.4 NUMERICAL SIMULATIONS OF DOWNHOLE SOLVENT DEASPHALTING

The numerical simulation of solvent deasphalting (SDA) at reservoir conditions was recently reviewed by Pourabdollah and Mokhtari [2013]. As mentioned, most of the articles found in the literature are concentrated on evaluating the benefit of solvent and solvent deasphalting on production rates and oil recovery [Nghiem et al. 2001, Vargas-Vasquez and Romero-Zerón 2007]. Several authors have studied different aspects of this technology such as well locations and fracture spacing [Rahnema et al. 2008, Azin et al. 2005], gas dispersion and solubility [Kapadia 2004], solvent diffusivity [Kapadia 2004, Upreti et al. 2007], solvent injection rate [Das 2005], fracture and matrix permeability [Azin et al. 2005], and asphaltene precipitation [Nghiem 1998, Pourabdollah et al. 2010]. However, except for viscosity, *no other* physical and chemical properties of the produced oils in comparison with the original crude oils are calculated or reported. Thus, the effect of subsurface upgrading (SSU) cannot be precisely determined. This section is concentrated on describing the numerical simulation studies with particular emphasis on determining the degree of upgrading of the produced crude oils, not only in terms of viscosity reductions but increases in the API gravity and decreases of asphaltene, heteroatom, and metal contents.

Pourabdollah *et al.* carried out the numerical simulation of the 2-D experiments of VAPEX described earlier [Pourabdollah *et al.* 2010]. The viscosity pattern in the two-dimensional cell was calculated using a CMG simulator (version 2006 from the Computer Modeling Group). Based on the fluid characteristics, the viscosity of bitumen was computed using a modified Pedersen's equation. Figure 5.21 shows the distribution pattern of the oil viscosity in the VAPEX cell for the swept zone and bitumen chamber. As seen, the viscosity of residual oil in the swept region was in the range of 0.019–0.033 Pa s, while the viscosity of the fluid in the bitumen chamber remained at 18 Pa s. These changes represent a reduction of three orders of magnitude which is due to the presence of solvent within the VAPEX chamber [Pourabdollah *et al.* 2010]. No further characterization of the produced oils was reported.

Haghighat *et al.* carried out the numerical simulation of VAPEX experiments using a cylindrical cell and discussed in the Section 5.2.2 [Haghighat *et al.* 2013]. As before, the GEM simulator from CMG was conducted. The complete characterization of Mackay River bitumen was performed using the oil characterization module of VMG-Sim, and the CMG's commercial PVT package, WINPROP, was used for the definition of the pseudo-components of bitumen, property calculations, and matching the PVT data through regression. The numerical model was history matched with the experimental results (22–60°C), and the verified simulation model was used to extrapolate to higher temperatures (70–90°C). The results showed that the validated model was successful in predicting the oil rates and vapor chamber growth of the conducted VAPEX experiments with a minor adjustment to relative permeability parameters [Haghighat *et al.* 2013].

Figure 5.22 shows the viscosity-distribution maps within the vapor chamber in the simulation models at the extrapolated temperatures of 70°C and 90°C after nine hours [Haghighat *et al.* 2013]. As before, the viscosity of the oil in the solvent chamber is one order of magnitude lower than the untouched areas of the model. When the temperature is increased from 70°C to 90°C, oil drainage rate is improved not only due to the lower viscosity of the draining diluted oil but also due to faster oil production by gravity drainage [Haghighat *et al.* 2013].

Ovalles *et al.* [2017] developed a new asphaltene precipitation model that included four pseudo-components and three pseudo-chemical reactions (Equation 5.16–5.18). The first one (Equation 5.16) considers the phase separation of asphaltenes in going from the liquid to the solid phase as it was experimentally shown in the PVT and propane flood experiments discussed earlier

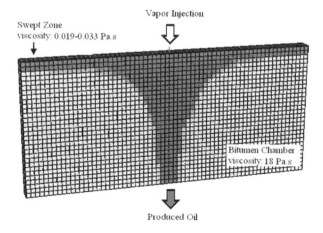

FIGURE 5.21 Distribution pattern of the oil viscosity in the VAPEX cell for the swept zone and bitumen chamber [Pourabdollah *et al.* 2010]. Reprinted with permission from Society of Petroleum Engineering.

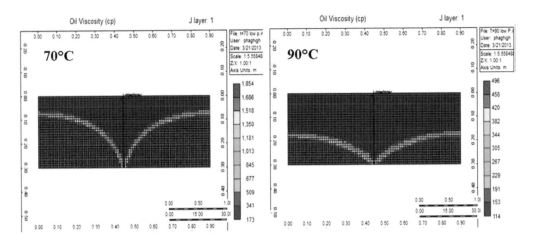

FIGURE 5.22 Viscosity-distribution maps within the vapor chamber in the simulation models at T = 70 and 90°C after nine hours [Haghighat *et al.* 2013]. Reprinted with permission from Society of Petroleum Engineering.

(Section 5.2.1). Equation 5.17 considers the reversibility behavior observed during light hydrocarbons and CO_2 asphaltene precipitation studies [Kohse and Nghiem 2004, Mirzabozorg *et al.* 2009]. Finally, Equation 5.18 describes the formation of a Heavy Fraction which was observed during the propane flood experiment. By using STARS (from Computer Modeling Group), this asphaltene precipitation model was validated by history match of the one-dimensional C3 flood experiments described earlier.

$$\text{Asphaltene-1 (1)} + nC3 \xrightarrow{k_1} \text{Asphaltene- 2 (s)} + nC3 \tag{5.16}$$

$$\text{Asphaltene-2 (s)} + mC3 \xrightarrow{k_2} \text{Asphaltene- 1(1)} + mC3 \tag{5.17}$$

$$\text{Asphaltene-2 (s)} + pC3 \xrightarrow{k_3} 1.39 \text{ Heavy F.(l)} + pC3 \tag{5.18}$$

Next, the results of Arrhenius pre-exponential factors A for pseudo-reactions (5.16 and 5.17), asphaltene deposition, and permeability reductions were transferred, with the appropriate units and scaling factors applied [Ovalles *et al.* 2017]. A one-well pair in an SAGD configuration was simulated using typical properties for a heavy oil reservoir (7.7°API gravity, viscosity = 67,678 cP at 40°C, porosity = 32.5%, permeability = 12D, 1,012 psi, and 120°F). Three cases were evaluated: Base Case (steam only, no propane injection), Technology Case 1 (10% liquid propane in cold water equivalent of steam), and Technology Case 2 (50% liquid propane in cold water equivalent steam) [Ovalles *et al.* 2017].

Figure 5.23 shows the effect of propane concentration on the cumulative production and oil rate obtained by numerical simulation of SSU-SDA in a one-well pair in an SAGD configuration [Ovalles *et al.* 2017]. As seen, accelerated oil production was achieved in both Technology Cases (dashed and gray lines) in comparison with Base Case (steam only, solid black line). As discussed in Section 3.1.3.3, these results are consistent with those reported in the literature in regard to the effects of solvent during SAGD operations [Akinboyewa *et al.* 2010, Duhaime and Moir 2010, Gupta *et al.* 2005, Jha 2012, Lin *et al.* 2014].

Figure 5.24 shows the effect of propane concentration on the cumulative API gravity and viscosity of the crude oil produced as calculated by numerical simulation of SSU-SDA in a one-well pair in a SAGD configuration [Ovalles *et al.* 2017]. By using 10% of propane (Figure 5.24, lighter

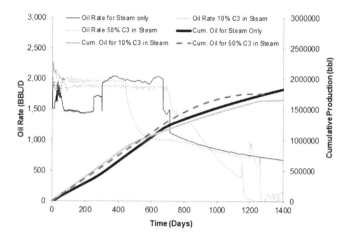

FIGURE 5.23 Effect of propane concentration on the cumulative production and oil rate obtained by numerical simulation of SSU-SDA in a one-well pair in an SAGD configuration. Data were taken from Ovalles [2017].

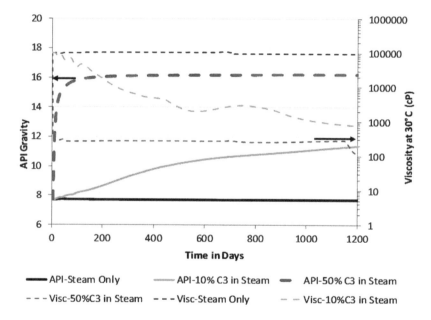

FIGURE 5.24 Effect of propane concentration on the cumulative API gravity (left-hand side y-axis) and viscosity (right-hand side y-axis) of the crude oil produced by numerical simulation of SSU-SDA in a one-well pair in an SAGD configuration.

solid line) a modest increase in the API gravity was obtained with values reaching 10°API (2°API increase) at the end of the simulation whereas no upgrading was obtained in the Base Case (black solid line, steam only). These results are consistent with the observed increases in the API of the produced crude oil as reported by Encana [Gupta *et al.* 2005] and Imperial Oil [Duhaime and Moir 2010] during their field tests of Solvent-SAGD and LASER tests, respectively (Section 5.5).

The use of a higher propane-to-steam ratio (1:1) led to an increase of ~8°API (up to 16°API) which was achieved at ~80 days into the simulations (Figure 5.24, thicker dashed line). This result is consistent with the 1-D [Vallejos *et al.* 2002, Oliveira *et al.* 2017, Ovalles *et al.* 2017, Moreno-Arciniegas and Babadagli 2014, Mamora *et al.* 2003] and 2-D [Mokrys and Butler 1993,

Pourabdollah and Mokhtari 2013, Haghighat and Maini 2010, Rezaei *et al.* 2010] experiments as discussed earlier. Furthermore, Figure 5.24 also shows (lighter dashed line, right-hand side y-axis) that the upgraded crude oil (16°API) has an estimated viscosity of < 300 cP at 30°C (surface conditions) which makes it transportable using conventional pipelines without the need for diluent or heated lines. In this case, an upgrader unit would not be needed which makes this case attractive from the business point of view and could offset the relatively high solvent-to-steam ratio used to achieve crude oil upgrading [Ovalles *et al.* 2017].

From the results presented in this section, it can be concluded that the downhole numerical simulation of subsurface upgrading of solvent deasphalting (SSU-SDA) has been effectively carried out in the literature and that the results are consistent with the physical simulation experiments for Canadian and Venezuelan HO/B. As before, more data is needed for predicting the properties of the produced oils (e.g., SimDis, elemental analysis, heteroatom content, asphaltene composition, and stability, etc.) at variable operating conditions to fully evaluate the application of SSU-SDA to different HO/B reservoirs across the globe.

5.5 FIELD TESTS

As discussed in Section 3.1.3, there have been several solvent–steam co-injection field tests that have found increases in the API gravity of the produced oil throughout the life of the project. As shown in Table 3.1, in the EnCana Solvent Aided Process (Semlac), butane was added to steam to increase oil rate by 1,100 B/d and to reduce steam-to-oil ratio (SOR) from 2.6 to 1.6. They observed that the API of the produced oil increased from 12.7° to 13.7°API [Gupta *et al.* 2005]. Similarly, the Cristina Lake project, 15 wt. % pf C4 was used to increase the oil rate from 630 B/d to 1,900 B/d and to reduce the SOR from five to one. As before, they also found an increase of the API gravity of the produced oil by 1°API [Gupta and Gittins 2006]. For both cases, the authors attributed the increase in API gravity to subsurface upgrading via solvent deasphalting [Gupta *et al.* 2005, Gupta and Gittins 2006].

In the same order of ideas (see Section 3.1.3.1), Imperial Oil reported the use of a process called Liquid Addition to Steam for Enhanced Recovery (LASER) to produce Clearwater and Cold Lake reservoirs in Alberta, Canada [Leaute 2002, Leaute and Carey 2007, Stark 2013]. This process involves adding a small amount of C_5-condensate solvent, or diluent, to the steam under cyclic steam stimulation. Their commercial expectations were to increase 3–5% the recovery factor, reduction of the SOR, and small increases (~1°API) in the API gravity of the produced oil [Duhaime and Moir 2010].

In this section, three field tests are described aimed to increase the oil rates, cumulative oil production but also to upgrade the HO/B subsurface. The fields trials are VAPEX, Cyclic Propane Injection, and N-Solv.

5.5.1 VAPEX

The VAPEX involves the injection of solvent vapor (typical light hydrocarbons in the C3 to the C5 range) into an upper horizontal well in a conventional SAGD configuration pair (see Figure 5.16 and 5.25) [Mokrys and Butler 1993, Das 1998, 2005]. As discussed in Section 5.2.2, the hydrocarbon dissolves into the viscous oil making it mobile enough to drain down to the production well. Laboratory experiments have shown that asphaltene precipitates *in-situ* which results in a 6–7°API increase.

The VAPEX process was field tested at the Dovap site in Canada. This project was a Can $30 MM, well-instrumented, seven-year field pilot. Its design consisted of one cold-start-up well pair and one hot-start-up well pair. Figure 5.25 shows the well array and surface process diagram for the VAPEX field test. As seen, the surface facilities were equipped to inject downhole a methane/propane gas mixture and to separate the produced fluids into gas (sent to flare) and solvent-free oil. Steam facilities were included as well.

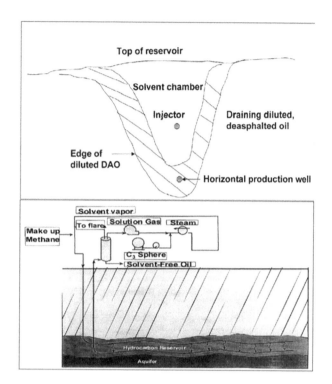

FIGURE 5.25 Well array and surface process diagram for the VAPEX field test.

Figure 5.26 shows the gas (methane and propane) and oil rates vs. time for the cold-start-up well pair of the VAPEX field test. Unfortunately, the cold-start-up well pair suffered from lack of communication and failed to give the expected results. From September 2005 to July 2006, the oil rate was less than ~100 B/d (~15 m³/d) even though upgrading from 10.5°API to 16.8°API was observed. Reductions in the percentages of asphaltenes (from 17 wt. % to 1 wt. %) and the viscosity were also found.

Figure 5.27 shows the temperature of the well pair (A), and steam, oil, and solvent rates (B) vs. time for the hot test start-up well pair of the VAPEX field test. A seen, the SAGD portion (from July 2004 to January 2005) performed as expected, i.e., temperatures of ~300°F from toe to heel (Figure 5.27a), up to 600 B/d (100 m³/d) of oil production with SOR ≈ 2.6 (Figure 5.27b). Unfortunately, the transition from SAGD to VAPEX was unsuccessful, and oil production fell to less than 120 B/d (~20 m³/d). As before, upgrading from 10.5°API to 12°API was observed with

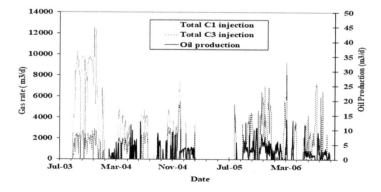

FIGURE 5.26 Gas and oil rates vs. time for the VAPEX field test (*Cold pair*).

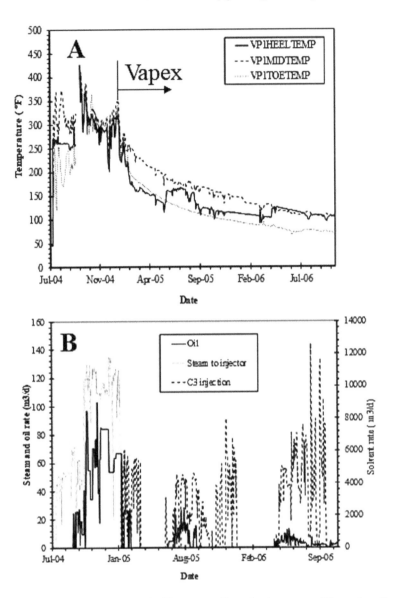

FIGURE 5.27   Temperature of the well pair (**A**), steam, oil, and solvent rates (**B**) vs. time for the VAPEX field test (*Hot pair*).

reductions in the percentage of asphaltenes (from 17 wt. % to 5 wt. %) and the viscosity. The project was discontinued in 2008 [Pourabdollah and Mokhtari 2013].

5.5.2 Cyclic Propane Injection

Cyclic propane injection was field tested in an eastern Venezuela reservoir (< 10°API) in 2000, and it is based on the downhole injection of a light solvent (propane) at reservoir conditions (130°F, 1,000 psi, no steam was used) in a conventional "Huff and Puff" production process (see Section 3.1). Following the injection step, the same well is then operated under production conditions, and after separation of the light solvent, the upgraded hydrocarbon product is obtained.

In the field test, 10,000 barrels of LNG (> 80% propane) were injected downhole, and the well was put into production without soaking time. The API gravity and the percentages of the heavy

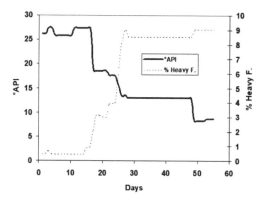

FIGURE 5.28 API gravity and percentage of the heavy fraction of the upgraded crude oil produced during the field test of cyclic propane injection.

fraction (defined as insolubles in paraffins) in the produced oil were monitored, and the results are shown in Figure 5.28. As can be seen, 26–27°API crude (darker trace) with a minimal amount of heavy fraction (lighter line) were produced during the first 15 days of the test. After that, a reduction in the API gravity and an increase in the heavy fraction of the produced oil were observed reaching its original values (8.7°API and 9%) after approximately 50 days. The overall API gravity of the cumulative oil was 20°API.

Also, a significant increase in the oil rate from 50 to 300 B/d was observed with a cumulative production of approximately 10,000 barrels of 20°API crude oil at the end of the 50-day period. From these results, it can be concluded that the solvent to upgraded crude oil ratio is approximately 1:1 v/v with 95% solvent recovery. Furthermore, during the field test, a reduction of the water cut was observed along with 97 wt. % a metal removal (V + Ni). At the end of the test, no formation damage due to the asphaltene precipitation was observed because the original oil rate was obtained (50 B/d).

5.5.3 Nsolv Technology

As discussed in Section 5.2.3, this process pumps downhole warm light alkanes (propane or butane) at approximately 60°C into an SAGD well configuration (see Figure 5.29) [Nenninger and Nenninger 2005, Nenninger and Dunn 2008, Nenninger 2009]. When the alkane solvent interacts with the sands underground (temperature is between 6 and 13°C year-round) it condenses, and the bitumen dissolves into the liquefied alkane. By operating at slightly elevated temperatures, the light alkane warms the bitumen, thus accelerating the extraction rate [Nenninger and Nenninger 2005, Nenninger and Dunn 2008, Nenninger 2009].

Due to the high solvent-to-bitumen ratio (4:1), *in-situ* precipitation of asphaltenes takes place leaving the sulfur and metal contaminants behind. It is claimed that the produced oil contains less sulfur, heavy metals (zinc, vanadium, iron), and carbon residue. This process partially upgrades the oil to 13–16°API from a value of approximately 8°API for the raw bitumen. The produced oil is also less viscous and thus, requires less diluent for pipeline transportation to the refinery [Nenninger and Nenninger 2005, Nenninger and Dunn 2008, Nenninger 2009].

Nsolv Technology is also claimed to be less vulnerable to non-condensable gas contaminations (methane, ethane, CO_2, etc.) in comparison to VAPEX. If the non-condensable gas is not adequately vented, accumulation of non-condensable gases within the chamber may interfere in mass and heat transfer mechanisms at the chamber boundary. Condensation of the solvent at the oil–solvent interface during Nsolv is said to provide the advantage of carrying most of the non-condensable gas to the production well and preventing its buildup within the chamber [Nenninger and Nenninger 2005, Nenninger and Dunn 2008, Nenninger 2009].

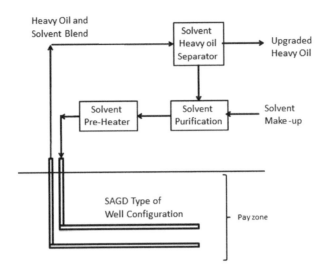

FIGURE 5.29   General scheme of the Nsolv process.

Nsolv began field operations with the launch of its pilot plant at Suncor's Dover lease in January 2014. The Bitumen Extraction Solvent Technology (BEST) test facility is a 500 barrel-per-day unit, comprising a 300 m horizontal well pair placed in a relatively shallow, thin occurrence of the McMurray formation (15 m pay zone with 75% oil saturation, 35% porosity, and 5 Darcy of permeability). The BEST pilot has a surface plant for processing the produced hydrocarbons and seven observation wells for monitoring and control [Nsolv 2017].

In March 2015, the BEST pilot plant started the injection of heated C4 into an oil sands reservoir leading to a partially upgraded oil (from 8–9° to 13–14°API). After two years of operation, the pilot plant produced over 125,000 barrels of upgraded crude oil, while meeting all the following key performance indicators: Extraction rate greater than SAGD with zero water usage, reduction of GHG Emissions by 75–80% relative to SAGD, produced oil with nickel and vanadium contents less than 125 ppm, and Micro-carbon residue of ~5 wt. %. The solvent recovery is expected to be greater than 90% after blow-down [Krawchuk 2016].

It was reported that no foaming or crude oil emulsification issues were observed at the dehydration/dewatered unit and no de-emulsification chemical was required. Similarly, no asphaltene deposits were found at the surface facilities. It seemed that the asphaltenes were sequestered by the matrix, and no sign of permeability degradation was seen. However, the mobility of the C4-asphaltenes continued to be "a key question" at reservoir conditions [Eichorn 2016].

At the 2016 World Heavy Oil Congress, a study of the development of the solvent chamber was presented using data collected from the field trial after one year of operation (March 2015–March 2016). The data included the following information: 4-D seismic imaging, observation well thermocouples and piezometers, observation well wireline logging, and horizontal well fiber-optic temperature sensors. Results showed good agreements between the thermocouple and seismic data in terms of solvent chamber dimension (50–60 m at the top and 40–50 m at the middle) and growth (~2.5–2.7 cm per day horizontally and 3.2 cm/day vertically). The data also showed how the solvent chamber interacted with the geological heterogeneities. The numerical model was history matched which allowed prediction of the test performance [Eichorn 2016].

In early 2017, Nsolv began shutting down the plant, marking the end of a highly successful project. It was claimed that Nsolv was a game-changing technology with both economic and environmental advantages. Currently, the company is looking for a variety of partnerships for developing heavy oil reserves around the globe [Nsolv 2017].

5.6 ASPHALTENE PRECIPITANTS

As discussed in the preceding sections, all the SSU concepts and processes that use solvent deasphalting require a relatively high solvent-to-crude ratio (SvOR from 1:1 to 10:1 v/v) to induce subsurface asphaltene precipitation, increase oil production, and generate an upgraded crude oil. Figure 5.30 shows the operating temperature versus the SvOR for various steam and solvent–steam co-injection processes. As seen, the SSU-SDA processes require a higher solvent-to-crude ratios than SAGD and ES-SAGD. Also, the first two processes need to recycle the solvent to improve the economic prospects. Therefore, costly surface facilities are required. In this order of ideas, the use of additives that can act as asphaltene precipitants to reduce the SvOR represents a feasible alternative to decrease capital and operating expenditures and improve the economic benefits of the process [Ovalles *et al.* 2016, Ovalles and Rogel 2017, Rogel *et al.* 2017].

A potential application of asphaltene precipitants is depicted in Figure 5.31 [Ovalles *et al.* 2016, Ovalles and Rogel 2017, Rogel *et al.* 2017]. As seen, these additives can be mixed with the solvent and injected downhole to induce subsurface asphaltene precipitation with the concomitant reduction of the SvOR. This process can be carried out using vertical and horizontal wells or in SAGD configuration to generate an upgraded crude oil. The asphaltene precipitants are not expected to be recovered because they remain associated with the asphaltenes downhole. Also, Figure 5.31 shows a solvent recovery unit to improve the economic prospect of the process [Ovalles *et al.* 2016, Ovalles and Rogel 2017, Rogel *et al.* 2017].

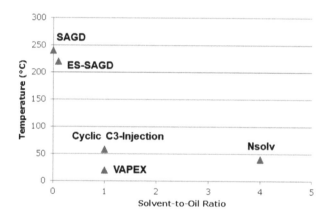

FIGURE 5.30 Operating temperature vs. solvent-to-oil ratio for various steam and solvent–steam co-injection processes.

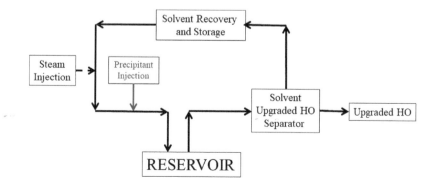

FIGURE 5.31 Diagram for the process for subsurface upgrading of HO/B via solvent deasphalting using asphaltene precipitant additives.

Lab scale experiments have shown that benzoyl peroxide, 4-vinyl pyridine methacrylate, 4-vinyl phenol methacrylate, poly (maleic anhydride), and Fe_2O_3 and NiO nanoparticles are active asphaltene precipitants for Venezuelan and Canadian HO/B. Figure 5.32 shows four optical microscopy photographs of the effect of asphaltene precipitants in a 4:1 w/w n-heptane/heavy crude oil blend with no asphaltene precipitant (**A**), 100 ppm of benzoyl peroxide (**B**), Fe_2O_3 nanoparticles (**C**), and 100 ppm of poly (maleic anhydride) (**D**) [Ovalles *et al.* 2016, Ovalles and Rogel 2017]. As seen, in the absence of asphaltene precipitant additive, a minimal amount of asphaltene particles with dimensions lower than 25 μm was observed (Figure 5.32A). The addition of 100 ppm of the benzoyl peroxide (Figure 5.32B), Fe_2O_3 nanoparticles (Figure 5.32C), or poly (maleic anhydride) (Figure 5.32D) led to the formation of a larger amount of asphaltenes particles. These experiments clearly demonstrated that the addition of asphaltene precipitants (peroxide and metal nanoparticles) increased the amount of precipitated asphaltenes in hydrocarbon solvent/heavy crude oil blends [Ovalles *et al.* 2016, Ovalles and Rogel 2017, Rogel *et al.* 2017].

Using benzoyl peroxide and nickel and iron nanoparticles, the percentages of C7-asphaltenes were measured. Up to ~20 wt. % of increase in asphaltene content was observed in comparison with the case without the additive. Spectroscopic and mechanistic studies using benzoyl peroxide as precipitant indicated a free radical mechanism. Whereas in the presence of nickel and iron-containing precipitants, most of these metals are found in the asphaltenes, showing that the nanoparticles are acting as nucleation sites [Rogel *et al.* 2017].

In the same order of ideas, Pourabdollah *et al.* used montmorillonite nanoclay particles as additives to enhance heavy oil recovery during the VAPEX process [Pourabdollah *et al.* 2011]. The lab experiments were carried out using Iranian heavy oil and propane. The setup consisted of two sand-packed cells, one packed only with glass beads as the oil matrix and the other with glass beads and modified with the montmorillonite nanoclay. Both cells had similar porosity and permeability. The results showed an increase in oil recovery from 25% to 31% of OOIP and a reduction of one order of magnitude of the produced oil viscosity in the nano-assisted VAPEX compared to those found in the conventional VAPEX. It was proposed that the montmorillonite nanoclay particles adsorbed asphaltenes and asphaltene micelles into their interlayer spaces, which in turn led to viscosity reduction and higher oil rates [Pourabdollah *et al.* 2011].

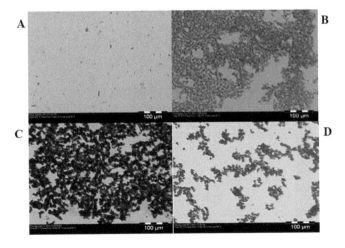

FIGURE 5.32 Optical microscopy photographs of the effect of asphaltene precipitants in a 4:1 w/w n-heptane/heavy crude oil blend. (A) No asphaltene precipitant, (B) 100 ppm of benzoyl peroxide, (C) 100 ppm of Fe_2O_3 nanoparticles, and (D) 100 ppm of poly (maleic anhydride).

REFERENCES

Akinboyewa, J., Das, S. K. Wu, Y.-S., Kazemi, H., 2010, "Simulation of Expanding Solvent - Steam Assisted Gravity Drainage in a Field Case Study of a Bitumen Oil Reservoir", SPE No. 129963, presented at the SPE Oil Recovery Symp., Tulsa, OK, 24–28 April.

Ali, M. A., Islam, M. R., 1998, "The Effect of Asphaltene Precipitation on Carbonate-Rock Permeability: An Experimental and Numerical Approach", SPE No. 50963, *SPE Prod. Facil.*, 13, 178–184.

Al-Murayri, M. T., Maini, B. B., Harding, T. G., Javad Oskouei, J., 2016, "Multicomponent Solvent Co-Injection with Steam in Heavy and Extra-Heavy Oil Reservoirs", *Energy Fuels*, 30, 2604–2616.

Al-Qasim, A., Bubshait, A., 2017, "Asphaltenes: What Do We Know so Far", SPE No. 185690, presented at SPE Western Regional Meeting, Bakersfield, California, 23–27 April.

Al-Qasim, A. S., 2011, "Simulation of Asphaltene Deposition during CO_2 Flooding", Master Thesis, The University of Texas at Austin.

Andersen, S. I., 1995, "Effect of Precipitation Temperature on the Composition of n-Heptane Asphaltenes, Part 2", *Fuel Sci. Technol. Int.*, 13, 579 and references therein.

Azin, R, Kharrat, R, Ghotbi, C, Vossoughi, S., 2005, "Applicability of the VAPEX process to Iranian Heavy Oil Reservoirs", SPE No. 97766, presented at SPE International Thermal Operations and Heavy Oil Symposium, Calgary, Alberta, Canada, 1–3 November.

Banchio, A. J., Nägele, G., 1999, "Viscoelasticity and Generalized Stokes–Einstein Relations of Colloidal Dispersions, *J. Chem. Phys.*, 111, 8721.

Bayestehparvin, B., Ali, S. M. F., Abedi, J., 2016, "Use of Solvents with Steam – State-of-the-Art and Limitations", SPE No. 179879, present at SPE EOR Conference at Oil and Gas West Asia, 21–23 March and references therein.

Bayliss, P., Levinson, A. A., 1976, "Mineralogical Review of the Alberta Oil Sand Deposits (lower Cretaceous, Mannville group)", *Can. Pet. Geol.*, 24(2), 211–224.

Benson, I. P., Ovalles, C. F., 2017, "Method for Upgrading In Situ Heavy Oil", US. Patent No. 9,739,125, 22 August.

Blevins, T. R., Duerksen, J. H., Ault, J. W., 1984, "Light-Oil Steamflooding–An Emerging Technology", SPE 10928, *J. Pet. Technol.*, 36, 1115, July.

Boek, E. S., Fadili, A., Williams, M. J., Padding, J., 2011, "Prediction of Asphaltene Deposition in Porous Media by Systematic Upscaling from a Colloidal Pore Scale Model to a Deep Bed Filtration Model", SPE No. 147539, presented at SPE Annual Technical Conference and Exhibition, Denver, CO, 30 October–2 November.

Butler, R.M., Yee, C.T., 2002, "Progress in the In-Situ Recovery of Heavy Oils and Bitumen", *J. Can. Pet. Technol.*, 41, January 31.

Choiri, M., 2010, "Study of CO_2 Effect on Asphaltene Precipitation and Compositional Simulation of Asphaltenic Oil Reservoir", Master Thesis, University of Stavanger, Norway.

Cimino, R., Correra, S., Del Bianco, A., Lockhart, T. P., 1995, "Solubility and Phase Behavior of Asphaltenes in Hydrocarbon Media", In: *Asphaltenes: Fundamentals and Applications*, E. Y. Sheu, O. C Mullins, Ed. Plenum Press, New York, and references therein.

Coelho, R.S.C., Ovalles, C., Benson, I.P., Hascakir, B., 2017, "Effect of Clay Presence and Solvent Dose on Hybrid Solvent-Steam Performance", *J. Pet. Sci. Eng.*, 150, 203–207.

Coelho, R.S.C., Ovalles, C., Hascakir, B., 2016, "Clay-Asphaltene Interaction during Hybrid Solvent-Steam Injection into Bitumen Reservoirs", SPE-180723 presented at SPE Canada Heavy Oil Technical Conference, Calgary, Alberta, Canada, 7–9 June.

Das S., 2005, "Diffusion and Dispersion in the Simulation of VAPEX Process", SPE No. 97924, presented at SPE International Thermal Operation and Heavy Oil Symposium, Calgary, Alberta, Canada, 1–3 November and references therein.

Das, S. K., 1998, "VAPEX: An Efficient Process for the Recovery of Heavy Oil and Bitumen", Paper SPE 50941, *SPE J.*, 3, 6, 22 September.

Das, S. K., Butler, R. M., 1994, "Effect of Asphaltene Deposition on the VAPEX Process: A Preliminary Investigation Using a Hele-Shaw Cell", *J. Can. Pet. Technol.*, 33(6), 39.

Duerksen, J. H., Hsueh, L., 1983, "Steam Distillation of Crude Oil", *Soc. Pet. Eng. J.*, 23, 265, April.

Duhaime, C., Moir, R., 2010, "ERCB Annual LASER Review" on 16 April, in www.ercb.ca/docs/products/osp rogressreports/2010/2010ImperialOilAnnualLASERReview.pdf (Dec. 14, 2011).

Eichorn, M., 2016, "Observations and Predictions on Field-Scale Solvent Chamber Development", Paper No. WHOC16-134, presented at the 2016 World Heavy Oil Congress, Calgary, Alberta, Canada, 6–9 September.

Eskin, D., Mohammadzadeh, O., Akbarzadeh, K., Taylor, S. D., Ratulowski, J., 2016, "Reservoir Impairment by Asphaltenes: A Critical Review", *Can. J. Chem. Eng.*, 94, 1202–1217.

Figuera, L., Marin, M., Lopez, G. L. R., Marin, E., Gammiero, A., Granado, C., 2010, "Characterization and Modelling of Asphaltene Precipitation and Deposition in a Compositional Reservoir", SPE No. 133180, presented at SPE Annual Technical Conference and Exhibition, Florence, Italy, 19–22 September.

Foo, Y. Y., Chee, S. C., Zain, Z. M., Mamora, D. D., 2011, "Recovery Processes of Extra Heavy Oil - Mechanistic Modelling and Simulation Approach", SPE No. 143390, presented at SPE Enhanced Oil Recovery Conference, Kuala Lumpur, Malaysia, 19–21 July.

Frauenfeld, T., Jossy, C., Wang, X., 2005, "Experimental Studies of Thermal Solvent Oil Recovery Process for Live Heavy Oil", presented at the Canadian International Petroleum Conference, Calgary, Alberta, Canada, PETSOC-2005-151, 7–9 June.

Guo, K., Li, H., Yu, Z., 2016, "In-Situ Heavy and Extra-Heavy Oil Recovery: A Review", *Fuel*, 185, 886–902, and references therein.

Gupta, S., Gittins, S., Picherack, P., 2005, "Field Implementation of Solvent Aided Process", *J. Can. Pet. Technol.*, 44, 8.

Gupta, S. C., Gittins, S. D., 2006, 'Christina Lake Solvent Aided Process Pilot", *J. Can. Pet. Technol.*, 45(9), 15–18.

Gupta, S. C., Gittins, S., Canas, C., 2012, "Methodology for Estimating Recovered Solvent in Solvent-Aided Process", SPE-136402, *J. Can. Pet. Technol.*, 51(5), 339.

Haghighat, P., Maini, B. B., 2010, "Role of Asphaltene Precipitation in VAPEX Process", *J. Can. Pet. Technol.*, 49(3), 14.

Haghighat, P., Maini, B. B., Abedi, J., 2013, "Experimental and Numerical Study of VAPEX at Elevated Temperatures", SPE No. 165467, presented at SPE Heavy Oil Conference-Canada, Calgary, Alberta, Canada, 11–13 June.

Hammami, A., Phelps, C. H., Monger-McClure, T., Little, T. M., 2000, "Asphaltene Precipitation from Live Oils: An Experimental Investigation of Onset Conditions and Reversibility", *Energy Fuels*, 14, 14–18.

Hong, K. C., 1994, *Steamflood Reservoir Management*, PennWell, Tulsa, Chapter 1–3 and references cited therein.

Hosseininejad-Mohebati, M., Maini, B. B., Harding, T. G., 2012, "Experimental Investigation of the Effect of Hexane on SAGD Performance at Different Operating Pressures", SPE No. 158498, presented at SPE Heavy Oil Conference Canada, Calgary, Alberta, Canada, 12–14 June.

Huc, A.-Y., 2011, *Heavy Crude Oils: From Geology to Upgrading: An Overview*, Technip, Chapter 23 and references therein.

Ivory, J., Frauenfeld, T., Jossy, C., 2010, "Thermal Solvent Reflux and Thermal Solvent Hybrid Experiments", SPE- No. 133202, *J. Can. Pet. Technol.*, 49(2), 23.

Jamaloei, B. Y., Dong, M., Mahinpey, N., Maini, B. B., 2012, "Enhanced Cyclic Solvent Process (ECSP) for Heavy Oil and Bitumen Recovery in Thin Reservoirs", *Energy Fuels*, 26(5), 2865–2874.

Jha, R. K., Kumar, M., Benson, I., Hanzlik, E., 2012, "New Insights into Steam-Solvent Co-Injection Process Mechanism", SPE No. 159277, presented at the SPE Annual Technical Conf. and Exhibition held in San Antonio, TX, 8–10 October.

Kapadia R.A., 2004, "Dispersion Determination in VAPEX: Experimental Design, Modeling and Simulation", *Masters Abstr. Int.*, 47(1).

Kar, T., Mukhametshina, A., Unal, Y., Hascakir, B., 2015, "The Effect of Clay Type on Steam-Assisted-Gravity-Drainage Performance", *J. Can. Pet. Technol.*, 54(6), 412–423.

Kariznovi, M., Jamialahmadi, M., Shahrabadi, A., 2008, "Optimization of Asphaltene Deposition and Adsorption Parameter in Porous Media by Using Genetic Algorithm and Direct Search", SPE No. 114037 presented at the SPE Western Regional and Pacific Section-AAPG Join Meeting, Bakersfield, CA, 31 March–2 April.

Kelland, M. A., 2009, *"Production Chemicals for the Oil and Gas Industry"*, CRC Press, Boca Raton, FL, pp 291–312 and references therein.

Kohse, B. F., Nghiem, L. X., 2004, "Modelling Asphaltene Precipitation and Deposition in a Compositional Reservoir Simulator", SPE No. 89437, presented at the SPE/DOE 14th Symp. on Improved Oil Recovery, Tulsa, OK.

Krawchuk P., 2016, "Development and Operation of the Nsolv BEST Demonstration Facility", Paper No. WHOC16-135, presented at the 2016 World Heavy Oil Congress, Calgary, Alberta, Canada, September 6–9.

Leaute, R. P., 2002, "Liquid Addition to Steam for Enhancing Recovery (LASER) of Bitumen with CSS: Evolution of Technology from Research Concept to a Field Pilot at Cold Lake", CIM/CHOA 79011, presented at Int. Therm. Oper. Symp., held in Calgary, Alberta, Canada, 4–7 November.

Leaute, R. P., Carey, B. S., 2007, "Liquid Addition to Steam for Enhancing Recovery (LASER) of Bitumen with CSS: Results from the First Pilot Cycle", *J. Can. Pet. Technol.*, 46(9), 22–30.

Leontaritis K.J., Mansoori G.A., 1987, "Asphaltene Flocculation during Oil Production and Processing: A Thermodynamic Colloidal Model", SPE No. 16258, presented at SPE International Symposium on Oilfield Chemistry, 1987, San Antonio, TX, February 4–6.

Leontaritis, K., Amaefule, J, Charles, R. E., 1994, "A Systematic Approach for the Prevention and Treatment of Formation Damage Caused by Asphaltene Deposition", *SPE Prod Facil.*, 9(3), 157–164.

Leontaritis, K. J., 1998, "Asphaltene Near-Wellbore Formation Damage Modeling", SPE No. 39446, presented at SPE Formation Damage Control Conf., Lafayette, LA, 18–19 February.

Leyva-Gomez, H., Babadagli, T., 2016, "Efficiency of Heavy Oil/Bitumen Recovery from Fractured Carbonates by Hot-Solvent Injection", SPE No. 184095, presented at SPE Heavy Oil Conference and Exhibition, Kuwait City, Kuwait, 6–8 December.

Lin, L., Ma, H., Zeng, F., Gu, Y., 2014, "A Critical Review of the Solvent-Based Heavy Oil Recovery Methods", SPE No. 170098, presented at SPE Heavy Oil Conf. Calgary, Alberta, Canada, 10–12 June.

Liu, P., Yuan, Z., Zhang, S., Xu, Z., Li, X, 2018, "Experimental Study of the Steam Distillation Mechanism during the Steam Injection Process for Heavy Oil Recovery", *J. Pet. Sci. Eng.*, 166, 561–567.

Mamora, D. D., Rivero, J. A., Hendroyono, A., 2003, "Experimental and Simulation Studies of Steam-Propane Injection for the Hamaca and Duri Fields", SPE No. 84201, presented at SPE Annual Technical Conference and Exhibition, Denver, Colorado, 5–8 October.

Mirzabozorg, A., Bagheri, M.B., Kharrat, R., Abedi, J., Ghotbi, C., 2009, "Simulation Study of Permeability Impairment Due to Asphaltene Deposition in one of the Iranian Oil Fractured Reservoirs", 2009-088 PETSOC Conf. Paper, Canadian International Petroleum Conference, Calgary, Alberta, Canada, 16–18 June.

Mitchell, D. L., Speight, J. G., 1973, "The Solubility of Asphaltenes in Hydrocarbon Solvents", *Fuel*, 52, 149.

Mokrys, I. J., Butler, R. M., 1993, "In-Situ Upgrading of Heavy Oils and Bitumen by Propane Deasphalting: The VAPEX Process", SPE 25452, presented in Productions Operations Symp. Oklahoma City, OK, 21–23 March and references therein.

Moreno-Arciniegas, L., Babadagli, T., 2013, "Optimal Application Conditions of Solvent Injection into Oilsands to Minimize the Effect of Asphaltene Deposition: An Experimental Investigation", SPE No. 165531, presented at SPE Heavy Oil Conference-Canada, Calgary, Alberta, Canada, 11–13 June. Also published in *SPE Res. Eval. Eng.*, November 2014, 530.

Moreno-Arciniegas, L., Babadagli, T., 2014, "Asphaltene Precipitation, Flocculation and Deposition during Solvent Injection at Elevated Temperatures for Heavy Oil Recovery", *Fuel*, 124, 202–211.

Mukhametshina, A., Kar, T., Hascakir, B., 2016, "Bitumen Extraction by Expanding Solvent Steam Assisted Gravity Drainage (ES-SAGD) with Asphaltene Solvents and Non-Solvents", *SPE J.*, 21(2), 380–392.

Muñoz, J. A. D., Ancheyta, J., Castañeda, L., 2016, "Required Viscosity Values to Ensure Proper Transportation of Crude Oil by Pipeline", *Energy Fuels*, 30, 8850–8854 and references therein.

Neasham, J. W., 1977, "The Morphology of Dispersed Clay in Sandstone Reservoirs and Its Effect on Sandstone Shaliness, Pore Space and Fluid Flow Properties", SPE No. 6858, presented at SPE Annual Fall Technical Conference and Exhibition, Denver, Colorado, 9–12 October.

Nenninger, J., Nenninger, E., 2005, "Method and Apparatus for Stimulating Heavy Oil Production", US Patent 6,883,607.

Nenninger, J. E., Dunn, S. G. 2008, "How Fast Is Solvent Based Gravity Drainage?", Can. Int. Pet. Conf. (CIPC)/SPE Gas Tech. Symp. Joint Conf., 59th Annual Tech. l Meeting, Calgary, Alberta, Canada, June 17–19, Paper 2008-139.

Nenninger, J. E., Gunnewiek, L., 2009, "Dew Point vs Bubble Point: A Misunderstood Constraint on Gravity Drainage Processes", presented at the Can. Int. Pet. Conf. (CIPC), Calgary, Alberta, Canada, 16–18 June, Paper 2009-065.

Nghiem, L. X., 1998, "Compositional Simulation of Asphaltene Deposition and Plugging", SPE No. 48996, presented at SPE Annual Technical Conference and Exhibition, New Orleans, Louisiana, 27–30 September.

Nghiem, L. X., Sammon, P. H., Kohse, B. F., 2001, "Modeling Asphaltene Precipitation and Dispersive Mixing in the VAPEX Process", SPE No. 66361, presented at SPE Reservoir Simulation Symposium, Houston, TX, 11–14 February.

Nourozieh, H., Kariznovi, M., Abedi, J., 2012, "Liquid-Liquid Equilibria of Solvent/Heavy Crude Systems: In Situ Upgrading and Measurements of Physical Properties", SPE 152319, presented at SPE Western Regional Meeting, Bakersfield, CA, 21–23 March.

Nsolv website, 2017, www.nsolv.ca/ retrieved on Dec. 30.

Oliveira, M. C. K., Lopes, H., Teixeira, C. S., Silvino, L., Junior, C., Gonzalez, G., Altoé, R., 2017, "Liquid Extract Separated by Propane-Induced Crude Oil Fractionation", *Energy & Fuel*, 31, 13198–13214.

Ovalles, C., Angel, R., Perez-Perez, A., Hernandez, M., Guzman, N., Anselmi, L., Manrique, E. 2002, "Physical and Numerical Simulations of Steamflooding in a Medium Crude Oil Reservoir, Lake Maracaibo, Venezuela", SPE No. 75131-MS, presented at SPE/DOE Improved Oil Recovery Symposium, Tulsa, OK, 13–17 April.

Ovalles, C., Rogel, E., Moir, M. E., Thomas, L., Pradhan, A., 2012, "Characterization of Heavy Crude Oils, Their Fractions, and Hydrovisbroken Products by the Asphaltene Solubility Fraction Method", *Energy Fuel*, 26, 549.

Ovalles, C., Rogel, E. Chilton, E., Hussein Alboudwarej, H., Inouye, A., Vaca, P., 2014, unpublished results.

Ovalles, C., Rogel, E., Moir, M. E., Morazan H., 2015, "Effect of Temperature on the Analysis of Asphaltenes by the On-Column Filtration/Redissolution Method", *Fuel*, 146, 20–27 and references therein.

Ovalles, C., Rogel, E., Vien, J., Morazan, H., Benson, I., Carbognani Ortega, L., 2016, "Subsurface Upgrading of Heavy Oils via Solvent Deasphalting in the Presence of Asphaltene Precipitants", presented at the World Heavy Oil Congress, held in Calgary, Alberta, Canada, 7–9 September.

Ovalles, C., Rogel, E., Alboudwarej, H., Inouye, A., Benson, I. P., Vaca, P., 2017, "Physical and Numerical Simulations of Subsurface Upgrading Using Solvent Deasphalting in a Heavy Crude Oil Reservoir", SPE-183636, *SPE Res. Eval. Eng.*, 20(3), 654–668.

Ovalles, C., Rogel, E., 2017, "Process for *In Situ* Upgrading of Heavy Hydrocarbons Using Asphaltene Precipitant Additives", US Patent No. 9,670,760, 6 June.

Papadimitriou, N. I., Romanos, G. E., Charalambopoulou, G. Ch., Kainourgiakis M. E., Katsaros, F. K., Stubos, A. K., 2007, "Experimental Investigation of Asphaltene Deposition Mechanism during Oil Flow in Core Samples", *J. Pet. Sci. Eng.*, 57, 281–293.

Pan, H. Q., Firoozabadi, A., 1998, "A Thermodynamic Micellization Model for Asphaltene Precipitation: Part I: Micellar Size and Growth", *SPE Prod. Facil.*, 13, 118–127.

Pathak, V., Babadagli, T., Edmunds, N., 2012, "Mechanics of Heavy-Oil and Bitumen Recovery by Hot Solvent Injection", SPE-144546, *SPE Res. Eval. Eng.*, 15(2), 182–194.

Peramanu, S., Singh, C., Agrawala, M., Yarranton, H. W., 2001, "Investigation on the Reversibility of Asphaltene Precipitation", *Energy Fuels*, 15, 910–917.

Pourabdollah, K., Moghaddam, A. Z., Kharrat, R., Mokhtari, B., 2010, "Study of Asphaltene and Metal Upgrading in VAPEX Process", *Energy Fuels*, 24(8), 4396–4401.

Pourabdollah, K., Moghaddam, A. Z., Kharrat, R., Mokhtari, B., 2011, "Improvement of Heavy Oil Recovery in the VAPEX Process using Montmorillonite Nanoclays", *Oil Gas Sci. Technol.*, 66(6), 1005–1016.

Pourabdollah, K., Mokhtari, B., 2013, "The VAPEX Process, from Beginning Up to Date", *Fuel*, 107, 1–33, and references therein.

Ramirez-Corredores, M. M., 2017, *"The Science and Technology of Unconventional Oils: Finding Refining Opportunities"*, Elsevier, 1st Ed. London, Chapter 2, p 2 and references therein.

Rahnema, H, Kharrat, R, Rostami, B., 2008, "Experimental and Numerical Study of Vapor Extraction Process (VAPEX) in Heavy Oil Fractured Reservoir", presented at Canadian International Petroleum Conference, Calgary, Alberta, PETSOC-2008-116, 17–19 June.

Rezaei, N., Mohammadzadeh, O., Chatzis, I., 2010, "Warm VAPEX: A Thermally Improved Vapor Extraction Process for Recovery of Heavy Oil and Bitumen", *Energy Fuels*, 24(11), 5934–5946.

Redford, D. A., McKay, A. S., 1980, "Hydrocarbon-Steam Processes for Recovery of Bitumen from Oil Sands", SPE No. 8823 presented at SPE/DOE Enhanced Oil Recovery Symposium, Tulsa, OK, 20–23 April.

Rogel, E., 2012, private communication.

Rogel, E., Vien, J., Morazan, H., Lopez-Linares, F., Liang, J., Benson, I., Carbognani Ortega, L. C., Ovalles, C., 2017, "Subsurface Upgrading of Heavy Oils via Solvent Deasphalting Using Asphaltene Precipitants. Preparative Separations and Mechanism of Asphaltene Precipitation Using Benzoyl Peroxide as Precipitant", *Energy Fuels*, 31(9), 9213–9222.

Sammon, P. H., 2003, "Dynamic Grid Refinement and Amalgamation for Compositional Simulation", SPE No. 79683, presented at SPE Reservoir Simulation Symposium, Houston, TX, 3–5 February.

Sharpe, H. N., Richardson, W. C., Lolley, C. S., 1995, "Representation of Steam Distillation and In-situ Upgrading Processes in a Heavy Oil Simulation", SPE 30301, presented in Int. Heavy Oil Symp., Calgary, Alberta, Canada, 19–21 June, p 551.

Smith, D. G., Hubbard, S. M., Leckie, D. A., Fustic, M., 2009, "Counter Point Bar Deposits: Lithofacies and Reservoir Significance in the Meandering Modern Peace River and Ancient McMurray Formation", Alberta, Canada. *Sedimentology*, 56(6), 1655–1669.

Stark, S. D., 2013, "Cold Lake Commercialization of the Liquid Addition to Steam for Enhancing Recovery (LASER) Process", IPTC 16795, presented at the Int. Pet. Tech. Conf., Beijing, China, 26–28 March.

Telmadarreie, A., Trivedi, J., 2017, "Dynamic Behavior of Asphaltene Deposition and Distribution Pattern in Fractured Porous Media during Hydrocarbon Solvent Injection: Pore-Level Observations", *Energy Fuels*, 31, 9067–9079.

Unal, Y., Kar, T., Mukhametshina, A., Hascakir, B., 2015, "The Impact of Clay Type on the Asphaltene Deposition during Bitumen Extraction with Steam Assisted Gravity Drainage", SPE No. 173795, presented at the Int. Symp. Oil Field Chem., The Woodlands, TX, 13–15 April.

Upreti, S. R., Lohi, A., Kapadia, R.A., El-Haj, R., 2007, "Vapor Extraction of Heavy Oil and Bitumen: A Review", *Energy Fuels*, 21, 1562–1574.

Vallejos, C., Vasquez, T., Siachoque, G., Layrisse, I., 2002, "Process for In Situ Upgrading of Heavy Hydrocarbon", US Patent No. 6,405,799, 18 June.

Vargas-Vasquez, S. M., Romero-Zerón, L. B., 2007, "The Vapor Extraction Process: Review", *Pet. Sci. Technol.*, 25, 1447–1463.

Victorov, A.I., Firoozabadi, A., 1996, "Prediction of Asphaltene Precipitation Using Thermodynamic Micellization Model", *AIChE J.*, 42, 1753–1764.

Vogel, A. I., 1974, *Practical Organic Chemistry*, 3rd. Ed., Longman, London, UK, p 12.

Wang, S., Civan, F., 2001, "Productivity Decline of Vertical and Horizontal Wells by Asphaltene Deposition in Petroleum Reservoirs", SPE No. 64991, presented at SPE International Symposium on Oilfield Chemistry, Houston, TX, 13–16 February.

Wildemuth, C. R., Williams, M. C., 1984, "Viscosity of Suspensions Modeled with a Shear-Dependent Maximum Packing Fraction", *Rheol. Acta*, 23(6), 627–635.

Wu, C. H., 1977, "Critical Review of Steamflood Mechanisms", SPE 6550, Pres. in 47th Annual California Regional Meeting of Pet. Eng. of AIME, Bakersfield, CA, 13–15 April.

Wu, C. H., Elder, R. B., 1983, "Correlation of Crude Oil Steam Distillation Yields with Basic Crude Oil Properties", *Soc. Pet. Eng. J.*, 23, 937, December.

Wu, J. Z., Prausnitz J. M., Firoozabadi, A., 1998, "Modeling Asphaltene Precipitation by n-Alkanes from Heavy Oils and Bitumen Using Cubic-Plus-Association Equation of State", *AIChE J.* 44, 1188–1199.

Yakubov, M., Yakubova, S., Borisov, D., Romanov, G., Yakubson, K., 2014, "Asphaltene Precipitation Inhibitors and Phase Behavior Control for Bitumen Recovery by Solvent Injection", SPE No. 170165, presented at SPE Heavy Oil Conference-Canada, Calgary, Alberta, Canada, 10–12 June.

6 Thermal Conversion

In this subsurface upgrading route of HO/B, reservoir conditions are increased up to or above cracking temperature (> 200°C) so oil-containing chemical bonds (carbon–carbon, sulfur–carbon, etc.) are broken with the concomitant permanent reduction in the viscosity and increase in distillable materials in the produced oil. Thermal-only processes have the advantage of being relatively simple and in general, do not require complicated surface installations. However, they suffer several disadvantages such as lack of low cost and abundant energy sources, the absence of hydrogen which limits the value of the upgraded products, instability due to asphaltene precipitation, the formation of olefins, and potential for development of scales through dissolution/transport of formation minerals.

This chapter is concentrated on describing low-temperature cracking and Visbreaking as well as downhole pyrolysis. These subsurface upgrading concepts are carried out in the *absence* of hydrogen or hydrogen sources. However, the use of catalysts during thermal conversion routes (i.e., catalytic cracking) is within the scope of this chapter. Whereas thermal or catalytic hydrocracking and hydrotreatment pathways are described in Chapters 7 and 8, respectively. The chemistry of thermal cracking and Visbreaking was discussed in Section 4.1.2.

Several energy sources have been reported to heat the reservoir such as steam, electricity, electromagnetic energy (Section 10.1), and sonication (Section 10.3). However, this chapter does not describe the use of steam because, for reasons that are discussed later, steam is an active hydrogen source via the Aquathermolysis reaction. This route is covered in Chapters 7 (in the presence of hydrogen donors) and 8 (using catalysts). To this date, Shell's *In-Situ* Upgrading Process (IUP) is the only thermal conversion concept that has been field tested and is described in Section 6.3. Finally, it is important to mention that the application of thermal methods for the upgrading of oil shale reservoirs is out of the scope of this chapter. A recent review in the area can be found elsewhere [Youtsos *et al.* 2012].

6.1 LOW-TEMPERATURE CRACKING AND MILD VISBREAKING

As described in Section 4.1.2, thermal cracking involves the breaking of chemical bonds to generate lighter hydrocarbons via decomposition at high temperatures (> 330°C). However, this phenomenon can occur at temperatures as low as 200°C. At these low severity conditions, the conversion of the petroleum-containing compounds may take extended periods of time from days/months to thousands or millions of years [Huc 2011]. As mentioned in Section 4.1.2, the chemistry of low-, medium-, and high-temperature cracking is fundamentally the same, i.e., the reaction mechanism involves the generation, propagation, and termination of free radicals [Gray 2015].

In this section (6.1.1), low severity thermal reactions are discussed with particular emphasis on mild Visbreaking (see Section 4.1.2.2). Lab experiments, as well as kinetic studies, are described with the objective of extrapolating the learnings to subsurface conditions. Specifically, some of these concepts were numerically simulated, and the results are presented in Section 6.1.2. Similarly, the notion of asphaltenes as unreactive species is not supported in the literature [Naghizada *et al.* 2017]. In fact, there are several examples that prove otherwise.

6.1.1 Low Severity Thermal Conversion

Visbreaking is a relatively mild form of thermal cracking and is operating at relatively low pressure (Section 4.1.2). The thermal reactions are not allowed to proceed to completion, and little or no coke formation takes placed [Joshi *et al.* 2008, Huc 2011, Speight 2012, Gray 2015]. Visbreaking leads to permanent changes in the physical and chemical properties such as reductions of viscosity, pour point, and density as well as increases in the amount of distillable materials in comparison with the original oil. During Visbreaking, most of the chemical reactions are thermally driven so there is no selectivity in bond cleavage. The area of pyrolysis of Athabasca Bitumen has been recently reviewed by Gates and coworkers covering temperatures up to 800°C [Kapadia et al. 2015].

In 1965 Henderson and Weber studied the thermal upgrading of heavy crude oils [Henderson and Weber 1965]. These authors carried out lab experiments using seven different crudes in a batch stainless-steel reactor in the absence of sand and water at 264–430°C from 1 h to 1,100 h. For an Athabasca bitumen, results showed increases in the API gravity from 6.9 to 12.8°API and percentages of conversion of the 427°F⁺ residue (see Equation 4.44) up to 24% [Henderson and Weber 1965].

Henderson and Weber also studied the kinetics of the cracking process by following the changes in the weight fractions of the 427°F⁺ residue measured in the products by vacuum distillations in comparison with the original oils. They assumed pseudo-first-order kinetics as shown in Equation 6.1 [Henderson and Weber 1965]:

$$\text{Rate of Upgrading } = dC/dt = k\,C\,t \tag{6.1}$$

Where C = weight fraction of unreacted crude oil, k = Arrhenius rate constant (in this case in s⁻¹), t = time (s), and dC/dt is the derivative of the concentration vs. time, i.e., the rate of upgrading.

If the crude oil concentration is changing from the initial value C_0 to the value of C at time t, the integrated form of Equation 6.1 is Equation 6.2 [Henderson and Weber 1965]:

$$\text{Ln}\left(C_0/C\right) = k\,t \tag{6.2}$$

As expected, the rate of reaction k follows the Arrhenius equation that describes the dependency with the temperature, Equation 6.3 [Henderson and Weber 1965]:

$$k = A\,e\left(Ea/RT\right) \tag{6.3}$$

Where A = Arrhenius pre-exponential factor, Ea = Arrhenius is the activation energy in kJ/Mol, R = ideal gas constant, 8.3143 J/K Mol, and T = temperature in K.

Henderson and Weber assumed that C and C_0 are related to the percentage of conversion of the 427°F⁺ residue as shown in Equation 6.4:

$$\%\,\text{Conversion of } 427°\text{F + Residue} = \left[\left(C_0 - C\right)/C_0\right] \times 100 \tag{6.4}$$

Solving for C and substituting into Equation 6.2 allowed obtaining of the rate constants at different temperatures, pressures, and times. Henderson and Weber warned that, during the determination of rate constants, the percentage of conversion should be kept small because high conversions resulted in the onset of parallel and consecutive secondary reactions. This situation changed the nature of the material being heated and tended to diminish the rate constant calculated using the assumption of a first-order reaction. The effects of secondary reactions were detected and minimized by extrapolating to "zero conversion" a graph of the rate constant vs. conversion [Henderson and Weber 1965].

Table 6.1 shows the typical times and temperatures required to upgrade different crude oils [Henderson and Weber 1965]. As seen, Athabasca bitumen as well crudes B and F are remarkably stable in the absence of sand and water at a temperature lower than 316°C (600°F). At these lower temperatures, it requires days to years to achieve a 15% conversion of the distillation residue. However, caution should be taken in extrapolating Henderson's and Weber's results outside the temperature range used (264–430°C) because the mechanism of thermal cracking could change substantially.

On the other hand, at temperatures near 430°C (800°F), the times required are much shorter (< 1 h) so it is not surprising that refinery-based Visbreaking units operate at these conditions

TABLE 6.1

Typical Times and Temperatures Required to Upgrade Crude Oils in the Absence of Sand and Water

Crude Oil	% Residue Conv.[b]	Time Required[a] 204°C (400°F)	260°C (500°F)	316°C (600°F)	371°C (700°F)	427°C (800°F)
Athabasca bitumen	15%	450 years	1.9 years	8.8 days	0.26 days	–
Crude B	15%	3,000 years	8.3 years	24 days	0.46 days	0.39 h
Crude F	30%	–	53.8 years	87 days	0.96 days	0.57 h
Crude F	40%	–	76.9 years	123 days	14 days	0.81 h

Data taken from Henderson and Weber [1965].

[a] Time (hours, days, or years) required to achieve the degree of upgrading as measured by the percentage of residue conversion at a given temperature.

[b] Percentage of 427°F+ residue conversion using Equation 4.4.

(see Section 4.1.2.2). Thus, higher temperature or longer times, or both, are needed to thermally upgrade the crude oil under subsurface conditions.

Shu and Venkatesan studied the kinetics of Visbreaking of a Cold Lake bitumen at 260–325°C with residence times up to one month [Shu and Venkatesan 1984]. A model was developed in which the oil was first cracked to a less viscous fraction, and the latter imposed a solvent effect on the former. Thus, the viscosity of the mixture is calculated using Equation 6.5:

$$Ln\ \mu\ =\ X\ ln\ \mu o\ +\ (1-X)\ Ln\ \mu t \tag{6.5}$$

Where μ = viscosity of the mixture of original crude and visbroken products, μ_0 = viscosity of the original Cold Lake bitumen, μ_t = viscosity of the visbroken product at a time t, and X is a parameter that depends on the degree of viscosity reduction. Figure 6.1 shows the pseudo-first-order rate

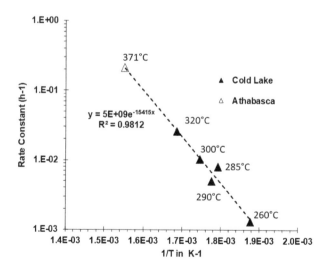

FIGURE 6.1 Pseudo-first-order rate constant for the low severity Visbreaking of Cold Lake and Athabasca bitumens vs. the inverse of temperature (K^{-1}). Data taken from Shu and Venkatesan [1984] and Henderson and Weber [1965].

constants for the low severity Visbreaking of Cold Lake and Athabasca bitumens. As seen, both crude oils fall on the same line. This finding was attributed to the similar chemical composition of these Canadian crudes [Shu and Venkatesan 1984]. Also, the results showed that the first-order rate constant for the visbroken Cold Lake bitumen has an activation energy of 31 Kcal/mol. This value is within those reported in the literature [Joshi *et al.* 2008].

In 1986, Venkatesan and Shu extended their low-severity Visbreaking studies by carrying out isothermal batch experiments at 572°F (300°C) of heavy crude oils from San Ardo (California) and Celtic (Saskatchewan) reservoirs for up to 480 hours [Venkatesan and Shu 1986]. As shown in Figure 6.2, the experimental results showed a permanent reduction in oil viscosity by a factor as high as ten, without significant formation of coke after 120 hours of soaking. Extending the heating time up to 480 h (20 days) did not lead to further viscosity reduction [Venkatesan and Shu 1986].

Venkatesan and Shu also reported the reduction of asphaltene content from 8% to 15% (wt./wt.) during mild Visbreaking (300°C) [Venkatesan and Shu 1986]. These results are in contrast with those reported by Carbognani *et al.* using Athabasca VR [Carbognani *et al.* 2007]. These researchers carried out preparative medium severity Visbreaking experiments by keeping the feed at a constant temperature (380°C) for different residence times. Figure 6.3 shows the percentage of Athabasca VR conversion vs. the asphaltene stability measured as P-value [Carbognani *et al.* 2007], ΔPS [Rogel *et al.* 2010], and SARA group-type distribution [Carbognani *et al.* 2007]. As seen, as the VR conversion increases from 8 to 30%, and the contents of asphaltenes also increase from 15 to 30 wt. %. At the same time, the visbroken product became intrinsically unstable, i.e., lower P-value and higher ΔPS (see Section 4.3). The authors attributed the higher asphaltenes instability to the cracking of alkyl appendages from the asphaltene compounds with the concomitant increases in their aromaticity. These alkyl moieties created paraffinic light ends that are poorly suited solvent media for the asphaltenes. The loss of alkyl appendages from aromatic compounds facilitated further stacking and recombination of free radicals [Carbognani *et al.* 2007, 2008].

Another significant outcome of Figure 6.3B is the decrease of resins that occurred in parallel with increasing asphaltene content when the conversion increased. This finding confirms the importance of the resins/asphaltenes (R/A) ratio for stabilization of the asphaltene components. Also, saturates were observed to increase as a function of Visbreaking conversion. These facts support the mentioned instability caused by improper oil media properties for the dispersion of asphaltenes [Carbognani *et al.* 2007].

It is important to point out that the tendency to bond breaking reaction, also known as crackability, tends to increase with the molecular weight (or boiling range) of the starting material. Larger

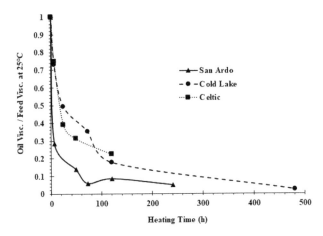

FIGURE 6.2 Viscosity reduction for the low severity Visbreaking of HO/B as a function of heating time at 300°C. Data taken from Venkatesan and Shu [1986]. Lines were drawn to show tendency.

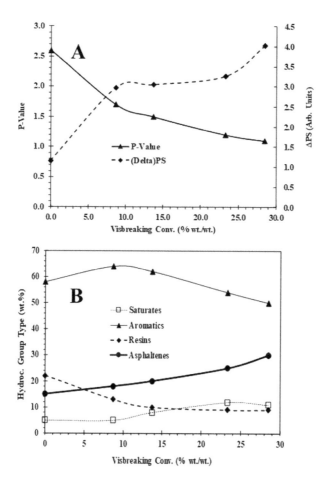

FIGURE 6.3 Percentage of conversion of Athabasca VR Visbreaking vs. A) Asphaltene stability measured as P-value and ΔPS, and B) SARA group-type distribution. Data taken from Carbognani *et al.* [2007] and Rogel *et al.* [2010]. (Lines were drawn to show tendency.)

molecules have more bonds which can easily rupture which, in turn, increase the probability of breaking [Gray 1994 and Blanchard 1997]. For these reasons, it is not surprising that asphaltenes are not very thermally stable. In fact, de Klerk *et al.* studied the reactivity of C5-asphaltenes from Athabasca oil sands bitumen over the temperature range 100–250°C [Naghizada *et al.* 2017]. The authors found that, on heating the asphaltenes to 150°C, the aromatic hydrogen content increased relative to the feed by a factor of 1.12. It was also found that the n-heptane insoluble fraction increased from 67 to 75%. The latter finding was tentatively attributed to the free radical combination. Despite clear evidence of cracking reactions were taking place, almost no gas phase products were produced. Unfortunately, the nature of the reactive species in the asphaltenes was not conclusively determined [Naghizada *et al.* 2017].

Strausz and coworkers studied the composition of produced gas during pyrolysis at low temperature (70–210°C) of oil sand, asphaltenes, and maltenes. The authors found trace quantities of C2–C5 hydrocarbons, CO_2, CO, H_2S, COS, CS_2, and SO_2 [Strausz *et al.* 1976]. These results agree with those reported by Ovalles *et al.* [1998]. These authors studied the thermal stability of Cerro Negro crude oil from the Orinoco Basin by heating in a batch reactor at 250°C for 8 h. The products were characterized by gas chromatography, FTIR in the v(C=O) region, and by H-NMR. The results were consistent with a decarboxylation reaction with the subsequent formation of olefins and CO_2, as shown in Equation 6.6 [Ovalles *et al.* 1998].

$$\underset{\substack{|\\ \text{R-CH-CH}_2\text{COOH}}}{\overset{\text{L}}{}} \xrightarrow[\text{for 8 h}]{250°C} \text{R-CH=CH}_2 + \text{HL} + CO_2 \qquad\qquad (6.6)$$

Where R = alkyl chain and L = $-NR_3$, $-NHR_2$, $-NH_2R$, $-CN$, etc.

Additionally, Strausz and coworkers found that the Arrhenius activation energies of thermal cracking products had lower values (56.9 kJ/mol) for the evolution of methane from bitumen in the presence of mineral matter as compared to that from the asphaltene and maltene fractions in the absence of the solid (117 and 130 kJ/mol, respectively). These results suggested that there were catalytic effects of the mineral formation in the cracking of Athabasca bitumen at low severity conditions [Strausz *et al.* 1976].

The thermal reactivity of asphaltenes can be enhanced by using solid catalysts [Adams 2014]. For example, Montoya *et al.* studied the effect of the Ni-Pd nanocatalysts supported on fumed silica nanoparticles on post-adsorption catalytic thermal cracking of n-C7 asphaltenes at temperatures as low as *100°C* [Montoya *et al.* 2016]. The results showed reductions in the activation energy (Ea) by using the metal supported nanoparticles. For the virgin n-C7 asphaltenes, the activation energy varied between 117 and 195 kJ/mol, whereas for silica supported Ni-Pd nanocatalysts, Ea ranged between 55 and 170 kJ/mol [Montoya *et al.* 2016]. These findings showed the potentiality of using metal-containing catalysts to enhance thermal cracking processes under low severity conditions.

All the experimental results described in this section lead to the conclusion that it is plausible to thermally crack the HO/B at subsurface conditions by using mild Visbreaking (< 280°C). These processes lead to permanent viscosity reductions of the produced oils and depending on the severity and the presence of mineral formation, reductions in asphaltene content. Unfortunately, the time required to achieve the necessary upgrade of the HO/B could be between days to months. Based on these findings, the numerical simulation of low and medium severity thermal processes has been reported, and the results are shown in the next section.

6.1.2 Numerical Simulations of Downhole Mild Visbreaking

Shu and Venkatesan [1984] and Kasraie and Farouq Ali [Kasraie 1989] carried out the numerical simulation of mild Visbreaking during HO/B recovery. Even though steam was used to heat the reservoirs to the cracking temperatures, the authors only modeled the viscosity reduction of the produced oils by using the kinetic data previously discussed [Shu and Venkatesan 1984]. In the late 1980s, the role of water as hydrogen source via Aquathermolysis was not completely understood and was not included in the simulations. As mentioned, this topic is presented in Section 7.5. However, this modeling by Shu and Venkatesan [1986] and Kasraie and Ali [Kasraie 1989] represents one of the earliest attempts to numerically simulate subsurface upgrading processes.

Shu and Venkatesan [1986] modeled a typical Cold Lake reservoir (10.9°API) located in Alberta, Canada. They used unconsolidated sand with a porosity of 35% and a permeability of 6.4 Darcy. The depth was 400 m (1,445 ft), and the pay zone was 45.7 m (150 ft) thick. The initial temperature was 12.8°C (55°F) and pressure 3.4 MPa (495 psi). Steam was injected at 313°C (596°F) with 78% quality at a rate of 300 B/d. Using the lab kinetic data discussed in the previous section three steam injection strategies were studied. Those are cyclic steam stimulation (CSS), steamflooding, and steam slug process [Shu and Hartman 1986].

Figure 6.4 shows the distribution of the visbroken oil after the end of steam injection for the numerical simulation of CSS. The authors defined the steam zone as the regions where gas saturation was higher than 0.08. The part next to the steam contained high water saturation and was designated as condensation zone. Ahead and on top, the heavy oil zone was identified and corresponded to the "untouched" or original oil. The authors defined the mobile visbroken area as the region where the visbroken oil saturation was higher than 0.02. For all of Shu's and Venkatesan's numerical simulations, the total oil saturation was the sum of the original oil plus the visbroken oil

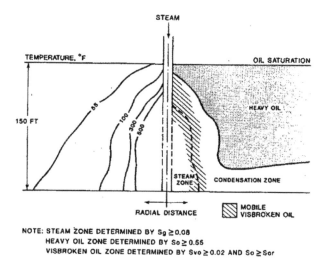

FIGURE 6.4 Distribution of the visbroken oil after the end of the injection for the numerical simulation of cyclic steam stimulation [Shu and Hartman 1986]. Reprinted with permission of Society of Petroleum Engineers.

[Shu and Hartman 1986]. As seen, at the end of the steam injection, the highest saturation of the thermally cracked oil was the region peripheral to the steam zone (shaded area in Figure 6.4). This area corresponded to a temperature range of 204 to 260°C (400–500°F) as shown in the left-hand side of Figure. 6.4.

Shu and Venkatesan simulated a case *without* Visbreaking (Base Case) with the same amount of steam injected, and the results were compared with the upgrading case. For the latter, they found that the cumulative oil production was higher by 5% than that observed for the Base Case. The reason for this finding was attributed to a viscosity reduction mechanism [Shu and Venkatesan 1986].

Next, the same authors carried out the simulation of steamflooding and the distribution of the visbroken oil and temperature is shown in Figure 6.5. As seen, about two-thirds of the reservoir was never significantly heated by steam. Under Shu's and Venkatesan's conditions, continuous steamflood encouraged gravity override and did not allow enough time for heat transfer between the steam and oil (Figure 6.5a). Also, a hot area (400°F) is found (Figure 6.5b) near the condensation zone in which little oil is present. Thus, the cumulative oil production was similar for the Visbreaking and Base Cases. Also, the cumulative visbroken oil produced was minimal (6% with respect to the total oil produced).

Finally, the authors decided to evaluate a steam slug process [Shu and Hartman 1986]. In this case, steam was injected first for 222 days. Then, the injection was terminated, and the fluids were produced by pressure drawdown. Figure 6.6 shows the distribution of the visbroken oil after 500, 600, and 700 days. As seen in Figure 6.6a, there is a significant amount of visbroken oil present in front of the steam zone. During the pressure drawdown phase, and because of water condensation, the oil displacement came through the bottom in the direction of the arrows toward the producing well (Figure 6.6b and c). As before, the Base and Visbreaking cases gave similar cumulative oil production. However, it is important to mention that the amount of visbroken oil increased to 10% in comparison to the continuous steamflooding (6% vs. total oil produced). These results indicated that the location of visbroken oil with respect to the direction of steam flow determined the relative importance of the Visbreaking reactions [Shu and Hartman 1986].

Similar results were reported by Kasrale and Farouq Ali [1989]. These authors used the mild Visbreaking-viscosity reduction model reported by Shu and Venkatesan [1986] to numerically simulate thermal upgrading of Saskatchewan oil in a five-spot pattern with an area of 20 acres (pay zone thickness was 36 ft). The authors found up to 13% increase in oil recovery in comparison with

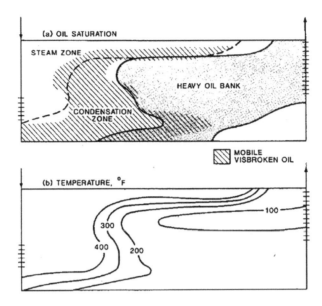

FIGURE 6.5 Distribution of the visbroken oil vs. (a) oil saturation and (b) temperature (°F) after 600 days for the numerical simulation of steamflooding [Shu and Hartman 1986]. Reprinted with permission of Society of Petroleum Engineers.

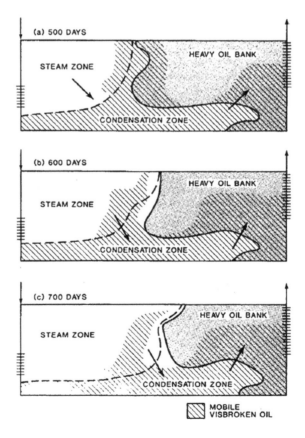

FIGURE 6.6 Distribution of the visbroken oil for the numerical simulation of steam slug process [Shu and Hartman 1986]. Reprinted with permission of Society of Petroleum Engineers.

the Base Case (no upgrading). The examination of the temperature profiles showed that the reason for the small upgrading effect on oil recovery was the fact that the mild Visbreaking process lowered the oil viscosity only in the parts of the reservoir with high temperatures, where the oil was mobile enough already. Upgrading did not affect oil viscosity in the cold parts of the formation [Kasrale and Farouq Ali 1989].

The authors concluded that thermal upgrading via mild Visbreaking is not likely to be significant in steamflood projects of short duration. However, in continuous steam injection, the oil around the producers could undergo thermal cracking leading to permanent viscosity reduction. Thus, the authors recommended that the thermal upgrading effect should be included in the numerical simulation of steam injection processes [Kasrale and Farouq Ali 1989].

6.1.3   ACID-CATALYZED CRACKING

As described in Section 4.2.2.1, solids with acid centers are active for hydrocarbon cracking at mild conditions. In this section, two types of catalysts are described. Those are heterogeneous and ionic liquid catalysts. The first types of materials are generally transition metals supported in acidic inorganic oxides (see the following paragraphs). Commonly practiced production engineering methodologies may be adaptable to injection and placement of heterogeneous catalysts downhole [Weissman 1997]. This topic is discussed in detail in Section 8.1.2. Conventional cased or openhole gravel pack completions or proppant injection are good alternatives for placing solid catalysts around a wellbore and into the oil-bearing formation. By this way, the crude oil could flow through the catalytic materials during the oil production with the subsequent upgrading.

Wang and coworkers compared the catalytic activity of impregnated and hydrothermally prepared tungsten oxide (WO_3) on zirconia catalysts for the upgrading of Liaohe crude oil (16°API) at 220°C for 6 h [Wang et al. 2012]. These experiments were run under thermal conditions, so no hydrogen source was present. Wang and coworkers found that the catalytic activity mainly depends on the acidity of the catalyst and that the materials prepared via the hydrothermal method exhibit higher viscosity reduction (82%) than those synthesized by impregnation (72%). Similar results were obtained for sulfur removal (22% vs. 10%) and asphaltene conversion (22% vs. 15%). The authors attributed these findings to a better dispersion of WO_3 in the former catalyst than in the latter [Wang et al. 2012].

Furthermore, Wang and coworkers conducted a blank test (no catalyst) and found that the viscosity reduction was only 18.7%, which demonstrated that the use of the WO_3/ZrO_2 catalyst was necessary to promote the viscosity reduction of the heavy oil. After the upgrading reaction, the catalyst was recovered, and its XRD pattern showed that its crystalline structure was almost unchanged. This result indicated that this acid catalytic system was stable under the upgrading conditions [Wang et al. 2012].

Wang and coworkers also studied the composition of the produced oils by SARA analysis and found that the contents of resins and asphaltenes decreased whereas the saturates and aromatics increased [Wang et al. 2012]. The authors attributed these findings to acid catalyzed cracking reactions occurring during the upgrading runs. It is important to point out that these experiments were carried out *in the absence* of a hydrogen source. Thus, the production of lower molecular weight liquids with higher H/C molar ratios was accompanied by the generation of coke with a lower H/C ratio. The latter material was found deposited on the catalyst surface and was observed by the color change of the catalysts. Indeed, the coke content measured on the recovered WO_3/ZrO_2 solid was ~0.04 wt. % [Wang et al. 2012].

Nowadays, the utilization of ionic liquids (IL) has gained popularity in solvent extraction, desulfurization, and scale removal due to their excellent solubility and catalytic properties under various operating conditions [Fan et al. 2007]. These compounds have attracted attention from both academia and industry since the late 1990s due to a series of exciting features such as excellent chemical and thermal stabilities, the ability to solvate a broad range of compounds, and the capability of tailoring their properties by a reasonable selection of the cation–anion pairs. The use of IL as catalysts for the subsurface upgrading of HO/B has excellent potential applications because they are environmentally friendly, have low toxicity, and are catalytically active, i.e., they have Brønsted acidity for cracking reactions.

Fan *et al.* carried out the thermal upgrading of Liaohe crude oil in a batch reactor at *very* mild conditions (90°C) using the ionic liquid [(Et)$_3$NH] [AlCl$_4$] (IL) with and without transition metals, e.g., Ni, Cu, or Fe [Fan *et al.* 2007] for periods of time compatible with EOR processes (days). As before, these experiments were carried out in the absence of a hydrogen source. The use of the IL led to a 44% viscosity reduction, 49% sulfur removal, and 18% asphaltene conversion [Fan *et al.* 2007]. Consistent with an acid catalyzed process, after upgrading reaction, the contents of saturates, aromatics, and resins increased, while the content of asphaltenes decreased. This change in composition led to the decrease of the average molecular weight and the reduction of viscosity of the heavy oil [Fan *et al.* 2007].

Fan *et al.* found improved results using a nickel-containing ionic liquid catalyst. They reported a 65% viscosity reduction, 85% desulfurization, and 43% asphaltene conversion. The authors attributed the previous findings to a complex formed between the IL-Ni and the organic sulfur compounds present in the heavy crude oil. This complex weakens the C–S bonds with the concomitant sulfur removal [Fan *et al.* 2007]. However, no evidence was presented to support this mechanism.

In 2009, Fan and coworkers carried out the synthesis of the ionic liquid [BMIM] [AlCl$_4$] by the reaction of 1-methyl-imidazole with chloro-n-butane and aluminum chloride at 90°C for 20 h [Fan *et al.* 2007]. The authors conducted the thermal upgrading of Xinjiang crude oil (21.7°API) using [BMIM] [AlCl$_4$] at 90°C, and 50% viscosity reduction, 66% desulfurization, and 59% asphaltene conversion were reported. Fan and coworkers found that the mixture of ionic liquid and transition metal salts (NiSO$_4$, FeSO$_4$, Ni-Naphthenate, and Fe-Naphthenate) gave higher viscosity reductions and lower asphaltene contents than the samples treated with only [BMIM] [AlCl$_4$] [Fan *et al.* 2009].

Thus, based on the evidence presented in this section it can be concluded that the thermal upgrading of HO/B can be effectively catalyzed by acid-containing heterogenous catalysts or ionic liquids. The use of very low temperatures in the presence of IL is a fascinating route that deserves further scrutiny.

6.2 DOWNHOLE MEDIUM VISBREAKING OR PYROLYSIS

In the previous section, the effect of mild Visbreaking (150–280°C) on the oil recovery and upgrading of HO/B was physically and numerically modeled. In this section, a medium Visbreaking or downhole pyrolysis is discussed. In these concepts, the temperature was increased to the 325–380°C range to achieve higher viscosity reductions, residue conversion, and API gravities and at the same time, reduce the time of reaction.

Phillips *et al.* [1985] thermally cracked Athabasca bitumen and bitumen-derived products with and without sand at 360°C, 400°C, and 420°C. The products were separated into six pseudo-components: coke, asphaltenes, heavy oils, middle oils, light oils, and gases. The results showed that the yields of coke and gases from cracking bitumen–sand mixtures were higher than those from cracking bitumen alone. The presence of sand also affected the production of the other pseudo-components. The authors proposed two kinetic models to describe the results. The activation energies involved in cracking bitumen–sand mixtures were lower than those included in cracking bitumen alone, indicating the possibility of a catalytic effect of the mineral formation [Phillips *et al.* 1985].

Millour and coworkers [1985] studied the thermal cracking for three oils (8°API, 13°API, and 24°API) at temperatures of 360°C, 397°C, and 420°C for periods of time up to 48 h. The tests were performed on oxidized and unoxidized samples in the presence of water and rock matrix. Four pseudo-components were separated: Coke, asphaltenes, maltenes, and non-condensable gas. A simple implicit model with a small number of parameters was developed and gave a good match of their experimental runs [Millour *et al.* 1985].

Murugan *et al.* [2012] carried out a kinetic study of the thermal decomposition of Athabasca oil sand (Bitumen/reservoir sand ratio of 1:4) using thermogravimetric analysis (TGA). TGA experiments were conducted at multiple heating rates of 5, 10, 20°C/min from room temperature up to 800°C to obtain the pyrolysis characteristics of bitumen. The differential method was used for determining the kinetic parameters and the best fit for the order of reaction. Kinetics results confirmed the presence of two different stages in bitumen pyrolysis with different values. The first one from ~100°C to ~400°C (~30% weight loss) and the second from that temperature up to 550°C (~70% weight loss). The average activation energy for the first and second stage was 29 and 60 kJ mol^{-1}, and the average order of the reaction was 1.5 and 0.25, respectively. The presence of sand in bitumen led to an increase in coke formation with lower activation energy compared to the pyrolysis of bitumen alone [Murugan *et al.* 2012].

Kumar and coworkers [2011] performed the physical simulation of subsurface pyrolysis of bitumen (8.6°API) using an unconsolidated core (0.05 m diameter, 0.15 m length, 30% of porosity, and 4,158 mD permeability) for 97 hr [Kumar *et al.* 2011]. The initial gas, water, and oil saturations were 0.1, 0.10, and 0.80 v/v, respectively. The core was mounted vertically in a steel core holder divided into four equal zones for electrical heating. The sections were heated progressively from top to bottom at 8°C/min until each section reached 375°C. The producer well was placed on the cell top at a constant pressure of 1.5 MPa (218 psi).

Figure 6.7 shows the oil rate and API gravity of the produced oil vs. time of the physical simulation of subsurface pyrolysis of bitumen [Kumar *et al.* 2011]. As seen, the oil rate (g/hr) increased with time reaching a maximum after 2.4 h. Then, it decreased until the end of the experiment. The total recovery was 48% of the OOIP. The authors found a significant amount of gases was generated. At the same time, they observed increases of the API gravity of the produced oil from 8.6 to ~40°API (Figure 6.7). These results were attributed to the cracking of the bitumen molecules to generate light molecular weight materials [Kumar *et al.* 2011]. This lab experiment mimicked the thermal front of downhole pyrolysis and gave valuable information about the reservoir behaviors under subsurface upgrading. Unfortunately, Kumar and coworkers [2011] did not disclose viscosity data of their upgraded oils. However, the changes in API were drastic, so very low-viscosity values are expected.

Perez-Perez *et al.* carried out [Perez-Perez *et al.* 2014] the numerical simulation of the experiment performed by Kumar and coworkers [2011]. Perez-Perez and coworkers coupled a reactive model with a thermodynamic model to represent the phase distribution and decomposition reactions of the bitumen fractions. The kinetics of the pyrolysis process was described by a total number of 29 pseudo-components and 24 chemical reactions [Perez-Perez *et al.* 2014]. The thermo-kinetic model plus a flow formulation was implemented in a commercial reservoir simulator (STARS from CMG).

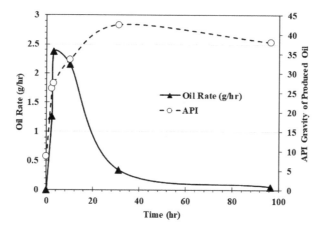

FIGURE 6.7 Oil rate and API gravity of the produced oil vs. time of the physical simulation of subsurface pyrolysis of bitumen. Data taken from Kumar *et al.* [2011]. (Lines were drawn to show tendency.)

Figure 6.8 shows a comparison between the experimental and simulated cumulative oil and water production for the physical simulation of downhole pyrolysis of bitumen [Perez-Perez *et al.* 2014]. As seen, a good match between the experimental and calculated values was obtained. Three stages were identified for the pyrolysis experiment. The first stage, from the start of the run to ~0.2 days, was controlled by the thermal expansion of fluids and water production. The second step was driven by the pyrolysis reactions in which much of the upgraded bitumen was generated. According to their simulation results, a smooth slope of hydrocarbon production (light to intermediate oil fractions) and gases were produced. The third step corresponded to a gas injection process with the purpose of evacuating the liquids that remained in the core [Perez-Perez *et al.* 2014].

Figure 6.9 shows the comparison between the experimental and simulated API gravity of the produced oil during the physical simulation of downhole pyrolysis of bitumen [Perez-Perez *et al.* 2014]. As mentioned, the first two samples corresponded to the thermal expansion process while the third to fifth samples were from the second step, i.e., *in-situ* upgrading process (from 10 to 98 hours). As seen, the API gravity increased from 8.6°API of the initial bitumen to the 26–40°API range. The reduction in the oil production quality for the last sample (No. 6) was due to a large heavy fraction production retained in the core at the end of the pyrolysis experiment [Perez-Perez *et al.* 2014].

Based on the physical and numerical simulations presented in this section, it can be concluded that subsurface upgrading of HO/B is technically feasible at medium severity Visbreaking or pyrolysis. Unfortunately, the stability of the visbroken product could be compromised, especially at high conversion levels. However, more work is needed to characterize the upgraded oil fractions to understand further the chemical changes occurring under downhole conditions in the presence

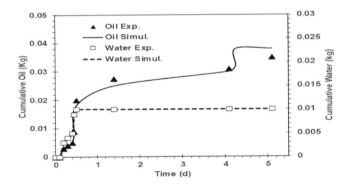

FIGURE 6.8 Comparison between the experimental and simulated cumulative oil and water production for the physical simulation of subsurface pyrolysis of bitumen. Data taken from Perez-Perez *et al.* [2014].

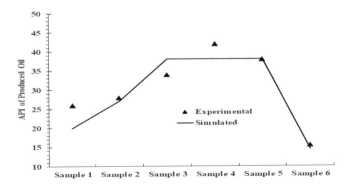

FIGURE 6.9 Comparison between the experimental and simulated API gravity of the produced oil during the physical simulation of subsurface pyrolysis of bitumen. Data taken from Perez-Perez *et al.* [2014].

of the mineral formation. Neither Kumar *et al.* [2011] nor Perez-Perez and coworkers [2014] mentioned how they envision the heating of the reservoir up to pyrolysis temperatures (> 325°C) for several months. One possibility is the use of Shell downhole heating technology. This topic is described next.

6.3 SHELL'S TECHNOLOGY AND FIELD TESTS

After exhaustive lab tests, the Shell *In-Situ* Upgrading Process (IUP) was piloted in five locations around the world in oil shale, heavy oils, and bitumen reservoirs. According to the available literature, the timeline for the development of this technology is the following [Karanikas 2016]:

- Mahogany Field Test and Demonstration Project, [1996–1998 and 2003–2005, respectively], Colorado, USA (Oil Shales)
- Viking Heavy Oil Pilot, [2004–2009], Peace River, Canada (HO/B)
- Grosmont Heavy Oil Pilot, [2012–2014], Grosmont, Canada (HO/B)
- Jordan Field Experiment, 2014–2021, Jordan (Oil Shales)

In this section, a description of Shell's IUP process is presented as well as the main characteristics and results of the four field tests mentioned above.

6.3.1 DESCRIPTION AND FUNDAMENTALS

Shell's IUP is a thermal recovery method based on thermal conduction. It has been reported that this technology has reached the stage of commercial pilots with potential applications to oil shales, heavy oils, and bitumens. It is formally a carbon rejection upgrading process as described in Section 4.1.3 (Coking and Delayed Coking). It uses downhole electrical heaters to achieve subsurface temperatures higher than 300°C (generally in the 500–700°C range) and very long residence times (months to years). The electrically heated sections have lengths more than 1,000 m, have high output power, and are designed to have a life expectancy of at least five years. Figure 6.10 shows the schematic of IUP using horizontal wells [Karanikas 2012]. As seen, the well pair is placed in an SAGD configuration with the heater in the bottom and the producer at the top.

To maximize reservoir contact, Shell uses tightly spaced heater wells to deliver heat directly to the oil-containing zone. For example, Figure 6.11 shows the schematic of IUP using vertical producers, but slanted heaters could be used as well [Vinegar *et al.* 2002]. It is reported that well spacing of 10–20 m is typical. As expected, this design is capital intensive but makes sure that the heat is effectively delivered to minimize the time required for conversion and upgrading. Additionally, Shell IUP can use a seven-spot array having six electrical heaters with one producer in the center [Biglarbigi *et al.* 2007, 2009].

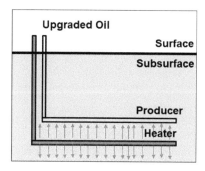

FIGURE 6.10 Schematic of Shell *In-Situ* Upgrading Process using horizontal wells.

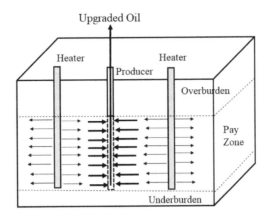

FIGURE 6.11   Schematic of Shell *In-Situ* Upgrading Process using vertical wells.

By using a thermal conduction mechanism, the heat input is not affected by flow barriers such as those found in heterogeneous or fractured reservoirs. Also, due to the downhole operating conditions (T > 325°C, P ~7 atm), the upgraded crude oil exists mostly in the vapor phase. Thus, oil recovery is less vulnerable to subsurface complexities such as reservoir connectivity that can impact the performance of other recovery processes where downhole transport of hydrocarbons occurs primarily in the liquid phase [Karanikas 2012].

Furthermore, gas evolution and HO/B thermal expansion are the two primary fluid production mechanisms which drive the upgraded crude oil and any unconverted material toward the production wells [Karanikas 2012]. Alpak *et al.* carried out the numerical simulation of the IUP stages [Alpak *et al.* 2013]. At the beginning of the simulation, the self-generated gases and the bitumen expansion are the main products. Later, once the cracking temperatures are achieved, the upgraded oil and the coke are produced [Alpak *et al.* 2013].

Depending on the operating conditions, the quality of the upgraded products varied. For example, Karanikas reported for an initial heavy oil of 6.5–7.5°API, the API gravity of the cumulative oil at the end of the test was close to 30°API. He also reported that the instantaneous value of the produced oil reached 19°API. This oil density (0.9465 g/cc) meets the acceptance specifications of Canadian pipelines without the need for mixing with the diluent. Finally, he mentioned that the sulfur content of the oil drops to about 2.5 wt. % from an initial value of 6–7 wt. %. Toward the end of the IUP treatment [Karanikas 2012]. As discussed in Section 4.1.3 (Coking and Delayed Coking), carbon rejection processes are known to produce asphaltenes precipitation and olefin-containing materials. Unfortunately, no data has been released regarding the stability of the upgraded crude oil as well as its olefin contents.

6.3.2   Mahogany Field Test and Demonstration Projects

Shell has carried out seven field tests of the *In-Situ* Conversion Process (ICP) in the Green River oil shales of Colorado, USA. Figure 6.12 shows the geographical location of those tests, and Table 6.2 presents a summary of the objectives, years, depth and heaters, producers, and other wells used [Fowler and Vinegar 2009]. Four pilots were aimed to demonstrate the ICP technology, one to test the heaters used during such pilot tests, and two to study the formation and robustness of the freeze wall methodology [Vinegar 2006].

The Red Pinnacle Thermal Conduction Test (Table 6.2) was the first project of the ICP technology and was carried out by Shell in the early 1980s. The first phase was aimed to confirm the laboratory measurements of thermal properties. The second part used a seven-well hexagonal configuration, and the formation was heated at 300 W/ft for two months. It was reported that the total production was ~30 gal of oil, 2,000 SCF of gas, and 20 gal of water [Fowler and Vinegar 2009].

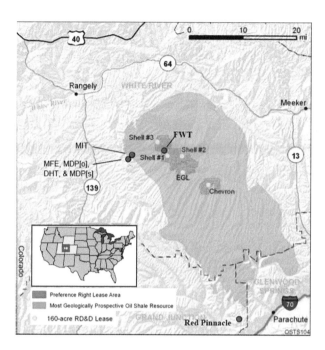

FIGURE 6.12   Geographical location of the seven-field test of Shell's *In-Situ* Upgrading Process [Fowler and Vinegar 2009]. Reprinted with permission of Society of Petroleum Engineers.

The Mahogany Field Experiment (Table 6.2) was initiated in 1996, but the heating began in mid-1997 until early 1998. It consists of six heaters and one producer in the center of a hexagonal array. The total production was 230 BBL and $1,300 \times 10^3$ SCF of gas. The produced oil was in the naphtha/jet/diesel boiling range as predicted by their lab experiments [Fowler and Vinegar 2009].

Shell reported that the primary purpose of the Mahogany Demonstration Project (Table 6.2) was to determine the ICP recovery efficiency and energy balance with a small uncertainty range. The other objective was to confirm the product quality at greater depth (600 ft). Unfortunately, the original heaters failed due to a breaching of the whole canister, hot spots, temperature failure, foreign debris, and geo-mechanical forces. The average temperature was well below the 650°F target, and the total oil production was less than 2% of the expected volume [Fowler and Vinegar 2009].

The Mahogany Demonstration Project South (Table 6.2) was aimed to reduce uncertainty around recovery efficiency by producing a more significant amount of oil than the previous tests. Because of prior experience, the project was designed with redundancy and other protective measures to maximize the probability of success. Twenty-five heaters and two producers at the center were used. Heating began in spring 2004 and continued for 15 months. It was reported that the total production was 1,860 BBL which corresponded to a liquid efficiency of 60% [Fowler and Vinegar 2009]. Fan and coworkers carried out the numerical simulation of this *in-situ* upgrading field test by using a thermal/compositional model that included chemical reaction and flow formulation [Fan *et al.* 2010]. The results showed that oil and gas productions were highly dependent on the heater temperature. Their simulations showed that the energy output-to-energy input ratio was in the range of 6.2–6.9 for all sensitivity cases considered. This ratio is mostly affected by heat loss to the over-burden and under-burden [Fan *et al.* 2010].

The Deep Heater Test goal (Table 6.2) was to field-test of a variety of heaters in near commercial operations (up to 410 ft in length). The energy delivery tools were subjected to a myriad of extreme conditions such as mechanical and thermal stresses, water influx, wellbore stability, etc. Twenty-one heaters with 30 ft spacing and two producers were used for a total of 45 wells. This test helped

TABLE 6.2
Field Tests of Shell's *In-Situ* Conversion Process for Oil Shales

Name	Objective	Years	Depth (ft)	Heaters	Producers	Other Wells	Total Wells
Red Pinnacle Thermal Conduction Test	ICP Demo	1981–1982	20	3	3	Observers	14
Mahogany Field Experiment	ICP Demo	1996–1998	130	6	1 (center)	Dewatering Observers Monitoring	26
Mahogany Demonstration Project	ICP Demo, Recovery	1998–2005	600	46	11	Water Prod. Observers Monitor	101
Mahogany Demonstration Project South	ICP Demo, Recovery	2003–2005	400	25	2 (center)	Temp. Observers Water Prod.	27
Deep Heater Test	Test Heaters	2001–2005	700	21	2	Observers/Geophone Water Prod.	45
Mahogany Isolation Test	Evaluate Freeze Wall	2002–2004	1,400	2	1	Freeze Hole Monitor Temp. Monitor	53
Freeze Wall Test	Evaluate Freeze Wall	2005–2010	1,700	0	0	Freeze Hole Monitor Temp. Monitor	233

Data taken from Fowler and Vinegar [2009].

Shell to identify many technical challenges including hot spots, geo-mechanical stress, non-uniform gripping of the heaters, and several others [Fowler and Vinegar 2009].

The Mahogany Isolation Test (Table 6.2) had the objective to technically demonstrate that a freeze wall could be formed using conduction cooling and that pyrolysis-derived fluids could be contained within the wall during the *in-situ* upgrading process. It was pointed out that the use of freeze walls was a relatively common practice in some underground operations and could also prevent water influx at the point of heating. For this purpose, 18 wells (8.7 ft spacing on 50 ft diameter ring) were drilled. The result showed that the freeze wall could be constructed within 2 ft radial tolerance for the entire depth (1,400 ft). For this test, minimal oil (3 BBL) and gas (70×10^3 SSF) productions were obtained [Fan *et al.* 2010].

The Freeze Wall Test (Table 6.2) was designed to investigate the viability of using freeze walls containment for a commercial scale oil shale project. Two hundred and thirty-three holes were drilled for this purpose, and the walls were formed by using chilled aqua-ammonia (29% anhydrous ammonia in water). A fiber-optic system was used to monitor temperatures profiles. Shell's results showed that freeze walls were formed across the entire commercial intervals and could withstand the expected pressure differential (200–430 psi) and over 150 psi above the full hydrostatic head [Deeg *et al.* 2011].

6.3.3   VIKING HEAVY OIL PILOT

The Viking Pilot was a multi-year pilot project designed to determine the commercial viability of the *In-Situ* Upgrading Process (IUP) in heavy oils. The test was conducted at the Peace River lease in Alberta, Canada, where it ran from December 2004 to June 2008. The key properties of the heated interval in their pilot area were [Karanikas 2012]:

- Depth: ~600 m.
- Net pay: ~25–30 m.
- Porosity: ~0.25–0.30.
- Oil saturation: ~0.75–0.80.
- Oil viscosity and API: ~50,000–900,000 centipoise with 6–8°API.
- Depth-dependent viscosity is consistent with observations at other deposits in Alberta.

The pilot consisted of 18 horizontal heater wells, three horizontal producers, and a mix of eight horizontal and vertical observation wells [Karanikas 2016]. The observation wells were used to measure temperature, pressure, and micro-seismic, and to run well logs. Karanikas reported that the pilot produced well over 150,000 barrels of oil with an average density of 30°API. Including the produced hydrocarbon gas, the cumulative volume exceeded 200,000 barrels of oil equivalent [Karanikas 2012].

The ratio of the heating value of produced hydrocarbons to the cumulative amount of heat injected into the heated reservoir was slightly over four. This value is consistent with a value of slightly over six for commercial-scale operations [Karanikas 2012]: The author reported that the amount of injected heat, volumes of produced fluids, oil quality, and operating history were fundamental as predicted by reservoir simulations including chemical reactions derived from laboratory data. Based on the available information, the main accomplishments were the following:

- 450,000 labor-hours with no Lost Time Incidents
- No reportable oil spills
- 1,732 days of heating
- Recovery Efficiency ~77%
- Liquid Oil Recovery Efficiency ~60%
- Highest Average Reservoir Temperature ~320°C
- Highest Average Reservoir Pressure ~12.2 MPa

6.3.4   GROSMONT HEAVY OIL PILOT

This field test was focused on the confirmation of commercial viability of MI cable heater [Sandberg *et al.* 2016] or Long Horizontal Heater Test (LHT). Its objective was to de-risk the application of the *In-Situ* Upgrading Process (IUP) technology in the Grosmont area.

A 610 m horizontal MI cable heater drilled in an open hole test in the Upper Ireton formation was used [Karanikas 2016]. The characteristics of this reservoir were as follows:

- Dolomite with thin silt laminations, brecciated zones, corrosion intervals
- Depth 300 m
- Temperature 11°C @ 320 m
- Pressure 1.4 MPa @ 320 m
- Bitumen saturation 80–90%
- Porosity 22–35% (Matrix)
- Horizontal matrix permeability range 20–200 mD
- Fracture permeability range 2–20 D
- Bitumen viscosity 1,800,000+ cP

After some initial issues, the pilot achieved an average and maximum temperatures of 500–600°C and 650°C, respectively, with pressures in the 1,000–2,000 kPa range for the approximatively four-month period [Alberta 2014]. As mentioned before, these temperatures are sufficiently high to crack and coke the bitumen molecules into liquid and gases leading to the *in-situ* upgrading of the HO/B.

Shell reported that the Grosmont Heavy Oil Pilot was completed on 16 December 2013 and the results exceeded expectations. The test confirmed the commercial heater design to be used in the IUP technology for the recovery of extra-heavy oil and bitumen resources.

6.3.5   JORDAN FIELD EXPERIMENT

Jordan has one of the most abundant oil shale resources in the world, and although previous attempts to harness this energy source have been made, none have resulted in large-scale production [Meijssen *et al.* 2014]. In May 2009, Royal Dutch Shell plc ("Shell") signed an Oil Shale Concession Agreement to explore and evaluate the commercial potential of the deeper layers of Jordanian oil shale. Current activities are focused on demonstrating the technical feasibility of ICP technology.

This project is aimed to validate the subsurface understanding of the ICP process. It consists of seven vertical heaters in a close-spaced hexagonal array with a producer and two observer wells inside [Meijssen *et al.* 2014]. Additionally, seven monitor wells are included to examine possible product migrations and to safeguard the environment. Even though no detailed account of the results is known to date, the information available showed that more than 2,000 BBL of oil had been produced [Karanikas 2016].

REFERENCES

Adams, J. J., 2014, "Asphaltene Adsorption, a Literature Review", *Energy Fuels*, 28, 2831–2856 and references therein.

Alberta Energy Regulator, 2014, "Field Production and Heater Test Project" www.aer.ca/documents/oilsands/insitu-presentations/2014AthabascaShellGrosmont11487.pdf, retrieved 11 June 2017.

Alpak, F. O., Vink, J. C., Gao, G., Mo, W., 2013, "Techniques for Effective Simulation, Optimization, and Uncertainty Quantification of the In-Situ Upgrading Process", *J. Unconv. Oil Gas Res.*, 3–4, 1–14.

Biglarbigi, K., Dammer, A., Cusimano, J., Mohan, H., 2007, "Potential for Oil Shale Development in the United States", SPE No. 110590 presented at SPE Annual Technical Conference and Exhibition, Anaheim, CA, 11–14 November.

Biglarbigi, K., Mohan, H., Carolus, M., 2009, "Potential for Oil Shale Development in the United States", presented at the Oilsands Heavy Oil Technology Conf., Calgary, Canada, 15 July.

Carbognani, L., Gonzalez, M.F., Pereira-Almao, P., 2007, "Characterization of Athabasca Vacuum Residue and Its Visbroken Products. Stability and Fast Hydrocarbon Group-Type Distributions", *Energy Fuels*, 21, 1631–1639.

Carbognani, L., González, M. F., Lopez-Linares, F., Sosa-Stull, C., Pereira-Almao, P., 2008, "Selective Adsorption of Thermal Cracked Heavy Molecules", *Energy Fuels*, 22, 1739–1746.

Deeg, W., Arbabi, S., Crump, L., Hansen, E., Lin, M., 2011, "Shell's Colorado Oil Shale Freeze Wall Tets", presented at 31st Oil Shale Symp. Golden, CO, 17–21 October.

Fan, Y., Durlofsky, L., Tchelepi, H. A., 2010, "Numerical Simulation of the In-Situ Upgrading of Oil Shale", SPE No. 118958, *SPE J.*, 15(2), 368–381.

Fan, H.-F., Li, Z.-B., Liang, T., 2007, "Experimental Study on Using Ionic Liquids to Upgrade Heavy Oil", *J. Fuel Chem. Technol.*, 35(1), 32–35.

Fan, Z.-X., Wang, T.-F., He, Y.-H., 2009, "Upgrading and viscosity reducing of heavy oils by [BMIM][AlCl4] ionic liquid", *J. Fuel Chem. Technol.*, 37(6), 690–693.

Fowler, T. D., Vinegar, H. J., 2009, "Oil Shale ICP – Colorado Field Pilots", SPE No. 121164, presented at SPE Western Regional Meeting, San Jose, CA, 24–26 March.

Gray, M. R., 2015, *Upgrading Oilsands Bitumen and Heavy Oil*, The University of Alberta Press, Edmonton, and references therein.

Henderson, J. H., Weber, L., 1965, "Physical Upgrading of Heavy Crude Oils by the Applications of Heat", *J. Can. Pet. Technol.*

Huc, A.-Y., Ed. 2011, *Heavy Crude Oils. From Geology to Upgrading. An Overview*, Technip, Paris, France, and references therein.

Joshi, J. B., Pandit, A. B., Kataria, K. L., Kulkarni, R. P., Sawarkar, A. N., Tandon, D., Ram, Y., Kumar, M. M., 2008, "Petroleum Residue Upgradation via Visbreaking: A Review", *Ind. Eng. Chem. Res.*, 47, 8960–8988.

Kapadia, P. R., Kallos, M. S., Gates, I. D., 2015, "A Review of Pyrolysis, Aquathermolysis, and Oxidation of Athabasca Bitumen", *Fuel Proc. Technol.*, 131, 270–289.

Karanikas, J. M., 2012, "Unconventional Resources: Cracking the Hydrocarbon Molecules In Situ", *J. Pet. Technol.*, 65, 68–75.

Karanikas, J. M., 2016, "Improved Recovery in Heavy Oil Plays: Near Wellbore Heating and In-Situ Upgrading Process", presentation to Chevron. Gasmer Road Site, Houston, TX, 11 April.

Kasrale, M., Farouq Ali, S. M., 1989, "Role of Foam, Non-Newtonian Flow and Thermal Upgrading in Steam Injection", SPE No. 18784 presented at SPE California Regional Meeting, Bakersfield, CA, 5–7 April, 389.

Kumar, J., Fusetti, L., Corre, B., 2011, "Modeling In-Situ Upgrading of Extraheavy Oils/Tar Sands by Subsurface Pyrolysis", SPE No. 149217 presented at Canadian Unconventional Resources Conference, Calgary, Alberta, Canada, 15–17 November.

Meijssen, T. E. M., Emmen, J., Fowler, T. D., 2014, "In-Situ Oil Shale Development in Jordan through ICP Technology", SPE No. 172135, presented at Abu Dhabi International Petroleum Exhibition and Conference, Abu Dhabi, UAE, 10–13 November.

Millour, J. P., Moore, R. G., Bennion, D. W., Ursenbach, M. G., Gie, D. N., 1985, "A Simple Implicit Model for Thermal Cracking of Crude Oils", SPE No. 14226 presented at SPE Annual Technical Conference and Exhibition, Las Vegas, NV, 22–26 September.

Montoya, T., Argel, B. L., Nassar, N. N., Franco, C. A., Corte, F. B., 2016, "Kinetics and Mechanisms of the Catalytic Thermal Cracking of Asphaltenes Adsorbed on Supported Nanoparticles", *Pet. Sci.*, 13, 561.

Murugan, P., Mani, T., Mahinpey, N., Dong, M., 2012, "Pyrolysis Kinetics of Athabasca Bitumen Using a TGA under the Influence of Reservoir Sand", *Can. J. Chem. Eng.*, 90, 315–319.

Naghizada, N., Prado, G. H. C., de Klerk, A., 2017, "Uncatalyzed Hydrogen Transfer during 100–250°C Conversion of Asphaltenes", *Energy Fuels*, 31(7), 6800–6811.

Ovalles, C., Garcia, M. C., Lujano, E., Aular, W., Bermudez, R., Cotte, E., 1998, "Interfacial and Thermal Properties of Acid, Basic and Neutral Fractions Derived from Orinoco Belt Crude Oil", *Fuel*, 77, 121–126.

Perez-Perez, A., Mujica, M., Bogdanov, L., Corre, B., 2014, "Modeling In-Situ Upgrading of Heavy Oils by Subsurface Pyrolysis", presented at the 2014 Heavy Oil Congress, New Orleans, LA, 05–07 March, Paper No. WHOC14-232.

Phillips, C., R., Haidar, N. I., Poon, Y. C., 1985, "Kinetic Models for the Thermal Cracking of Athabasca bitumen. The Effect of the Sand Matrix", *Fuel*, 64, 678.

Rogel, E., Ovalles, C., Moir, M. E., 2010, "Asphaltene Stability in Crude Oil and Petroleum Materials by Solubility Profile Analysis", *Energy Fuel*, 24 (8), 4369–4374.

Sandberg, C., Thomas, K., Penny, S., 2016, "The Use of Coiled Tubing for Deployment of Electrical Heaters in Downhole Applications", SPE No. 179095, presented at the SPE/ICoTA Coiled Tubing and Well Intervention Conf. and Exh., Houston, TX, 22–23 March.

Shu, W. R., Hartman, K. J., 1986 "Thermal Visbreaking of Heavy Oil During Steam Recovery Processes", *SPE Res. Eng.*, 1(05), 474.

Shu, W. R., Venkatesan, V. N., 1984, "Kinetics of Thermal Visbreaking of a Cold Lake Bitumen", *J. Can. Pet. Technol.*, 23(02).

Speight, J. G., 2012, "Visbreaking: A Technology of the Past and the Future", *Scientia Iranica*, 19, 569–573.

Strausz, O. P., Jha, K. N., Montgomery, D. S., 1976, "Chemical Composition of Gases in Athabasca Bitumen and in Low-Temperature Thermolysis of Oil Sand, Asphaltene and Maltene", *Fuel*, 56, 114–120.

Venkatesan, V. N., Shu, W. R., 1986, "Alteration in Heavy Oil Characteristics during Thermal Recovery", *J. Can. Pet. Technol.*, 25(4), 66.

Vinegar, H. J., 2006, "Shell In-situ Conversion Process", presented at the Colorado Energy Research Institute 26th Oil Shale Symposium, Golden, CO, 16–18 October, downloaded from www.ceri-mines.org/documents/R05a-HaroldVinegar.pdf on 17 January 2018.

Vinegar, H. J., de Rouffignac, E. P., Karanikas, J. M., Maher, K. A., Sumnu-Dindoruk, M. D., Wellington, S. L., Crane, S. D., Messier, M.t A., Roberts, B. E., 2002, "In Situ Thermal Processing of a Tar Sands Formation", US Patent No. 7,066,254.

Wang, H., Wu, Y., He, L., Liu, Z., 2012, "Supporting Tungsten Oxide on Zirconia by Hydrothermal and Impregnation Methods and Its Use as a Catalyst To Reduce the Viscosity of Heavy Crude Oil", *Energy Fuels*, 26, 6518–6527.

Weissman, J. G., 1997, "Review of Processes for Downhole Catalytic Upgrading of Heavy Crude Oil", *Fuel Proc. Technol.*, 50, 199–213.

Youtsos, M., Mastorakos, E., Cant, R. S., 2012, "Simulations of Thermal Upgrading Methods for Oil Shale Reservoirs", SPE No. 163338 presented at SPE Kuwait International Petroleum Conference and Exhibition, Kuwait City, Kuwait, 10–12 December.

7 Thermal Hydrogen Addition

In this route (Figure 1.9), hydrogen gas or hydrogen-donating compounds are used in the presence of an energy source for subsurface upgrading (SSU) of heavy crude oils and bitumens (HO/B) [Kapadia *et al.* 2015, Muraza and Galadima 2015]. Generally, most of the hydrocarbon conversion is thermal (> 230°C), and the presence of the hydrogen source improves the quality and stability of the upgraded products. As discussed previously, thermal hydrogen addition routes are relatively simple and do not require complicated surface installations. However, these processes suffer several disadvantages such as poor mixing between the hydrogen source and the heated heavy crude oil, and the lack of availability of inexpensive and abundant hydrogen and energy sources.

The chemistry and mechanistic pathways involved in thermal hydrogen addition to hydrocarbons were discussed in Section 4.2 [Noguchi 1991, Huc 2011, Gray 2015, Speight 2007, Ramirez-Corredores 2017]. In this chapter, several hydrogen sources for the subsurface upgrading of heavy crude oils and bitumens are described. Specifically, the use of hydrogen gas and hydrogen precursors is the first topic, followed by the utilization of steam during thermal Aquathermolysis reaction. The use of catalysts to enhance the hydrogen transfer from water to the upgraded HO/B is presented in Chapter 8. Next, the use of hydrogen donor solvents is discussed as well as the effect of methane and mineral formation. This chapter finishes with the use of refinery fractions as hydrogen donors as means of increasing the economic prospects of this SSU route.

As with thermal processes discussed in the previous chapter, several energy sources have been reported to heat the reservoir to the temperature necessary to achieve the thermal cracking of the hydrocarbon molecules. Those sources are steam, electricity, *in-situ* combustion, and electromagnetic energy. To date, only the use of steam has been field-tested. Because those pilot tests used catalysts, their discussion is presented in Chapter 8.

7.1 USE OF HYDROGEN AND HYDROGEN PRECURSORS

Table 7.1 shows a summary of the US patents and a report by the National Institute of Petroleum and Energy Research (NIPER) using hydrogen gas or hydrogen precursors for the subsurface upgrading of heavy crude oils and bitumens. These studies are discussed in the following sections.

7.1.1 USE OF HYDROGEN GAS

The utilization of hydrogen gas for the subsurface upgrading of hydrocarbons has been proposed as early as 1965. Dew and Martin disclosed a process for the recovery of hydrocarbons from underground formations by using *in-situ* combustion between two vertical wells and using hydrogen at temperatures between 600°F (315°C) and 700°F (371°C) [Dew and Martin 1965]. The concept involves the injection of air to initiate the hydrocarbon combustion and bring the reservoir to the cracking temperature. After the passage of the combustion front through the formation, a hydrogen-containing gas is injected at a pressure of at least 500 psig (3.4 MPa) for a period sufficient to convert the residual carbonaceous material. Unfortunately, no examples were given, but a graph showing the relation between the combustion front temperature and the amount of residual material left in the formation was disclosed [Dew and Martin 1965].

Hamrick and Rose reported the *in-situ* conversion of hydrocarbon-containing materials by using a gas generator in a borehole for burning a hydrogen-rich mixture of H_2 and O_2 [Hamrick 1977]. The authors claimed that the temperature of the exhaust gases is maintained sufficiently high to crack the hydrocarbons in the presence of an excess of hot hydrogen to form lighter and less viscous

TABLE 7.1

U.S. Patents and a Report from NIPER Using Hydrogen or Hydrogen Precursors for the Subsurface Upgrading of Heavy Crude Oils and Bitumens

First Author	Assignee	Hydrogen Source	Description	Comments	References
Dew	Continental Oil Comp.	Hydrogen gas	A method to recover hydrocarbons from underground formations using *in-situ* combustion and contacting the hydrocarbon with hydrogen at 500 psi and 400–950°F for a period sufficient to substantially lower the viscosity	A process scheme and a graph showing the temp. needed to achieve viscosity reduction were disclosed	Dew [1965]
Hamrick	World Energy System	Hydrogen gas	*In-situ* conversion of HO/B using a gas generator in a borehole for burning a hydrogen-rich mixture of H_2 and O_2. The temperature of the exhaust gases is maintained high enough to crack the hydrocarbons in with the excess hot hydrogen to form lighter and less viscous products	Diagrams are provided but no examples disclosed	Hamrick [1977]
Stine	UOP LLC	Hydrogen gas	Conversion of heavy crude oils containing a trace amount of vanadium, nickel, iron by using *in-situ* combustion in *one* or *two* vertical wells and contacting with hydrogen at a pressure from about 200–5,000 psig	Diagrams are provided, and one example using Orinoco crude oil is disclosed	Stine [1984a, 1984b]
Brunner	Kraftwerk Union Aktiengesell	CO, hydrogen, and steam	Method of extracting hydrocarbons from oil-containing sand, by adding hydrogen and carbonization at 350°C	Process diagram is shown but no examples disclosed	Brunner et al. [1985]
Ware	World Energy Systems	H_2 and steam	Hydrogen and steam are injected downhole to cause hydrogenation of the petroleum in the heated zone (350–900°F). The process can be used in a single (Huff and Puff) or two wells (steamflooding)	A few diagrams but no examples were disclosed	Ware et al. [1986]
Gregoli	World Energy Systems	Hydrogen gas	Hydrogen, air, and steam are fed to downhole combustion devices. Superheated steam and hot reducing gases are injected into the formation to fracture it and upgrade the HO/B into lighter hydrocarbons	Process diagrams, simulation, and comparison with steamflood were disclosed	Gregoli et al. [2000]
Stapp	IIT Research InstNIPER	Hydrogen and nitrogen gas	Lab batch experiments in the presence of sands and brine at 550–750°F for several days	Several examples were reported	Stapp [1989]
Hewgill	Union Oil Company of California	Formic acid and its salts	*In-situ* crude oil upgrading in a subterranean formation by adding a non-gaseous hydrogen precursor into steam to enhance oil recovery at 550°F (288°C) for at least one week	One control and one example were disclosed	Hewgill and Kalfayan [1992], Delgado et al. [2005]
Stine	UOP LLC	Ethers, ketones, alcohols, aldehydes, glycols, carbohydrates	Enhanced recovery by *in-situ* thermal cracking of an oxygenated organic compound to form hydrogen for the upgrading of the hydrocarbon using two horizontal wells at the temperature range 150–500°C	One diagram but no examples	Stine and Barger [2012]
Krumrine	Signa Chemistry, Inc	Alkali metal silicides	Enhanced oil recovery by downhole injection of alkali metal silicides to generate hydrogen gas, heat, and alkali metal silicate solutions *in situ* upon contact with water	Several plots disclosed but no specific examples were disclosed	Krumrine et al. [2017]

upgraded products [Hamrick and Rose 1977]. Several diagrams of the gas generator placed in the borehole were disclosed as well as the internal components of such a device. Unfortunately, no examples of changes in the crude oil were reported.

Brunner and coworkers used carbon monoxide, hydrogen, and steam for extracting and upgrading hydrocarbons from oil-containing sand at temperatures of 450–520°C and pressures of 5 to 15 MPa. As before, no examples were disclosed [Brunner *et al.* 1985].

Stine reported a method for the *in-situ* conversion and recovery of heavy crude oil containing indigenous trace metals from two adjacent non-communicating hydrocarbon reservoirs which are alternately pressured and recovered [Stine 1984a, 1984b]. Figure 7.1 shows the schematic diagram of the subsurface process which involves the following steps [Stine 1984b]:

1. Heating the heavy crude oil in a first reservoir (left-hand side of Figure 7.1) to a hydrocarbon conversion temperature (> 500°F or 260°C)
2. Contacting the first reservoir with hydrogen gas at a pressure at least 200 psi (1.4 MPa)
3. Heating the heavy crude oil in the second reservoir to a hydrocarbon conversion temperature (> 500°F or 260°C)
4. Depressurizing the first reservoir, and using the gas–liquid separator (Figure 7.1), separating the upgraded crude oil from the gaseous component
5. Contacting the second reservoir (right-hand side of Figure 7.1) with hydrogen gas recovered in step 4
6. Depressurizing the second reservoir to yield additional upgraded crude oil and unreacted hydrogen gas

The author mentioned that various techniques might be utilized for reservoir heating such as superheated steam, hot circulating oil, high-temperature nitrogen streams, electrical heating elements, and air injection [Stine 1984a, 1984b]. Stine disclosed an example in which a sample of Orinoco crude oil having the characteristics presented in Table 7.2 was utilized. A fire flood was started by injecting air and a source of ignition. A portion of the formation was consumed by fire to furnish enough heat to raise the surrounding reservoir to a temperature of about 850°F (427°C). Then the hot crude oil formation was pressured with hot hydrogen at a temperature of about 500°F (260°C) and a pressure of about 1,500 psig (10.3 MPa). The hydrocarbon zone (Figure 7.1) was permitted to remain in these conditions for 48 hours. During this period, the consumed hydrogen was replenished to maintain the desired reaction pressure [Stine 1984a, 1984b].

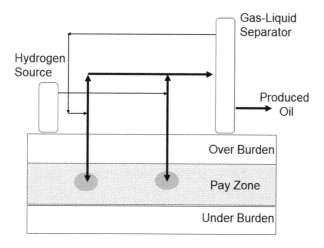

FIGURE 7.1   Schematic diagram of the *in-situ* conversion process of heavy hydrocarbons.

TABLE 7.2

Properties of the Orinoco Crude Oil Feed and the Upgraded Product as Disclosed by Stine Using *In-Situ* Combustion Followed by Hydrogen Gas Addition at 427°C and 10.3 MPa

Inspection	Orinoco Crude Oil	Upgraded Product
API gravity	9.9	14.0
Sulfur (wt. %)[a]	5.88	5.0 (15%)
Nitrogen (wt. %)[b]	0.635	0.6 (5.5%)
Heptane insolubles (wt. %)[c]	12.7	11.0 (13%)
Iron (ppm)	11	10
Nickel (ppm)	105	100
Vanadium (ppm)	1,260	1,200
Distillation		
Initial Boiling Point (°F)	187	170
10% off	572	550
30% off	840	820
43% off	1,000	–
50% off[d]		1,000 (12%)

Data taken from Stine [1984a, 1984b].

[a] Numbers in brackets indicate the percentage of desulfurization.

[b] Numbers in brackets indicate the percentage of nitrogen removal.

[c] Numbers in brackets indicate the percentage of reduction in asphaltene insolubles.

[d] Numbers in brackets indicate the percentage of conversion of 1000°F+ fraction (see Section 4.4).

Stine reported that, after the upgrading step was performed, the formation was depressurized and the resulting fluid was separated in the gas–liquid separator (Figure 7.1) at a temperature of 350°F (177°C) and a pressure of 400 psig (2.8 MPa). The properties of the upgraded crude oil are shown in Table 7.2. As discussed in Section 4.2.1.1., under these moderate conditions (427°C and 10.3 MPa, and 48 h), hydrovisbreaking occurs via a free radical mechanism, and an increase in the API gravity of ~5°API and a conversion of the 1000°F+ fraction of 12% were reported (Table 7.2). Similarly (Table 7.2), modest percentages of hydrodesulfurization (15%), nitrogen removal (5.5%), and heptane insolubles reduction (13%) were found [Stine 1984a, 1984b]. Even though the inventor claimed that the indigenous trace metals are acting as catalysts during the *in-situ* conversion [Stine 1984a, 1984b], the moderate upgrading results and the lack of evidence of hydrogenation reaction (i.e., no H/C increase, no aromatic fraction decrease, no metal removal, etc.) indicate otherwise. No more information about the development of this technology has been disclosed to date.

Ware *et al.* from World Energy Systems Inc. reported a method of recovering petroleum from underground formations by injecting hydrogen and steam under sufficient pressure to cause conversion of the hydrocarbon in the heated zone (350–900°F) [Ware *et al.* 1986]. The process can be used in a single well in a Huff and Puff mode (see Section 3.1.2.1) or in two wells in a steamflooding scheme (see Section 3.1.2.2.) [Ware *et al.* 1986]. A few diagrams were disclosed with the invention characteristics, but no examples were reported.

Gregoli and coworkers, also from World Energy Systems Inc., disclosed a similar process that injects hydrogen, water, and air but inserts a downhole combustion unit into the borehole. This unit generates superheated steam and hot reducing gases which are injected into the formation to fracture it and upgrade the HO/B into lighter hydrocarbons [Gregoli *et al.* 2000]. The inventors disclosed a simulation of the fracturing process to illustrate how formation fracturing makes possible the injection of superheated steam and a reducing gas into porous media which contains a very viscous hydrocarbon. Also, they discussed the advantages of the upgrading and increased recovery which occurs when a heavy hydrocarbon is produced by *in-situ* hydrovisbreaking rather than by steam drive [Gregoli *et al.* 2000]. As in the previous patent, a few diagrams were disclosed, but no examples were reported.

Stapp from NIPER conducted bench-scale experiments for thermal cracking of heavy hydrocarbons in the presence of hydrogen (Hydrovisbreaking) and nitrogen (Visbreaking) [Stapp 1989]. He also carried out experiments using sand, water, and brine which are discussed in the next section [Stapp 1989].

Table 7.3 shows the results of the upgrading of Cat Canyon crude oil in batch experiments at 14.5 MPa, 550°F (288°C) for ten days [Stapp 1989]. As seen, minimal changes were observed in the chemical properties (API, molecular weight, and elemental analysis) of the products vs. the feed. However, there was a significative reduction in viscosity in the upgraded crude oils in comparison with the original feedstock. These results are typical of a very mild thermal cracking process (see Section 4.1.2).

On the other hand, the Conradson carbon decreased whereas the aromatic carbon increased in the H_2-containing run vs. those found in the N_2 experiment (Table 7.3). Also, Stapp observed hydrogen consumption as the pressure diminished during the ten-day run. The results were attributed to a hydrovisbreaking reaction instead of hydrogenation [Stapp 1989]. In other words, hydrogen gas is involved in the radical capping of the intermediate radicals (see Section 4.2.1.3), rather than hydrogen addition to aromatic or unsaturated compounds (see Section 4.2.2.2). Stapp evaluated two other heavy crude oils, Saner Rach (–4°API) and Alaska North Slope (11.4°API), with similar results [Stapp 1989].

7.1.2 Use of Hydrogen Precursors

As shown in Table 7.1, Hewgill and Kalfayan disclosed the *in-situ* crude oil upgrading by adding a non-gaseous hydrogen precursor into steam to enhance oil recovery of HO/B [Hewgill and Kalfayan 1992]. The hydrogen precursors were selected from the group consisting of formic acid, organic salts of formic acid, alkali metal salts of formic acid, formate esters, formamide, N-substituted formamides, and mixtures thereof. The authors disclosed an example using Ventura crude oil (300 g;

TABLE 7.3

Upgrading of Cat Canyon Crude Oil in Batch Experiments at 14.5 MPa, 550°F (288°C) for Ten Days

Analysis	Feed	Hydrogen Experiment	Nitrogen Experiment
Gas	–	H_2	N_2
API gravity	10.7	10.0	9.8
Conradson carbon	14.3	13.0	14.3
Mol. weight (by VPO in g/mol)	594	599	589
Wt. % carbon	81.82	79.76	79.95
Wt. % hydrogen	10.28	10.68	10.70
H/C molar ratio	1.51	1.61	1.61
Wt. % sulfur	5.2	5.1	5.0
Viscosity at 54°C	2,955	1,900	1,091
Viscosity at 82°C	427	286	185
Aliphatic carbon by ^{13}C-NMR	75.1	73.8	71.7
Aromatic carbon by ^{13}C-NMR	24.9	26.2	28.3

Data taken from Stapp [1989].

16°API gravity), an aqueous formic acid solution (10 ml formic acid dissolved in ~290 g water), 1 g MoS_2, and a formation material (~100 g) heated at ~550°F (288°C) for a period of one week. They observed that the reactor pressure decreased over time which strongly suggested the uptake of hydrogen gas. This hypothesis was confirmed by the analysis of the gas phase at the end of the experiment. The analysis showed the presence of 16 mol. % of hydrogen, 0.4 mol. % of CO, and 35.5 mol. % of CO_2. These results were rationalized in terms of the decomposition of formic acid to form carbon monoxide (Equation 7.1) and carbon dioxide (Equation 7.2) under the reaction conditions [Hewgill and Kalfayan 1992]:

$$HCOOH \rightarrow H_2O + CO \tag{7.1}$$

$$HCOOH \rightarrow H_2 + CO_2 \tag{7.2}$$

In turn, the carbon monoxide reacts via water–gas-shift reaction to generate one additional mol of hydrogen gas and one of carbon dioxide according to Equation 7.3:

$$H_2 + CO \rightarrow H_2 + CO_2 \tag{7.3}$$

Hewgill and Kalfayan also found that the API of the upgraded oil increased whereas the asphaltene content decreased, but no actual data were disclosed. A control experiment was carried out in the absence of formic acid but failed to affect the API gravity and had a much smaller effect on the asphaltene content of the upgraded oil [Hewgill and Kalfayan 1992]. Even though the crude oil upgrading reaction was carried out in the presence of a catalyst (MoS_2) the important point is the use of HCOOH as hydrogen precursor.

These results were confirmed by Scott and coworkers several years later [Delgado *et al.* 2005]. These authors carried out laboratory physical simulations, using a batch reactor, at 280°C and 88–109 Bar (8.8 – 19.9 MPa). The reactor was fed with a mixture of crude oil and sand (99 wt. % SiO_2) having 10 wt. % crude oil (Hamaca C. from the Orinoco Belt with 9.1°API gravity), together with formic acid and water. They believed that these conditions more closely represent the actual situation found during crude oil production by steam stimulation processes [Delgado *et al.* 2005].

SARA analyses showed (see Table 7.4) that when the crude oil was treated in the presence of formic acid, reductions of the asphaltene and resin contents (10 wt. % and 32 wt. %) were obtained in comparison with the original Hamaca crude (13 wt. % and 45 wt. %, respectively). At the same time, the aromatics content increased (from 34 wt. % to 55 wt. %), but the saturates were close to the value in the original oil (~7–8 wt. %). The authors found that the best results were obtained when formic acid was used in conjunction with water. As shown, asphaltenes were reduced by 29.2% and

TABLE 7.4
SARA Fractions for Original and Upgraded Hamaca Crude Oil[a]

Feed and Reaction Products	Saturates wt. %	Aromatics wt. %	Resins wt. %	Asphaltenes wt. %
Hamaca crude oil	8.0 ± 0.8	34 ± 4	45 ± 2	13 ± 1
Formic acid	7 ± 1	51 ± 1	32 ± 1	10 ± 1
Formic acid + water	10.4 ± 0.5	55 ± 3	25 ± 1	2 ± 0.5
Water	11.9 ± 0.6	48 ± 2	29 ± 1	11.2 ± 0.6

Data taken from Delgado *et al.* [2005].

[a] Reactions carried batch wise, no stirring, at 88–109 Bar at 280°C for 24 h. Ratio sand/crude oil/formic acid/water = 10:1:1:1.

resins by 44.5% (Table 7.4). A control experiment using only water led to insignificant upgrading of the heavy crude. These results agreed with those reported by Hewgill and Kalfayan [1992] and were attributed to the chemical pathway similar to the one shown in Equations 7.1 to 7.3 [Delgado *et al.* 2005].

Scott *et al.* also reported the sulfur and nitrogen contents for the Hamaca crude oil, before and after upgrading [Delgado *et al.* 2005]. When formic acid alone was used, the percentage of desulfurization was less (12%) than that found when formic acid and water were used together (38%). On the other hand, the percentages of denitrogenation were about the same (27 ± 3%) no matter whether the reaction was carried out only with water or formic acid, or both. These results showed the potentiality of formic acid when used in conjunction with water, for generating H_2 under steam injection conditions [Delgado *et al.* 2005].

In 2012, Stine and Barger from the UOP disclosed the enhanced recovery via *in-situ* thermal cracking of oxygenated-organic compounds to form hydrogen for the upgrading of the hydrocarbons using two horizontal wells at the temperature range 150–500°C [Stine and Barger 2012]. The oxygen-containing compounds included ethers, ketones, alcohols, aldehydes, glycols, and carbohydrates. It was proposed that under reaction conditions the oxygenated-organic compounds decomposed with the production of hydrogen for the hydrocarbon upgrading reaction. Water may be optionally added to the *in-situ* reaction zone to promote steam reforming. As before, one process diagram was shown, but no examples were disclosed [Stine and Barger 2012].

Krumrine and coworkers disclosed the enhanced oil recovery by downhole injection of alkali metal silicides (NaSi, LiSi, or KSi) to generate hydrogen gas, heat, and alkali metal silicate solutions *in-situ* upon contact with water [Krumrine *et al.* 2017]. The author claimed that the heat could be used to reduce the viscosity and the hydrogen to upgrade the crude oil underground. The resulting alkaline silicate solution saponifies the acidic crude components to form surfactants which, in turn, emulsify the crude to improve its mobility toward the production well. As in the other patent cases, several process diagrams were shown, but no specific examples were disclosed [Krumrine *et al.* 2017].

As presented in this section, laboratory physical simulations have shown that hydrogen or hydrogen precursors could be effectively used to upgrade HO/B under subsurface conditions. However, none of these concepts have been numerically simulated nor field tested. Next, the use of steam as hydrogen source is presented.

7.2 THERMAL AQUATHERMOLYSIS

Aquathermolysis is a hydrogen addition route that involves the thermal cracking of heavy crude oils, bitumens, or oil-containing sands in the presence of water as a source of hydrogen [Clark *et al.* 1983, Viloria *et al.* 1985, Brons and Siskin 1994, Maity *et al.* 2010, Kapadia *et al.* 2015, Muraza and Galadima 2015]. This process can also be considered a hydrous pyrolysis [Montgomery *et al.* 2013] and has been studied for more than 30 years [Hyne *et al.* 1982]. During the HO/B production with steam or hot water, thermal energy breaks large molecules into smaller ones with the concomitant reduction of oil viscosity. As described in Section 4.1.2.2, thermal cracking involves the homolytic cleave of bonds with the formation of free radicals (Equation 4.1). It has been reported that the overall Aquathermolysis reaction is as follows [Hyne *et al.* 1982, Hyne 1986]:

$$\left.\begin{array}{c}\text{Heavy Oil \& Bitumens}\\+\\\text{Mineral Formation}\end{array}\right| \xrightarrow[\text{200--300°C}]{\text{Steam}} \begin{array}{c}\text{Upgraded}\\\text{Crude Oil}\end{array} + H_2S + CO_2 + CO + CH_4 + H_2 \qquad (7.4)$$

One of the relevant aspects of this chemical reaction (Equation 7.4) is the rupture of C–S bonds of the crude oil (bond energy in the range of 250–300 kJ/mol, see Table 4.2) with the concomitant generation of hydrogen sulfide and improvements in flow properties of heavy oils [Ovalles *et al.* 1995, Maity *et al.* 2010]. Also, some of the produced gases such as carbon dioxide and monoxide, methane, and hydrogen remain dissolved in the crude oil, reducing its viscosity even further at downhole conditions. Trace amounts of other sulfur-containing compounds are generated such as mercaptans, sulfides, thiophenes, etc. (see Section 7.2.1).

The effect of the porous media on the Aquathermolysis reaction has been studied in the literature [Stapp 1989, Maity *et al.* 2010, Monin and Audlbert 1988, Viloria *et al.* 1985, 1987] and is discussed in Section 7.3.2. The use of metal-containing catalysts to enhance the hydrogen transfer from steam to the HO/B is presented in Chapter 8. Several references reported in the literature are concentrated on the H_2S and CO_2 production during steam injection processes and the operational problems of this chemical phenomenon [Barroux and Lamoureux-Var 2013, Barroux *et al.* 2013, Goicetty 2010, Lamoureux-Var and Lorant 2005, Lin *et al.* 2016, Thimm 2007, 2014]. This section is focused on the upgrading of HO/B occurring during Aquathermolysis, the effect of the mineral formation, kinetics and mechanism of the reaction, and reservoir numerical simulations.

7.2.1 Upgrading During Steam Injection

Laboratory, field tests, and commercial steam and hot water operations have found that hydrogen sulfide and carbon dioxide are generated in the produced gases with the concomitant changes in the properties of the heavy oil and bitumens. As mentioned, this phenomenon is known as Aquathermolysis (Equation 7.4). Figure 7.2 shows the typical gases produced for the thermal treatment of Athabasca oilsands with steam under batch conditions at 300°C [Hyne *et al.* 1982, Hyne 1986]. As seen, significant volumes of H_2S, CH_4, H_2, and CO_2, as measured in mL/Kg of crude, were generated after the Aquathermolytic treatment for 30 days. The production of hydrogen and methane seemed to increase as time progressed. On the other hand, the generation of carbon dioxide and hydrogen sulfide seemed to level off at the end of the experiments. A detailed discussion of the Aquathermolysis mechanism is presented in Section 7.2.3.

In line with lab and field data, Figure. 7.3 presents the production of H_2S in function of time from the reaction of a Venezuelan heavy crude oil with steam in the 200°C–380°C range. The lab-scale experiments were carried out in a batch reactor using a water-to-oil ratio of 1:1 v/v in the absence of sand. As seen, below 300°C, the generation of hydrogen sulfide is minimal. When the temperature

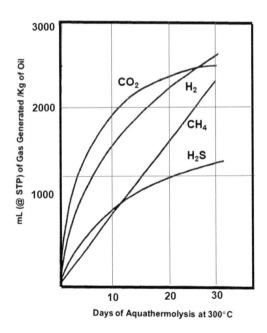

FIGURE 7.2 Typical gas production for the Aquathermolysis of Athabasca oilsands at 300°C. Data taken from Hyne *et al.* [1982].

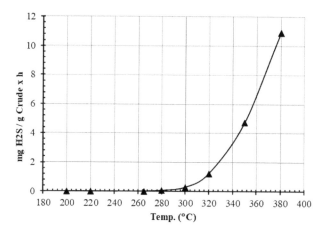

FIGURE 7.3 Generation of hydrogen sulfide vs. temperature for the thermal treatment of a Venezuelan heavy crude oil in the presence of steam (steam-to-oil ratio = 1:1, 280 psi of CH_4, 6 to 96 h lab experiments).

reaches 320°C, there is a significant increase in the rate of H_2S production. As mentioned, the production of hydrogen sulfide is due to the thermal decomposition of sulfur-containing organic compounds which are indigenous to the crude oils. As reported by Viloria and coworkers, similar results have been found for other Venezuelan crude oils such as Cerro Negro, Jobo, Pirital, Melones, and Tia Juana [Viloria et al. 1987].

Furthermore, Figure 7.4 shows moles of sulfur generated by the Aquathermolysis of a Cerro Negro core at 240°C for up to 20 days [Viloria et al. 1987]. As seen, the moles of sulfur as hydrogen sulfide and methyl mercaptan (CH_3SH) increased with time. Throughout the experiment, the last values were almost twice as high as the first one. These findings are significant because, during commercial thermal processes, the surface facilities must be able to handle H_2S as well as CH_3SH. Typically, hydrogen sulfide is oxidized at the production site to produce sulfur dioxide or recovery as elemental sulfur by using the Claus process.

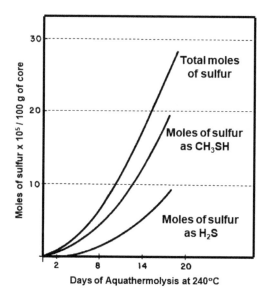

FIGURE 7.4 Moles of sulfur generated by the Aquathermolysis of a Cerro Negro core at 240°C. Data taken from Viloria et al. [1987].

Other organosulfur compounds are generated during the Aquathermolysis reaction. As shown in Figure 7.5., significant amounts of methyl mercaptan, dimethyl sulfide, and C2-thiophenes were detected in the gas phase during the thermal treatment of a Venezuelan crude. The concentration of CH_3SH increased with temperature whereas the other two compounds did not. These results are an indication of a complex reaction pathway as it is discussed in Section 7.2.2 [Clark 1983, Siskin *et al.* 1990]. Due to the small amount produced, these organosulfur compounds are normally removed at the surface by using commercial absorption processes.

Figure 7.6 presents the logarithm of the percentage of the overall thermal conversion of the original heavy oil or bitumen vs. the inverse of temperature in the presence of steam [Hyne 1986].

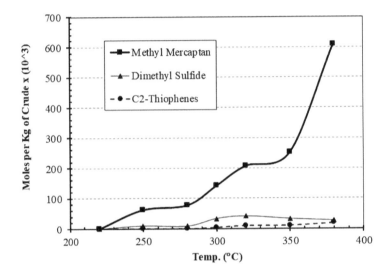

FIGURE 7.5 Generation of organosulfur compounds vs. temperature for the thermal treatment of a Venezuelan crude oil in the presence of steam (steam-to-oil ratio = 1:1, 280 psi of CH_4, 6 to 96 h lab experiments).

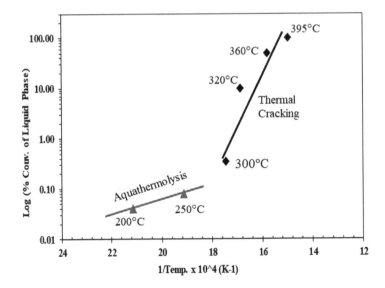

FIGURE 7.6 Log of the percentage of conversion of the liquid phase vs. the inverse of temperature for the thermal treatment of heavy crude oils in the presence of steam (lab). Data taken from Hyne [1986].

In this Arrhenius plot, the slope gives the energy of activation involved in the chemical conversions during the lab experiments. The substantially greater slope of the curve at a temperature higher than 300°C demonstrates a change of the chemical reactions with much higher energy requirements than the subtler, less destructive processes that characterize the chemistry of Aquathermolysis (< 300°C). During thermal cracking of the liquid phase (> 300°C), increasing amounts of gas and coke are produced even in the presence of water. This phenomenon indicates the predominant role of thermal cracking at much higher temperatures in comparison with the Aquathermolysis "window" (200–300°C) in which little conversion of the liquid phase is observed [Hyne 1986].

As mentioned earlier, the generation of hydrogen sulfide, carbon dioxide, and other gases led to changes in the properties of the heavy oil and bitumens. Figure 7.7 shows the changes in viscosity vs. residence time during Aquathermolysis experiments of Athabasca oil at three different temperatures, 200°C, 240°C, and 300°C [Hyne et al. 1982, Hyne 1986]. Hyne reported that at 200°C there was a steady increase in viscosity from ~1,550 cP to ~3,200 cP, whereas at 300°C, the opposite effect was observed (from ~1,550 cP to ~200 cP). At the intermediate temperature (240°C), the viscosity first increased (~200°C), but after prolonged Aquathermolysis (60 and 90 days), as would be characteristic of actual field operations, the viscosity of the recovered fluid decreased [Hyne 1986]. Similar observations have been reported by Viloria and coworkers for Venezuelan heavy crudes under steam stimulation conditions [Viloria et al. 1987].

Hyne studied the effect of Aquathermolysis on the SARA fractions (see Section 2.3.2.) on Canadian and Venezuelan heavy crude oils, and the results are presented in Figure 7.8 [Hyne 1986]. For both oils, the asphaltene content increased whereas the resin fraction decreased upon Aquathermolysis. This observation is particularly noticeable in the Canadian case compared to in the Venezuelan counterpart. As shown in Chapter 2 and reported in the literature [Speight 2007], the asphaltene and resin fractions have the highest amount of sulfur in comparison with those found in the saturates and aromatics. Thus, if indeed the organosulfur components of the oils are one of the most reactive species in the Aquathermolysis reaction, it might be expected that a substantial change would occur in those fractions [Hyne 1986]. Current understanding of the Petroleum Model using the Boduszynski's Continuum (Section 2.3.4) leads us also to consider that the observed Aquathermolytic changes are due to the cracking of the sulfur-containing hydrocarbon chains attached to the aromatic cores of the resins. This reaction, in turn, increases the aromaticity of this fraction, and more asphaltenes are obtained in the SARA analysis.

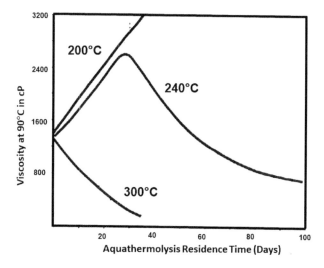

FIGURE 7.7 Changes in viscosity vs. residence time during Aquathermolysis of Athabasca oilsands. Data taken from Hyne et al. [1982].

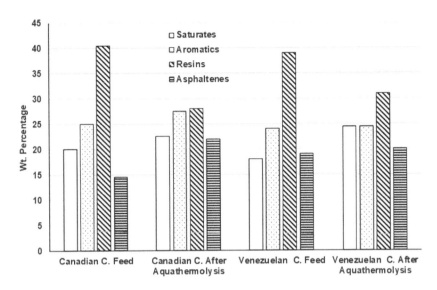

FIGURE 7.8   SARA composition for the feed and the Aquathermolysis products of Canadian and Venezuelan crudes. Data taken from Hyne [1986].

On the other hand, and as expected, the aromatic fractions (Figure 7.8) appeared to be the least affected via Aquathermolysis. There were changes, but these were relatively minor which indicated that the aromatic structures were less reactive at the temperatures characteristic of the Aquathermolysis (< 300°C).

Strausz and coworkers carried out laboratory batch experiments with oil sand samples from the Athabasca, Cold Lake, and Wabasca reservoirs with various amounts of water (0–20 wt. %) [Chen *et al.* 1991]. The oil sands were heated in sealed, evacuated quartz tubes at 250°C and 300°C for 3, 7, and 14 days. The authors observed the formation of gases, volatiles, and light oils, co-distilling with water (~13 wt. %). The amount of CO_2 released from the mineral component of the oil sands exhibited a marked water dependence [Chen *et al.* 1991].

Strausz *et al.* also reported decreases in the viscosity of the bitumens (~10–20%), and the asphaltene contents (18% for Athabasca, 44% for Cold Lake, and 18% for Wabasca). The latter findings are opposite to those reported by Hyne [Hyne 1986]. As seen in Figure 7.9, the asphaltene content of the Athabasca bitumen decreased with the time of the thermal treatment in the presence of water at temperatures of 250°C and 300°C (Aquathermolysis). Similar results were observed for the other two bitumen samples [Chen *et al.* 1991].

Furthermore, Strausz *et al.* reported that the oxygen and sulfur contents of Athabasca asphaltenes were somewhat lowered, but their molecular weight remained unchanged. The class composition of the steam-treated Athabasca maltenes showed an increase in the polyaromatic and saturate fractions and a decrease in the polar, diaromatic, and monoaromatic contents. The extent of these changes increased with the severity of thermal treatment applied [Chen *et al.* 1991].

Brons and Siskin carried out laboratory-scale simulations of steam-production processes using Cold Lake tar sand samples in neutral water and caustic brine at 250°C for up to 75 days [Brons and Siskin 1994]. The experiments were carried out in a batch reactor under a nitrogen atmosphere with a water-to-oil ratio of 0.25 v/v. The thermally treated oils were solvent extracted after the experiments to recover the upgraded bitumens. Similar to Strausz and coworkers, Brons and Siskin observed volatiles generation (~2.5 wt. %) and reduction of asphaltene content (from 16 wt. % to ~8 wt. %) with the concomitant decrease in the viscosity (~10%) of the upgraded crude oil. The authors observed maximum volatiles levels, and minimum viscosities and asphaltene concentrations in the 30- to 50-day time-frame.

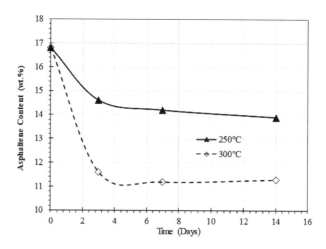

FIGURE 7.9   Changes of the asphaltene content of Athabasca bitumen in function of the reaction time in the presence of water. Average of two determinations. Data taken from Chen *et al.* [1991].

In contrast with Hynes' observations [Hyne 1986], longer thermal treatments showed an increase in the asphaltene content and viscosity of the products. The generated volatiles from the 50-day experiments were found to be mostly olefinic in nature. These olefins can then undergo addition reactions that would be consistent with the increase in viscosity and asphaltenes [Brons and Siskin 1994].

Hongfu and coworkers conducted the laboratory Aquathermolysis reactions using Liaohe heavy oils at 240°C for up to three days [Hongfu *et al.* 2002]. After reaction with steam, the viscosity of the heavy oil was reduced by 28–42% and the amount of the saturate and aromatic fractions increased, whereas the resins and asphaltenes decreased. The gas analyses showed that the accumulated amount of C20 carbon numbers increased from 13.30–20.92% to 38.79–53.92%. The authors attributed these results to a mild Visbreaking of the Liaohe heavy oils during steam drive and steam stimulation [Hongfu *et al.* 2002].

Montgomery *et al.* conducted laboratory Aquathermolysis experiments over a temperature range of 150–325°C for 24 h [Montgomery *et al.* 2013, Zeng 2013]. They treated a heavy oil extracted from Osmington Mills, Dorset, United Kingdom (18.4–18.9°API) to progressively higher temperatures and pressures in the presence of liquid water. The experiments revealed that at relatively mild conditions (250°C and 600 psig) the polar organic compounds present in the extracted oil led to the production of lighter materials richer in aliphatic hydrocarbons and a gas phase containing carbon dioxide. At higher temperatures and pressures (325°C and 1750 psig), the relative abundance of hydrocarbons over functionalized organic increased. At this relatively high temperature and pressure, cracking occurs, removing most solvent-soluble organic matter from the oil sand, generating a maximum amount of upgraded oil. The authors found that the upgraded crudes have lower asphaltenes (from ~14 wt. % to ~8 wt. %) and polar fractions (from ~45 wt. % to ~18 wt. %) than those found in the original crude [Montgomery *et al.* 2013]. These results are consistent with those reported by Strausz *et al.* [Chen *et al.* 1991] and Brons and Siskin [1994].

One point worth mentioning is the effect of steam distillation during thermal processes of HO/B via the Aquathermolysis reaction. As discussed in Section 5.1, steam distillation leads to the co-evaporation of fractions present in the crude oils at *much lower* temperatures than those obtained if the oil were alone. Thus, during steam recovery processes there is the possibility that indigenous organosulfur compounds are produced by steam distillations and not by the Aquathermolysis reaction [Viloria *et al.* 1987]. This phenomenon should be considered when analyzing HO/B samples generated by steam recovery processes.

Based on the previous observations and findings by different authors it can be safely concluded that during thermal operations of HO/B using steam or hot water, the generation of H_2S, CO_2, H_2, and CH_4 occurs via Aquathermolysis reaction (Equation 7.4). Depending on the nature of the crude oils, the subsurface upgrading can take place, and lower viscosity and asphaltene contents have been found in the produced oils.

7.2.2   EFFECT OF MINERAL FORMATION

Since the beginning of the studies of Aquathermolysis, it has been proposed that the mineral present naturally in the reservoirs can take part in the reaction at the steam injection conditions [Viloria *et al.* 1985, 1987, Stapp 1989, Maity *et al.* 2010]. Viloria *et al.* proposed that in order to explain the phenomena observed during Aquathermolysis, a catalytic effect of the formation could be considered [Viloria *et al.* 1985, 1987]. The use of catalysts to accelerate and enhance the properties of the upgraded crude oils is discussed in Chapter 8. However, several reports are worth mentioning at this moment.

Monin and Audibert carried out Aquathermolysis experiments using four crude oils with different geochemical compositions in the presence of water and a mineral formation representative of reservoir rocks at 350°C for 200 hours [Monin and Audlbert 1988]. Like other researchers, the authors reported that asphaltenes and, to a lesser extent, resins are the most reactive species. Gaseous hydrocarbons (mainly methane) were formed by pyrolysis. They found that the CO_2 evolution depended on matrix compositions. When Illite was associated with calcite or pyrite is present in the mineral formation, a partial matrix decomposition enhanced CO_2 and H_2S buildup [Monin and Audlbert 1988]. The change of the reaction equilibrium during Aquathermolysis in the presence of clays (Illite) has been previously reported by Siskin *et al.* [1990].

Fan and coworkers carried out laboratory reactions using Liaohe heavy oils at 240°C for up to 24 h [Fan *et al.* 2001]. When 10 wt. % of mineral formation was added to the reaction system, the saturates and aromatics increased, whereas the resins and asphaltenes decreased. After thermal treatment, VPO results showed that the average molecular weights of heavy oils and asphaltenes were reduced and the content of sulfur decreased. Similarly, the viscosity of heavy oils decreased to 23–26% when the mineral matrix was used in the presence of steam [Fan *et al.* 2001].

Montgomery and coworkers studied the effects of the porous media on gas and oil production [Montgomery *et al.* 2014, 2018]. They used an Alaskan oil-containing core and the isolated crude oil from that core. The thermal experiments were run at 250–300°C in the continuous presence of liquid water for 24 hours. All reaction products were studied with a variety of analytical techniques. Sulfur-rich oil sand sample produced H_2S at temperatures of 250°C and above. The principal finding was that the presence of mineral formation led to H_2S generated at 50°C lower than when the solid was not present (300°C). Also, the results indicated the existence of minerals that promote the CO_2 formation [Montgomery *et al.* 2014].

Furthermore, petroleum-containing reservoirs are large porous media consisting of sands, clay minerals, and non-clay minerals. A typical mineral formation contains a high percentage of rock (quartz, feldspar, etc.) and clays such as montmorillonite or Illite [Maity *et al.* 2010]. When high-temperature steam is injected into a reservoir, the generation of Brønsted acid centers (Al^{3+}) can be induced. These acid centers can catalyze the C–C and C–S bond cracking of organosulfur compounds with the concomitant acceleration of the Aquathermolysis reaction (Equation 7.4). In particular, Ovalles *et al.* conducted a lab study with and without mineral formation for the upgrading of Venezuelan crude oil under steam injection conditions at 280°C and 24 h [Ovalles *et al.* 2003a]. Figure 7.10 shows the effect of the presence of the mineral formation on the viscosity of the upgraded crude oil. As seen, the experiment carried out with porous media yielded a lower viscosity product than that found in the absence of the solid (e.g., 3,100 vs. 4,200 cP at 60°C).

Based on the results presented in this section, it can be concluded that depending on the mineral composition (quartz, clays, etc.) and the presence of metals (Fe, Zn, etc.), the catalytic activity of

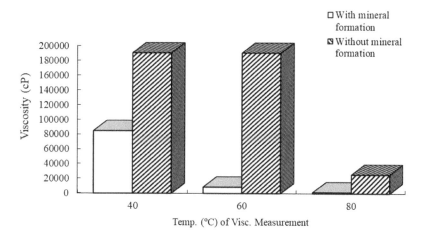

FIGURE 7.10 Effect of the presence of the mineral formation on the viscosity of the upgraded crude oils.

the mineral formation can be affected during the subsurface Aquathermolysis of the heavy oils and bitumens. In turn, the variability in chemical composition can lead to different rates for H_2S and CO_2 generation.

7.2.3 KINETICS AND MECHANISM

Over the years several kinetic and mechanistic studies on the Aquathermolysis reaction (Equation 7.4) have been carried out [Attar 1984, Clark *et al.* 1983, Hyne 1986, Brons and Siskin 1994, Belgrave *et al.* 1997, Chen *et al.* 1991, Lamoureux-Var and Lorant 2005, Barroux and Lamoureux-Var 2013, Barroux *et al.* 2013, Jiang 2005, Jia *et al.* 2016]. Recently, Gates and coworkers conducted a critical review on this subject with a very comprehensive list of the modeling and kinetic measurements reported in the literature [Kapadia *et al.* 2015].

As mentioned in the previous section, Hyne reported that the Aquathermolysis reaction involves the H_2S generation from organosulfur compounds which then reacted with water to yield smaller fragments [Hyne *et al.* 1982, Hyne 1986]. He also described various mechanisms to produce acid gases such as carbon dioxide and hydrogen sulfide. The effect of the water–gas-shift reaction (Equation 7.5) was also studied by changing the carbon monoxide concentration in the mixture of gases during Aquathermolysis [Hyne *et al.* 1982, Hyne 1986].

$$CO+H_2O \rightarrow CO_2+H_2 \tag{7.5}$$

In 1984, Viloria and coworkers carried out a sulfur speciation study during the H_2S generation of Cerro Negro (Orinoco) and Pirital cores in the presence of steam [Attar 1984]. The authors reported a complex reaction scheme (Figure 7.11) to describe the formation of hydrogen sulfide (H_2S) from alkyl thiophenes (a), thiophenes (b), sulfides (R-S-R), mercaptans (R-S-H), disulfides (R-S-S-R), and aryl sulfides (c) at 240°C at different times of reaction. As seen in Figure 7.11, the alkyl substituted thiophenes (a), and thiophenes (b) present in the crude oil react under a thermal condition to generate sulfides (R-S-R) and aryl sulfides (c). In turn, these intermediates suffer further transformation to mercaptans (R-S-H) with the final generation of hydrogen sulfide. Also, the disulfides (R-S-S-R) compounds present in the crude react with the production of H_2S [Attar 1984].

Using the reaction scheme shown in Figure 7.11, Viloria *et al.* solved the kinetic differential equations using a fourth-order Runge–Kutta algorithm, and the predicted values of the sulfur species were used to optimize the value of the rate constants. The calculated values of the H_2S

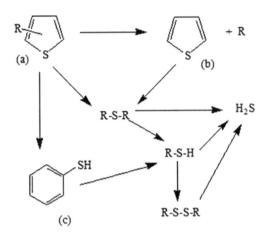

FIGURE 7.11 Reaction scheme proposed to account for the generation of H$_2$S during Aquathermolysis reactions.

concentrations were compared with the experimental data, and the results of the model matched reasonably well with the lab experiments [Attar 1984].

Lab studies using petroleum samples and model compounds have indicated that at steam injection conditions, the production of carbon dioxide might be from both the metal carbonates (e.g., Siderite, FeCO$_3$) present in the reservoir mineral and the heavy crude oils [Hyne 1986, Chen *et al.* 1991, Belgrade 1997, Maity *et al.* 2010]. As mentioned in Section 4.2.3.4, carboxylic acid groups are known to decompose under thermal conditions (200–400°C) to generate carbon dioxide and the alkanes, R-H (see Equation 7.9) [Ramirez-Corredores 2017]. In the presence of nitrogen-containing compounds, the reaction leads to olefin production and an alkyl substituted amine (Equation 7.7) [Ovalles *et al.* 1998a]:

$$RCOOH \rightarrow RH + CO_2 \qquad (7.6)$$

$$R\overset{L}{\underset{|}{-}}CH-CH_2-COOH \xrightarrow{\Delta} RCH_2{=}CH_2 + HL + CO_2 \qquad (7.7)$$

Where L = –NR$_3$, –NHR$_2$, –NH$_2$R, and R = alkyl chain.

It has been proposed that carbon monoxide in the products arises from the hydrolysis of thiophenes and to some extent of sulfides (R-S-H) as shown in Equation 7.8 [Hyne 1986, Chen *et al.* 1991]:

$$\text{(structures)} \xrightarrow[\ -H_2S\ ]{H_2O} \text{(OH, OH)} \rightarrow \text{(O, O)} \rightarrow 2CO + C_2H_6 \qquad (7.8)$$

Another possible source of carbon monoxide is the decomposition of the alcohols initially present in the HO/B via the production of aldehydes as intermediates. Hyne postulated that the following reaction sequence (Equation 7.9) might explain the CO generation during Aquathermolysis reaction [Hyne 1986]:

$$R - CH_2OH \rightarrow RCHO \rightarrow CO + H_2 \qquad (7.9)$$

Belgrade and coworkers developed a kinetic model that describes many experimental results and compositional changes during thermal cracking of Athabasca, North Bodo, Frisco Countess crude oils [Belgrade 1997]. These authors proposed that low-temperature oxidation increases the quantity

TABLE 7.5
Arrhenius Kinetic Data for the H_2S and CO_2 Production During Aquathermolysis Experiments

Reaction	Pre-exponential Factor (d^{-1})	Activation Energy (J/Mol)	Reference
H_2S Production	9.16	5.85×10^4	Jia et al. [2016]
	1.355	5.47×10^4	Kapadia et al. [2012]
CO_2 Production	1.67×10^6	1.129×10^5	Jia et al. [2016]

of molecular hydrogen generated by thermal cracking reactions, and H_2 generation was observed to correlate well with the decomposition of asphaltenes [Belgrade 1997].

Jia and coworkers carried out Aquathermolysis tests using different bitumen-to-water ratios at 225°C and 245°C for 3-, 10-, and 30-day reaction periods [Jia et al. 2016]. As found by other researchers, the results of the investigation showed that the amount of H_2S and CO_2 generated increased with the reaction time and temperature, and the CO_2 concentration stabilized over the test period. Data generated during the study were used to develop a kinetic model for predicting the time and temperature effects on H_2S and CO_2 production during the bitumen–steam/hot water Aquathermolysis [Jia et al. 2016]. Table 7.5 shows the Arrhenius pre-exponential factor and energy of activation for both reactions. As seen, Jia et al. calculated kinetic parameters [Jia et al. 2016] which matched reasonably well with the estimation by Kapadia et al. based on the experimental data by Hyne [Kapadia et al. 2012]. Significant deviations were seen between the estimated kinetic parameters and the literature data for CO_2 generation because the latter were based on tests that involved reservoir rock materials [Jia et al. 2016].

7.2.4 Numerical Simulations

There are very few publications focused on the numerical reservoir simulations of steam injection processes in which Aquathermolysis reactions are used for the subsurface upgrading of HO/B. Barroux and coworkers from the IFP Energies Nouvelle studied the H_2S production forecast during an SAGD process [Barroux et al. 2013]. A thermal and compositional model was used to simulate this process numerically and compare with experimental results in homogeneous [Barroux et al. 2013] and heterogeneous [Ayache et al. 2017] reservoirs. The authors developed a thermo-kinetic-based model using SARA fractions. The underground upgrading was followed by the decreasing of the resin and asphaltene contents [Barroux et al. 2013]. Figure 7.12 shows the comparison of the numerical simulation (lines) with the physical simulations (markers) for the Aquathermolysis of Athabasca crude oil. Panel A shows that the resins and asphaltenes decreased with time of reaction whereas saturates and aromatics increased. At the same time (Panel B), the H_2S generation increased [Barroux et al. 2013]. As seen, the results of the numerical simulations matched reasonably well with the experimental values. Even though the curve of the crude oil viscosity vs. temperature was presented in the article, no comparisons of the predicted and experimental values were reported.

Kapadia and coworkers included the Aquathermolysis reactions to a reservoir simulation model to understand the H_2S reactive zones in SAGD [Kapadia et al. 2014]. By using this model, the authors evaluated the potential of using an *in-situ* Claus process to scavenge the hydrogen sulfide from the steam chamber in the form of deposited sulfur as shown in Equation 7.10:

$$2H_2S + SO_2 \rightarrow (3/x)\, S_x + 2H_2O \qquad (7.10)$$

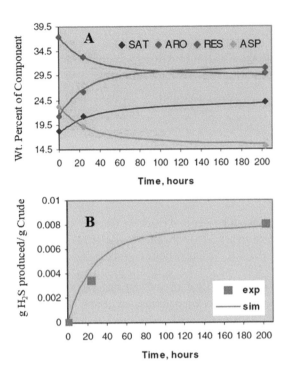

FIGURE 7.12 Comparison of the numerical simulation (lines) with the physical simulations (marks) for the Aquathermolysis of Athabasca Crude oil. (A) SARA fractions vs. time, where SAT = Saturates, ARO = Aromatics, RES = Resins, and Asp = Asphaltenes. (B) H_2S production vs time [Barroux *et al.* 2013]. Reprinted with permission of Society of Petroleum Engineers.

Where x is ~8 but S_6 and S_2 could also be produced.

The simulation results showed that injecting a small amount of SO_2, along with steam, initiated the Claus reaction (Equation 7.10) in the reservoir which converted H_2S into liquid sulfur. At 240°C, they found that hydrogen sulfide scavenging could be as high as 70 vol. %. In Equation 7.10, the liquid sulfur (S_x) generated has a considerably higher viscosity than that of bitumen. The author claimed that despite the deposition of sulfur in the pore space, pore blockage is negligible. These results show the potentiality of this concept for reducing H_2S emissions in Alberta SAGD processes [Kapadia *et al.* 2014].

Perez-Perez *et al.* proposed kinetic models for the evolution of H_2S from organosulfur compounds contained in resins and asphaltenes of bitumen and for the generation of CO_2 from the mineral formation [Perez-Perez *et al.* 2011]. They assumed that the Aquathermolysis reactions reached equilibrium after 90 days, and from laboratory data, kinetic models were constructed to predict gas generation plateaus vs. temperature. These models were integrated into homogeneous SAGD models to evaluate H_2S and CO_2 generation as a function of time. A 13-pseudo-component model was used but did not include the production of other gases like H_2, CH_4, CO, and C_2^+ gas compounds. The predictions showed a good match with the experimental and field data taken both from literature. Comparison of H_2S production results obtained by the reservoir model with field results indicated that predicted gas emissions (tons of SO_2 and CO_2 per BBL of bitumen) were of the same order of magnitude as those reported at the field scale.

Qian and coworkers reported a modified Aquathermolysis reaction scheme (ten pseudo-components) and coupled it to the commercial simulator CMG-STARS [Shijun *et al.* 2017]. Four different crude oils were studied, and the calculated values were compared with the field data with reasonable agreements. The authors found that the viscosity of heavy oil, and the asphaltene content decreased near the wellbore zone, was lower after the Aquathermolysis reaction [Shijun *et al.* 2017].

From the data presented in this section it can be concluded that the numerical simulation of SSU via Aquathermolysis of heavy crude oils and bitumens has been effectively carried out and that the results agree with lab and field data. However, more work is needed to develop new models that can predict API changes, amount of distillable materials, gas composition, and other properties of upgraded crude oils that are occurring due to Aquathermolysis during the subsurface upgrading under steam injection conditions.

7.3 USE OF HYDROGEN DONOR SOLVENTS

Hydrogen donors (H-donor) are chemical compounds that can easily transfer hydrogen to the HO/B during upgrading processes. The chemistry involved was discussed in Section 4.2.1.2. Under thermal conditions, the mechanism of hydrogen transfer from the H-donor to the accepting molecules follows a pathway known as *capping of free radicals* (Equations 4.20–4.21) and disproportionation reactions as shown in Equations 4.23–4.26. As mentioned in Section 4.2.1.3, by actively reacting with olefins, hydrogen donors can eliminate the pathway for polymerization and coke formation, therefore suppressing coking reactions (Equation 4.27) during thermal hydrogen addition processes. Also, in the presence of hydrogen donor and steam, the Aquathermolysis or hydrous pyrolysis reactions are affected due to the higher rate of hydrogen transfer of the H-donor in comparison with water [Ovalles *et al.* 2003]. Finally, hydrogen donor solvents have been used in the presence of catalysts to enhance the upgrading reactions further. The latter topic is discussed in Section 8.4.

For subsurface upgrading of HO/B, the selection of the hydrogen donor solvent is of paramount importance for the successful outcome of the process. As discussed in Section 3.1.3.3, in Expanded Solvent SAGD (ES-SAGD), the hydrocarbon solvent is selected so that it evaporates and condenses at the same conditions as the water phase at the boundary of the steam chamber. This phenomenon reduces the viscosity and the residual oil saturation with the concomitant increase in the oil rate and cumulative production (see Figure 3.13). Using the data from literature [Gupta *et al.* 2005], Jha and coworkers concluded that for a solvent to be successful for steam enhanced oil recovery (SAGD, CSS, etc.) the following conditions are necessary [Jha *et al.* 2012, Jha 2013]:

- The solvent or diluent must be injected in vapor phase along with the steam.
- For HO/B, C3 to C10, aromatic diluents and Syncrude are suitable candidates, operating temperatures of Canada and Venezuela yielding the highest cumulative oil productions as shown in Figure 7.13.
- Heavy solvents, like Syncrude, are not very efficient and offer little benefit as shown in Figure 7.13.

Similar conditions can be extrapolated to the case of subsurface upgrading of HO/B using steam/H-donor co-injection processes, i.e., the preferred way to transport the solvent to the oil interface is *in the vapor phase*. Also, solvent reduces oil viscosity at the o/w region, therefore minimizing the residence time of hot oil in the reservoir. It can be estimated that without solvent, the oil residence time could be 10–200 days at 400°F whereas, with solvent additives, the residence time is probably two to four times lower to carry out the upgrading reactions. Section 7.2.3 discusses the reservoir numerical simulations of hydrogen addition processes using hydrogen donors and steam.

In the next section, the use of naphtheno-aromatic compounds as hydrogen donors is also presented for the subsurface upgrading of HO/B using steam as a source of energy. Most of the work has been carried out at laboratory scale. However, a field test of this technology was conducted in the Liaohe Oilfield, China [Jiang *et al.* 2005]. Because this test also involved the use of catalysts, its discussion is presented in Chapter 8.

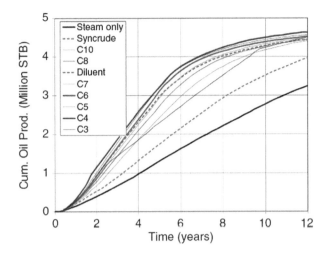

FIGURE 7.13 Cumulative oil production vs. time for steam–solvent co-injection process Jha *et al.* [2012]. Reprinted with permission of Society of Petroleum Engineers.

7.3.1 Use of Naphtheno-Aromatic Compounds

Naphtheno-aromatic compounds are one of most common and most studied of the hydrogen donors [Maity *et al.* 2010, Alemán-Vázquez *et al.* 2016]. This family of compounds are characterized for having at least one benzene ring and one cyclohexane ring (naphthene). They are also called hydroaromatic compounds because they are derived from the aromatic compounds by adding hydrogen to one of the rings. Their high rate of hydrogen donation is related to the favorable thermodynamics in going from naphtheno-aromatic to polycyclic aromatic with the concomitant generation of hydrogen. For example, the Gibbs' free energy for the dehydrogenation of tetralin to produce naphthalene (Equation 7.11) is −4.22 kJ/mol at 600K.

$$\text{(structure)} \longrightarrow \text{(structure)} + 2H_2 \qquad (7.11)$$

Reaction 7.11 can be carried out either thermally or catalytically. In this section, we discussed only the thermal hydrogen donation to upgrade HO/B. As mentioned, the use of catalysis in conjunction with H-donors is presented in Chapter 8. Other examples of hydroaromatic compounds are: 1,2,3,4-tetrahydroantracene, 9,10-dihydrophenanthrene, and 4,5-dihydropyrene. Their structures are shown below:

$$\text{(structures)} \qquad (7.12)$$

1,2,3,4-Tetrahydroanthracene   9,10-Dihydrophenanthrene   4,5-Dihydropyrene

In 1999, Ovalles and coworkers disclosed a process (Figure 7.14) for the subsurface upgrading of heavy and extra-heavy crude oils which involves the downhole addition of a hydrogen donor solvent under steam injection conditions (i.e., 280–315°C) and residence times of at least 24 h [Vallejos *et al.* 1999]. This concept could be operated under "Huff and Puff mode" similar to the cyclic steam stimulation with solvent co-injection as discussed in Section 3.1. The only difference is that the solvent was selected so it can donate hydrogen at subsurface conditions. The authors claimed that in

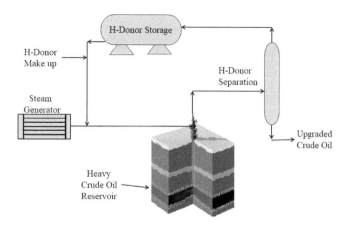

FIGURE 7.14 Concept proposed by Ovalles *et al.* for the upgrading of heavy crude oils using steam and hydrogen donors and operating in a "Huff and Puff" mode.

the presence of steam (heat source), the natural formation (catalysts), and methane (natural gas), the upgrading of extra-heavy crude oil takes place (Figure 7.14) with the recycling of ~70–80% of the hydrogen donor [Vallejos *et al.* 1999, Ovalles *et al.* 2001].

Table 7.6 shows the properties of the upgraded Hamaca (Orinoco) crude oil in the presence of tetralin as a hydrogen donor under steam injection conditions using the process described in Figure 7.14 [Ovalles *et al.* 2003]. Firstly, Ovalles *et al.* carried out a control experiment (0 h) to determine the effects of sample handling and experimental procedure at the batch conditions studied. Product characterization showed an increase of only 1°API, 7% reduction of asphaltenes, and 1,000 cP decrease in viscosity. These results are attributed to the time needed (20–30 min) to reach the temperature (300°C) in which tetralin and water were distilled from the oil/sand mixture. Then, at 280°C *for 24 h*, the presence of tetralin (hydrogen donor additive), water (steam), the natural formation, and methane led to an increase of 4° in the API gravity of the upgraded product and a two-fold reduction in its viscosity (Table 7.14). Further increases in the temperature led to products with improved properties reaching 15°API at 315°C [Ovalles *et al.* 2003].

The effect of tetralin concentration on the viscosity of the original and upgraded crude oils is shown in Figure 7.15. As seen, the viscosity of the product (measured at 60°C) decreased from 9,870 cP (9.8 Pa s) to 2,900 cP (2.9 Pa s) by adding hydrogen donor from 5 wt. % to 50 wt. %

TABLE 7.6

Properties of the Upgraded Hamaca Crude Oil Using Hydrogen Donor Under Steam Injection Conditions

Conditions	°API (±1°)	Viscosity at 60°C (cP)[b]	Asphaltenes (wt. %)[c]
Original crude oil	9.1	9,870	22.0
0 h, control reaction	10.2	8,310	20.7 (5.9%)
24 h, 280°C	13.2	5,500	18.9 (14.0%)
24 h, 300°C	14.0	5,070	19.1 (13.2%)
24 h, 315°C	14.7	4,610	18.0 (18.2%)

Data taken from Ovalles [2003].

[a] Experiments carried out under batch-wise conditions, no stirring, ratio sand/crude/tetralin/water = 10:1:1:1, 900 psi of CH_4 initial pressure.

[b] Results presented correspond to an average of at least two different reactions with 90% confidence using Student's value.

[c] Numbers in brackets indicate percentages of asphaltene conversions with respect to the original crude oil.

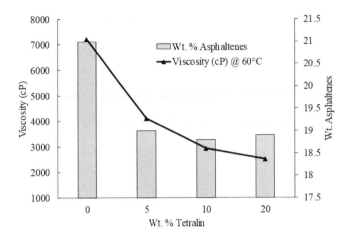

FIGURE 7.15 Effect of tetralin concentration on the viscosity (measured at 60°C) of the original and upgraded crude oils under steam injection conditions (280°C, 24 h, same conditions as Table 7.6).

[Ovalles 2003]. At the same time, the amount of asphaltenes present in the upgraded crude oil decreased from 21 wt. % in the original crude to ~19 wt. %. After removing the tetralin by vacuum distillation, 1,2-dihydronaphtalene (DHN) and naphthalene were detected in the products using gas chromatography. As discussed in Section 4.2.1., these findings are consistent with hydrogen transfer reactions from the H-donor to the heavy crude oil under mild hydrovisbreaking conditions. The overall reaction scheme can be seen in Equation 7.12:

$$\text{(7.13)}$$

On the other hand, the use of a lower capacity hydrogen donor, such as toluene, led to a product with properties similar to those of the original Hamaca crude oil (9.2°API and 23% asphaltenes) [Ovalles *et al.* 2001]. This result indicated that the presence of the hydrogen donor is of crucial importance for the downhole upgrading of heavy crude oil under steam stimulation conditions. Ovalles *et al.* extended their studies to other Venezuelan heavy crude oils. As shown in Figure 7.16, increases of 2–3°API were observed for Cerro Negro and Boscan crude oils under the same conditions as reported in Table 7.6. Also, Ovalles *et al.* reported significant reductions of the viscosities and asphaltene contents than those found the original crude oils [Ovalles *et al.* 2001].

Other authors have also studied the use of hydrogen donor for the subsurface upgrading of heavy crude oils under lab conditions [Liu and Fan 2002, Mohammad and Mamora 2008]. Liu and Fan reported the effect of a hydrogen donor additive (tetralin) on the viscosity of Liaohe heavy oil (China) in the presence of water at 240°C for up to 20 days [Liu and Fan 2002]. Figure 7.17 shows the changes in viscosity as a function of the tetralin concentration and time of reaction. As seen, the viscosity of the upgraded product decreased from 88.5 to 52.8 Pa s as the amount of the H-donor increased (from 0 to 1 wt. %) during the 20-day period. These results are consistent with those reported by Ovalles *et al.* and presented in Figure 7.15. Liu and Fan also studied the effect of V-, Ni-, and Fe-containing catalysts [Liu and Fan 2002], but these results are presented in Chapter 8.

Mohammad and Mamora conducted displacement experiments to verify the feasibility of sub-surface upgrading of a Venezuelan heavy crude oil by using a hydrogen donor (tetralin) under steam

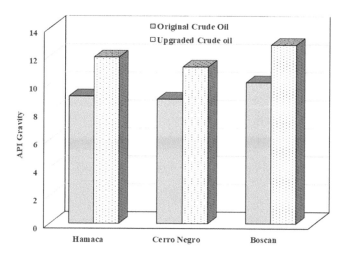

FIGURE 7.16 Upgrading of Venezuelan heavy crude oils in the presence of tetralin as hydrogen donors under steam injection conditions.

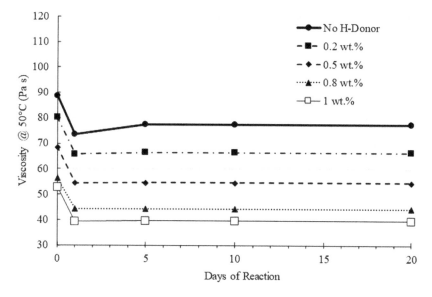

FIGURE 7.17 Effect of hydrogen donor (tetralin) on the viscosity of the products as a function of time after the upgrading reaction at 240°C (water content = 20 wt. %). Data taken from Lui [2002].

injection conditions [Mohammad and Mamora 2008]. The experiments were conducted in a vertical injection displacement cell utilizing a mixture of sand, water, and Jobo oil (12.4°API and a viscosity of 7.8 Pa s at 30°C). The experimental results showed that the addition of 5 wt. % of tetralin increased oil recovery by 15% above that with only steam injection. Mohammad and Mamora also observed API increases and viscosity reductions by using the H-donor and steam. The authors studied the effect of a Fe-containing soluble catalyst [Mohammad and Mamora 2008], but these results are presented in Chapter 8.

Hendraningrat and coworkers used decalin as hydrogen donors for the upgrading of Athabasca bitumen using batch experiments at 240°C for 12h [Hendraningrat et al. 2014]. Figure 7.18 shows the viscosity of the original bitumen, control reaction, and additives measured at 60°C. As seen, in the presence of decalin, a reduction of the viscosity of the upgrading product (~1,300 cP or 1.3 Pa s)

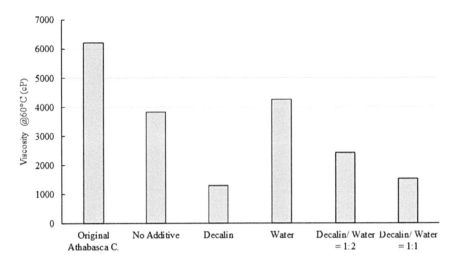

FIGURE 7.18   Effect of additives on the viscosity of products after upgrading at 240°C for 12h. Data taken from Hendraningrat *et al.* [2014].

was obtained vs. the no-additive and water-containing experiments—Aquathermolysis (~4,000 cP or 4.0 Pa s). Using decalin and water in 1:2 and 1:1 ratios led to decreases in the viscosity (~2,400 and 1,500 cP, respectively) with respect to the original bitumen and the control reaction, but no synergistic effect was observed [Hendraningrat *et al.* 2014].

The results presented in this section demonstrated that the upgrading of HO/B could be achieved by using hydrogen donor solvents under steam injection conditions (< 280°C, > 1,600 psi). Three elements, namely hydrogen donor, mineral formation, and natural gas (methane), were used to achieve crude oil improvement. The next section discusses the latter two in greater depth to gain knowledge regarding the chemistry associated with this upgrading process.

7.3.2   EFFECT OF MINERAL FORMATION

As mentioned in Section 7.2.2, there is a significant effect of the mineral formation on the properties of the upgraded HO/B during the subsurface upgrading of HO/B in the presence of hydrogen donors under steam injection conditions. Two possible reasons may explain the beneficial effect of the natural formation upon the heavy crude oil upgrading. As mentioned, the first one attributes catalytic activity to the mineral matrix, even at the low severity conditions of the Aquathermolysis reaction [Strausz *et al.* 1977, Stapp 1989, Phillips *et al.* 1985, Ovalles *et al.* 2003, Maity *et al.* 2010]. Actives sites on the mineral surface can be involved in the hydrogen transfer reactions by capping the free radicals (Equation 4.23–4.26) and improving the properties of the upgraded crude oil [Alemán-Vázquez *et al.* 2016].

Alternatively, the beneficial effect of the natural formation could be attributed to an improved heat transfer mechanism that diminishes the probability of the molecules to move away from the reaction zone. This extended contact period could increase the cracking reactions [Phillips *et al.* 1985]. To test this hypothesis, several upgrading runs were carried out varying the mineral formation/SiC ratio and keeping the total amount of solid constant. In these experiments, SiC was used as inert solid since it has excellent heat transfer properties. Figure 7.19 shows the effect of the concentration of mineral formation (sand) in the sand/SiC mixture on the viscosity of the upgraded product at 280°C for 24 h. As seen, the viscosity of the upgraded crude oil decreased as the proportion of mineral matrix (sand) increased in the mixture. For a 50 wt. % mineral formation, the viscosity of the upgraded crude oil was 55 Pa s (measured at 40°C), approximately, whereas for a 100% sand this value was reduced to 20 Pa s (see Figure 7.19).

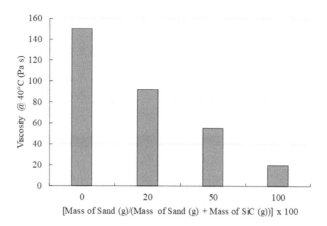

FIGURE 7.19 Effect of the concentration of mineral formation (sand) in the sand/SiC mixture on the viscosity of the upgraded product (measured at 40°C) under steam injection conditions at 280°C for 24 h.

These results strongly indicate that the mineral formation acts as a solid catalyst and not as a heat transfer matrix [Strausz *et al.* 1977, Stapp 1989, Phillips *et al.* 1985, Ovalles *et al.* 2003, Maity *et al.* 2010].

Furthermore, surface analysis of the Hamaca crude oil sand by X-ray photoelectron spectroscopy (XPS) showed a composition of 10% Al, 40% Si, and 0.6% Fe (balance is adventitious carbon from the organic phase). The presence of these elements on the surface of the sand is relevant because they can be catalytically active for the hydrogen transfer reactions and the concomitant upgrading of the HO/B. It is important to point out that the sand/crude oil ratio in the indigenous material is high (90%); consequently, there is a relatively high concentration of Fe, Si, and Al present that could potentially compensate for the low surface area of the material (< 1 m^2/g). Thus, different solids were evaluated for the upgrading of Hamaca crude under steam injection conditions (at 280°C for 24 h). As can be seen in Figure 7.20, the use of silica, iron oxide, and silica-alumina led to further reduction in the viscosity of the upgraded product in comparison with the mineral formation [Ovalles *et al.* 2003].

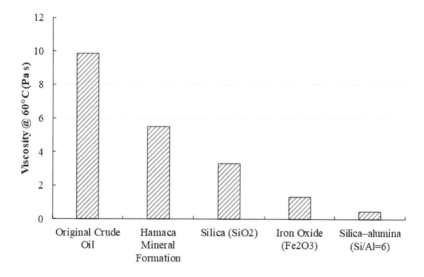

FIGURE 7.20 Effect of different solids used on the viscosity of the upgraded product (measured at 60°C) under steam injection conditions at 280°C for 24 h.

Based on the results shown in Figure 7.20, the effect of these materials on viscosity reduction was in the following order: Mineral formation < SiO_2 < Fe_2O_3 < silica-alumina [Ovalles *et al.* 2003]. These findings support the idea that the mineral matrix is acting as a catalyst, and more active catalysts could be potentially developed for subsurface upgrading via Aquathermolysis Ovalles *et al.* 2003, Maity *et al.* 2010]. This topic is discussed in Chapter 8.

7.3.3 EFFECT OF METHANE

As mentioned in Section 2.1.3, Canadian and Venezuela HO/B have gas-to-oil ratios (GOR) lower than 3 to 4 m^3 gas per m^3 of bitumen (~2–3 SCF/BBL) and ~19 m^3 gas per m^3 of crude (106 SCF/BBL), respectively. The main component of the dissolved gas is methane (> 90 vol. %) followed by carbon dioxide, nitrogen, and small amounts of C2 to C6 hydrocarbons.

Under Aquathermolysis conditions (240–315°C, residence times from 1–30 days), methane could be involved as a hydrogen donor in the upgrading reactions via a free radical pathway or catalyzed by the mineral formation. The effect of methane was evidenced by carrying out upgrading reactions of Hamaca crude oil under methane and nitrogen (control, no hydrogen donation). As seen in Figure 7.21, the viscosity (measured at 60°C) of the CH_4-containing reaction product was lower (5.5 Pa s) than that of the nitrogen-containing experiment (8.3 Pa s) and the original extra-heavy crude oil (9.87 Pa s). This result suggests that methane is most probably involved in the upgrading reactions as reported previously for higher temperature (380–410°C) processes [Steinberg 1986, Egiebor and Gray 1990, Ovalles *et al.* 1995, 1998b, He *et al.* 2017]. The mechanism associated with the thermal use of methane as hydrogen source proceeds via free radical process and was discussed in Section 4.2.1.5.

Furthermore, upgrading reactions carried out in the presence of CD_4 instead of CH_4 showed incorporation of deuterium into the crude oil via H–D exchange or CD_3 addition [Ovalles *et al.* 2003]. Also, experiments carried out in the presence of $^{13}CH_4$ gave further evidence for methane incorporation. By ^{13}C-NMR, a small signal near 20 ppm in the $^{13}CH_4$-containing experiment that is not present in the $^{12}CH_4$-counterpart. By comparison with the literature, this band was assigned to methyl groups in α-position to an aromatic ring. Therefore, the higher concentration of methyl groups in the run with $^{13}CH_4$ gives evidence that, most probably, methylation of aromatic rings was occurring as reported by Ovalles *et al.* [Ovalles *et al.* 1995, 1998b], Egiebor and Gray [1990], and Song and coworkers [He *et al.* 2017].

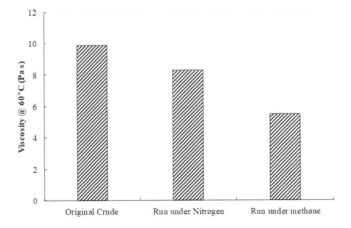

FIGURE 7.21 Effect of the gas used on the viscosity of the upgraded product (measured at 60°C) under steam injection conditions at 280°C for 24 h.

7.3.4 Numerical Simulations

As with thermal Aquathermolysis (Section 7.2.4), very few numerical simulation studies have been conducted using hydrogen donor solvents under steam injection conditions. Specifically, Ovalles and Rodriguez carried out the physical and numerical reservoir simulations of Hamaca crude oil (9.2°API) upgrading using tetralin in the presence of natural gas and sand under cyclic steam injection conditions [Ovalles and Rodriguez 2008]. A continuous bench scale plant was used at different temperatures (280–315°C) and residence times (24–64 h) for carrying out kinetic studies. In these experiments [Ovalles and Rodriguez 2008], live Hamaca crude oil (5.7 cm³ of CH₄/g crude) was mixed with fresh water and fed to heated sand-containing reactors with a sand/crude/water/tetralin ratio of 10:1:1:1 (similar to Hamaca reservoir). After the reaction zone, methane was separated by flash distillation at 300°C, and the reaction mixture was continuously distilled to remove water and tetralin and to generate the upgraded crude oil. The results were reported *at steady state conditions*, i.e., no changes in the properties of the upgraded crudes with time at a given residence time and temperature [Ovalles and Rodriguez 2008].

As previously found using a batch reactor [Ovalles *et al.* 2003], the continuous flow experiments showed significant improvement on the properties of the upgraded crude oil [Ovalles and Rodriguez 2008]. As shown in Figure 7.22, API increases in the 2–6°API range were obtained. Similarly, as seen in Table 7.7, reductions of the viscosity of the upgraded crude oils from 10% to 90% were measured with percentages of asphaltene and tetralin conversion in the 11–23% and 2–6% ranges, respectively. These results confirmed those obtained with the batch experiments and were used to carry out a kinetic study [Ovalles and Rodriguez 2008].

For the numerical simulation, a reaction model involving four pseudo-components (light, medium, heavy, and asphaltene fractions) was used, and the Arrhenius kinetic parameters (pre-exponential factors and activation energies) were determined for the forward and reversed reactions, as shown in Equation 7.14:

$$\boxed{\text{Asphaltene F.}} \underset{\text{Tetralin}}{\rightleftharpoons} \boxed{\text{Heavy F.}} \underset{\text{Tetralin}}{\rightleftharpoons} \boxed{\text{Medium F.}} \underset{\text{Tetralin}}{\rightleftharpoons} \boxed{\text{Light F.}} \qquad (7.14)$$

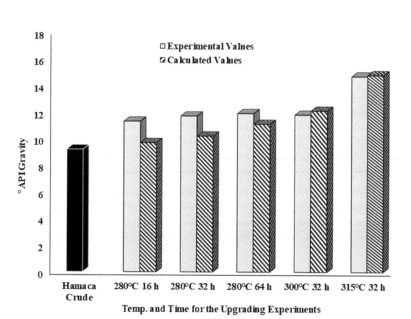

FIGURE 7.22 Comparison between experimental (±0.5°API) and calculated API gravities of the original Hamaca crude and the upgraded crude oils at different temperatures and residence times using a continuous bench scale plant.

TABLE 7.7

Comparison Between Experimental and Calculated Original and Upgraded Crude Oil Properties Obtained Using a Continuous Bench Scale Plant[a]

		HamacaCrude	Reaction Conditions				
			280°C 16 h	280°C 32 h	280°C 64 h	300°C 32 h	315°C 32 h
Viscosity	Exp.[b]	1,810	1,625	1,369	1,010	337	90
at 80°C (cP)[d]	Calc.[c]	1,972	1,461	1,183	800	538	199
%. Red. of	Exp.[b]	–	11	6	10	16	23
Asphaltenes[e]	Calc.[c]	–	2	4	8	12	23
% Conv. of	Exp.[b]	–	2	2	3	4	6
Tetralin[f]	Calc.[c]	–	1	1	2	3	7

[a] Experiments carried out using a continuous bench scale plant at the conditions described in [Ovalles 2008].

[b] Experimental values measured for the original and upgraded crude oils.

[c] Calculated values using compositional-thermal numerical simulations with an initial composition and kinetic parameters reported in [Ovalles 2008].

[d] Viscosity measured at 80°C.

[e] Percentage of reduction of asphaltenes with respect to the original crude oil.

[f] Percentage of tetralin conversion with respect to the original content.

Using these data, the compositional-thermal numerical simulations were carried out and validated using the continuous flow data. Figure 7.22 shows a comparison between experimental and calculated API gravities of the original Hamaca crude and the upgraded products at different temperatures and residence times. As seen, the results showed a good match between the calculated and experimental values with average relative errors ±0.5°API.

Similar matches were obtained for the viscosity of the upgraded products (measured at 80°C) and percentages of asphaltene and tetralin conversion [Ovalles and Rodriguez 2008]. As shown in Table 7.7, the average differences between the experimental and calculated values were ±87 cP, ±1%, and ±1%, respectively. All these results confirmed the validity of the model described in Equation 7.14 [Ovalles and Rodriguez 2008].

Using the previous model and under cyclic steam injection conditions (397.5 m³/d or 2,500 BBL/d of 1:1 steam/tetralin for 20 days with a ten-day soaking period), the subsurface upgrading process was numerically simulated in a typical well located in the Hamaca reservoir [Ovalles and Rodriguez 2008]. Figure 7.23 shows the API gravity of upgraded crude oil vs. time for the numerical simulation of three steam–solvent co-injection, soaking, and production cycles. In the first one, the cumulative API gravity (measured under surface conditions) of the upgraded crude oil reached a maximum value of ~35°API at the beginning of production time and decreased with time. This phenomenon was attributed to the generation of a lighter fraction closer to the well by chemical reactions with the hydrogen donor tetralin (as shown in Equation 7.14) in the presence of steam (temperature > 280°C). These lighter components represent the main fraction of the produced oil at the beginning of the simulation. At the end of the first cycle (~120 days in Figure 7.23), the cumulative API of the upgraded oil was found to be approximately 14°API [Ovalles and Rodriguez 2008].

However, the authors reported a reduction in the percentage of the calculated conversion of tetralin (0.8%) in comparison with the bench scale experiments (3%). This discrepancy was attributed to gravitational segregation of the steam coupled with low mixing efficiency of the hydrogen donor with the Hamaca crude oil at reservoir conditions. The numerical simulation showed that the higher temperatures were located on the top part of the reservoir (where steam was mostly found),

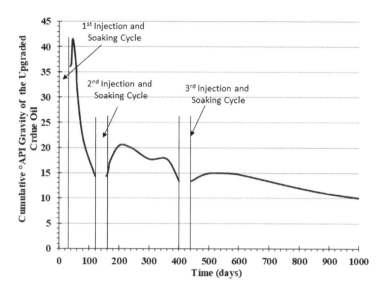

FIGURE 7.23 Cumulative API gravity of upgraded crude oil vs. time for the numerical simulation of cyclic steam–solvent co-injection.

whereas the heavy crude oil was displaced further away from the well due to steam and tetralin injection. In turn, the hydrogen donor remained in the bottom part of the grid mainly due to density difference with respect to steam. This situation, steam on top, tetralin on base, and crude oil pushed away from the well, is responsible for the reduced efficiency of the upgraded process with the concomitant lower hydrogen donor conversions and lower amount of produced upgraded crude oil in comparison with bench scale experiments [Ovalles 2008]. As it was pointed out in Section 1.5, the lack of control and low mixing efficiency downhole are two of the main disadvantages of subsurface upgrading processes. One possible improvement to the process could be the use of lower boiling point H-donors that evaporate and condense at the same conditions as the water phase at the boundary of the steam–crude interphase. As reported in the literature, the selection of the hydrogen donor solvent is of paramount importance for the successful outcome of the process.

Similarly, the second and third cycles were simulated, and increases in the API of the upgraded crude were obtained (Figure 7.23). For the second cycle, more extended production time was calculated (244 days) to reach the same cumulative API gravity of ~14° as the first cycle. Finally, as expected for a cyclic steam stimulation process (see Section 3.1.2.1.), the efficiency decreased for the third CSS cycle due to a reduction of the oil produced. At around ~1,000 days, the original API gravity of the Hamaca crude was obtained due to lack of mixing, as pointed out before.

7.4 USE OF REFINERY FRACTIONS AS HYDROGEN DONORS

Naphtheno-aromatic or hydroaromatic compounds are not widely commercially available, and sometimes their cost is higher than similar petroleum fractions [Sheng *et al.* 2016]. To improve the economic prospect of subsurface upgrading processes, the use of low-cost refinery fractions as H-donors can be proposed. Several precedents found in the literature involved coal liquefaction. Those experiments were conducted during the 1970s and 1980s [Kuhlmann *et al.* 1984, Yan and Espenscheid 1983, Bedel 1991, Shen and Iion 1994]. Other examples included the hydrovisbreaking of vacuum residues using a petroleum-derived H-donor [Gould and Wiehe 2007], a distilled fraction of a Venezuelan synthetic crude oil (Merey) with a boiling point less than 420°C [Wang *et al.* 2012], and narrow fractions from 180 to 500°C of coker gas oil [Chen *et al.* 2014].

In 1984, Kuhlmann and coworkers from Texaco studied the liquefaction of Illinois No. 6 coal using different naphtheno-aromatics as hydrogen donor solvents at 427°C, 4.13 MPa of hydrogen for 1 h [Kuhlmann *et al.* 1984]. They found that the order of hydrogen-donating reactivity was the following (Equation 7.15) where the number in brackets is the relative rate of reaction [Kuhlmann *et al.* 1984]:

$$ \qquad > \qquad \gg \qquad \tag{7.15} $$

4,5-Dihydropyrene (8)   9,10 Dihydrophenanthrene (7)   Tetralin (0.1)

Yan and Espenscheid used a heavy fraction coming from the fluid catalytic cracking unit for the liquefaction of Illinois No. 6 coal and North Dakota shale oil at relatively low severity conditions (315–427°C, 0–6.9 MPa, and residence times of 0.1–5 h) [Yan and Espenscheid 1983]. The authors obtained a maximum of 90% of coal dissolution using a refinery fraction as hydrogen donor with H/C molar ratio of 0.98. The characterization of the solvent by MS indicated the presence of the following naphtheno-aromatic compounds (Equation 7.16) [Yan and Espenscheid 1983]:

$$ \tag{7.16} $$

Fluorenes   Naphtheno-phenanthrene

Where R and R' are alkyl chains

Bedell and Curtis reported the liquefaction of Kentucky No. 9 coal in the presence a variety of naphtheno-aromatic and olefinic compounds using hexadecane as solvent at 380°C, 8.6 MPa for 1 h [Bedell and Curtis 1991]. As in Kuhlmann and coworkers [Kuhlmann 1984], the results indicated that the naphtheno-*poly*aromatic compounds are more active than the naphtheno-*mono*aromatic counterparts. For example (Equation 7.17), the authors reported higher conversion to liquid products using 5,6-dihydroanthracene (77%) than that found using tetralin (67%) at the same reaction conditions [Bedel 1991]:

$$ \qquad > \qquad \tag{7.17} $$

1,2-Dihydroanthracene (77%)   Tetralin (67%)

Shen and Iino conducted heat treatments of Zao Zhuang and Upper Freeport coals in several hydrogen-donating solvents under 0.1 MPa of N_2 at 175–300°C [Shen and Iion 1994]. They observed higher dissolution rates in 9,10-dihydroanthracene and 1,4,5,8,9,10-hexahydroanthracene than that measured in tetralin. The quantity of hydrogen transferred from the solvent to the coal was found to be well correlated with the extent of the dissolution reactions [Shen and Iion 1994].

Chen and coworkers studied three narrow fractions (180–350°C, 350–420°C, and 420–500°C) of coker gas oil as hydrogen donors in the Visbreaking of Karamay VR (China) at under batch conditions at 420°C, 0.4 MPa of initial nitrogen pressure for 30 min [Chen *et al.* 2014]. The results showed that the 350–420°C fraction was the most active hydrogen donor, achieving higher residue conversion and suppressing coke formation more than the other two analogs. No molecular characterization of distillation cuts was reported [Chen *et al.* 2014].

Wang and coworkers measured the hydrogen-donating ability of the narrow distillate fractions of coker gas oil (CGO), fluid catalytic cracking slurry (FCCS), and furfural extract oil (FEO) using an autoclave reactor, anthracene as a hydrogen acceptor probe at 380°C, 8 MPa of nitrogen for 8 min [Sheng *et al.* 2016]. Using spectroscopic data, the authors found that the H-donor activity of the fractions follows the order of FEO > FCCS > CGO. As expected, analysis of the hydrocarbon composition demonstrated that the total percentages of naphtheno benzenes, dinaphthene benzenes, and naphthene phenanthrenes play a significant role in the hydrogen-donating ability of the narrow distillate fractions [Sheng *et al.* 2016].

Based on the above-mentioned findings, it can be postulated that the naphtheno-*poly*aromatic compounds are more active hydrogen donors than the naphtheno-*mono*aromatic counterparts. Figure 7.24 shows the Gibbs' free energy ($\Delta G°$) vs. temperature for the dehydrogenation of tetralin and 9.10- dihydrophenanthrene. For all temperatures, the $\Delta G°$ is lower for the latter than those calculated for the first one. For example, at 573 K, the Gibbs' free energy for the dehydrogenation of dihydrophenanthrene is –64.85 kJ/mol whereas for tetralin it is almost 20-fold higher (3.3 kJ/mol). These results indicate that, from the thermodynamic point of view, the use of naphtheno-*poly*aromatic compounds as hydrogen donors is more favored than that of naphtheno-*mono*aromatic compounds as tetralin.

To test this hypothesis, a refinery fraction (LCO) was used as a hydrogen donor for the upgrading of Hamaca crude oil at steam injection conditions. This fraction was selected due to its relatively lower cost compared to conventional hydroaromatic compounds. Table 7.8 shows the characterization in terms of its boiling-point distribution and aliphatic and aromatic contents by mass spectrometry (MS). As seen, the LCO fraction has a boiling point range of 190–253°C, and its mid-boiling point (207°C) is similar to tetralin (208°C). It was found by MS that the concentration of naphtheno-benzenes is ~25 wt. %, which makes this fraction a potential hydrogen donor for the subsurface upgrading of HO/B.

As with the experiments with tetralin discussed in the previous section [Ovalles and Rodriguez 2008], a continuous bench scale plant was used to evaluate the LCO fraction as H-donor in the temperature range of 300–315°C and residence times of 32 and 64 h. As mentioned, live Hamaca crude oil was mixed with water and reacted with LCO in the presence of sand (sand/crude/water/LCO ratio of 10:1:1:1). After methane was separated by flash distillation, the reaction mixture was continuously distilled to remove water and LCO and to generate the upgraded crude oils. Table 7.9 shows the results *at steady state conditions* of the three upgrading experiments carried out using LCO at steam injection conditions. As seen, increases up to 4°API in the API gravity of the upgraded crude oils were observed in comparison with the starting Hamaca crude. Similarly, percentages of reduction of asphaltenes and Conradson carbons were in the 14–17% and 2–7% ranges, respectively.

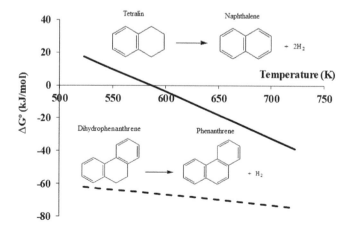

FIGURE 7.24 Change in the Gibbs' free energy vs. temperature for the dehydrogenation of tetralin and dihydrophenanthrene.

TABLE 7.8

Characterization of the Refinery Fraction (LCO) Used as Hydrogen Donor by Molecular Weight, Distillation, and Mass Spectrometry

Molecular Weight	145 g/mol	Mass Spectrometry	Wt. %[a]
Distillation ASTM D86	°C	% Aliphatic[b]	24.4
IBP	190	% Aromatics[c]	75.6
5 %v	198	% Monoaromatics[d]	62.9
10 %v	202	% Naphtheno-benzenes[e]	24.9
20 %v	206	% Diaromatics[f]	12.1
30 %v	206	% Triaromatics[f]	0
50 %v	207		
70 %v	213		
80 %v	214		
90 %v	223		
FBP	253		

[a] Weight percentage determined by mass spectrometry with respect to whole sample following the procedure reported by [Robinson 1971].

[b] Aliphatic weight percentage.

[c] Aromatic weight percentage.

[d] With respect to the amount of aromatics.

[e] Naphtheno-benzenes weight percentage with respect to the amount of monoaromatics.

[f] With respect to the amount of aromatics.

TABLE 7.9

Properties of the Upgraded Products of Hamaca Crude Oil Using a Refinery Fraction (LCO) as Hydrogen Donors at Steam Injection Conditions[a]

Conditions	Inc. in API[b]	% Red. Asphaltenes[c]	% Red Conradson C.[d]	% Conv. Of 500°C+ Fraction[e]
Hamaca crude oil[f]	(9.1)	(22%)	(17.9%)	(70.0%)
300°C, 32 h	1.1	13.9	2.2	5.7
315°C, 32 h	2.7	16.6	4.5	10.0
315°C, 64 h	3.8	17.2	6.7	12.9

[a] Experiments carried out using a continuous bench scale plant at a sand/live crude/LCO/water ratio of 10:1:1:1. Methane was separated by flash distillation at 300°C, and the reaction mixture was continuously distilled to remove water and underreacted LCO. The results were reported at steady state conditions.

[b] Increase in the API gravity with respect to the original crude oil.

[c] Percent of reduction of C7-asphaltenes with respect to the original crude oil.

[d] Percent of reduction of Conradson carbon with respect to the original crude oil.

[e] Percent of reduction of the 500°C+ fraction with respect to the original crude oil (Equation 1.2).

[f] Numbers in brackets indicate the characterization of the original crude oil used as feed.

Furthermore, the percentages of residue conversion (500°C⁺), as calculated by using Equation 1.2, ranged from 6 to 13%. In general, the properties of the upgraded products improved with the temperature and residence time (Table 7.8). These results indicated that the upgrading of Hamaca crude oil could be accomplished by using a refinery fraction containing naphtheno-benzene compounds. These preliminary findings agree with those found using tetralin as H-donor under batch [Ovalles *et al.* 2003] or continuous flow conditions [Ovalles and Rodriguez 2008] at steam injection conditions.

Figure 7.25 shows a comparison between tetralin and the refinery fraction (LCO) as hydrogen donors for the upgrading of Hamaca crude oil at 300–315°C and 32 h. As seen, the experiments using tetralin gave higher API gravities and percentages of reduction of asphaltenes than those using LCO. For example, at 300°C, the API of the upgraded crude was ~12°API using tetralin whereas a 10°API product was obtained in the presence of LCO. As shown in Figure 7.25, similar results were observed at 315°C (~15°API vs. ~12°API). These results could be attributed to the lower content of naphtheno-benzene compounds in the LCO (~25 wt. %) vs. tetralin (~100%). The use of *hydrogenated* fractions can be a potential path forward to increase the concentration of hydrogen donor compounds and thus maximize the properties of the upgraded products for the subsurface upgrading of HO/B.

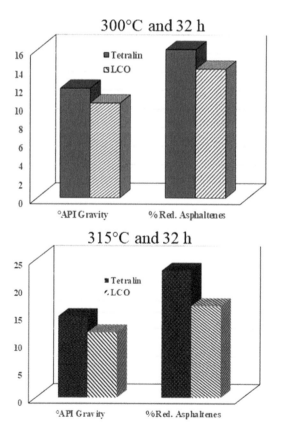

FIGURE 7.25 Comparison between tetralin and a refinery fraction (LCO) as hydrogen donors for the upgrading of Hamaca crude oil under steam injection conditions (300–315°C and 32h).

REFERENCES

Alemán-Vázquez, L. O., Torres-Mancera, P., Ancheyta, J., Ramírez-Salgado, J., 2016, "Use of Hydrogen Donors for Partial Upgrading of Heavy Petroleum", *Energy Fuels*, 30, 9050–9060.

Attar, A., Viloria, R. A., Verona, D. F., Parisi, S., 1984, "Sulfur Functional Groups in Heavy Oils and Their Transformations in Steam Injected Enhanced Oil Recovery." Preprint from the *Symp. of the Chem. of Enhanced Oil Recovery*, presented before the *Div. Pet. Chem., Amer. Chem. Soc. Philadelphia Meeting*, August 26–31, 1212–1222.

Ayache, S. V., Preux, C., Younes, N., Michel, P., Lamoureux-Var, V., 2017, "Numerical Prediction of H₂S Production in SAGD: Compositional Thermal-Reactive Reservoir Simulations", SPE No. 184998, presented at SPE Canada Heavy Oil Technical Conference, Calgary, Alberta, Canada, 15–16 February.

Barroux, C., Lamoureux-Var, V., 2013, "Using Geochemistry to Address H₂S Production Risk due to Steam Injection in Oil Sands", SPE No. 165437, presented at SPE Heavy Oil Conference-Canada, Calgary, Alberta, Canada, 11–13 June.

Barroux, C., Lamoureux-Var, V., Flauraud, E., 2013, "Forecasting of H2S Production Due to Aquathermolysis Reactions", SPE No. 164317, presented at SPE Middle East Oil and Gas Show and Conference, Manama, Bahrain, 10–13 March.

Bedell, M. W., Curtis, C. W., 1991, "Chemistry and Reactivity of Cyclic Olefins as Donors in Coal Liquefaction", *Energy Fuels*, 5, 469–476.

Belgrave, J. D. M., Moore, R. G., Ursenbach, M. G., 1997, "Comprehensive Kinetic Models for the Aquathermolysis of Heavy Oils", *J. Can. Pet. Technol.*, 36 (4), 38–44.

Brons, G., Siskin, M., 1994, "Bitumen Chemical Changes during Aquathermolytic Treatments of Cold Lake Tar Sands", *Fuel*, 73 (2), 183–191.

Brunner, G., Hoffman, R., Kunstle, K., 1985, "Method of Extracting Hydrocarbons from Oil-Containing Rock or Sand Through Hydrogenating Low Temperature Carbonization", US Patent No. 4,505,808.

Chen, H. H., Mojelsky, T. W., Payzant, J. D., Lown, E. M., Henry, D., Wallace, D., Strausz, O.P., 1991, "Chemical Changes in Alberta Oil Sands During – Steam Treatment", *AOSTRA J. Res.*, 7, 17–35.

Chen, Q., Gao, Y., Wang, Z., Guo, A., 2014, "Application of Coker Gas Oil Used as Industrial Hydrogen Donors in Visbreaking", *Pet. Sci. Technol.*, 32 (20), 2506–2511.

Clark, P. D., Hyne J. B., Tyrer, J. D., 1983, "Chemistry of Organosulphur Compound Types Occurring in Heavy Oil Sands", *Fuel*, 62, 959.

Delgado, O, Bolivar, C., Ovalles, C., Scott, C. E., 2005, "Hamaca Crude Oil Upgrading Using Formic Acid as Hydrogen Precursors under Steam Injection Conditions", Heavy Hydrocarbon Resources, *ACS Symp. Ser.*, 895, Chapter 10, 143–152.

Dew, J. N., Martin, W. L. 1965, "Recovery of Hydrocarbons by In Situ Hydrogenation", US Patent No. 3,208,514.

Egiebor, N. O., Gray, M. R., 1990, "Evidence for Methane Reactivity during Coal Pyrolysis and Liquefaction", *Fuel*, 69, 1276–1282.

Fan, H.-F., Liu, Y.-J., Zhong, L.G., 2001, "Studies on the Synergetic Effects of Mineral and Steam on the Composition Changes of Heavy Oils", *Energy Fuels*, 15, 1475–1479.

Goicetty, J. L. B., 2010, "Estimation of the H₂S Formation under Steam Injection Conditions for the Orinoco Oil Belt", SPE No. 141128, presented at SPE Annual Technical Conference and Exhibition, Florence, Italy, 19–22 September.

Gould, K. A., Wiehe, I. A., 2007, "Natural Hydrogen Donors in Petroleum Resids", *Energy Fuels*, 21 (3), 1199–1204.

Gray, M. R., 2015, *Upgrading Oilsands Bitumen and Heavy Oil*, The University of Alberta Press, Edmonton, and references therein.

Gregoli, A. A., Rimmer, D.P., Graue, D. J., 2000, "Upgrading and Recovery of Heavy Crude Oils and Natural Bitumens by In Situ Hydrovisbreaking", US Patent No. 6,016,867, 25 January.

Gupta, S., Gittins, S., Picherack, P., 2005. "Field Implementation of Solvent Aided Process", *J. Can. Pet. Technol.*, 44 (11), 8–13.

Hamrick, J. T., Rose, L. C. 1977, "In Situ Hydrogenation of Hydrocarbons in Underground formations", US Patent No. 4,050,515, 27 September.

He, P., Zhao, L., Song, H., 2017, "Bitumen Partial Upgrading over Mo/ZSM-5 under Methane Environment: Methane Participation Investigation", *Appl. Catal. B Environ.*, 201, 438–450.

Hendraningrat, L., Souraki, Y., Torsater, O., 2014, "Experimental Investigation of Decalin and Metal Nanoparticles-Assisted Bitumen Upgrading during Catalytic Aquathermolysis", SPE No. 167807, presented at SPE/EAGE European Unconventional Resources Conference and Exhibition, Vienna, Austria, 25–27 February.

Hewgill, G. S., Kalfayan, L. J., 1992, "Enhanced Oil Recovery Technique Using Hydrogen Precursors", US Patent No. 5,105,887, 21 April.

Hongfu, F., Yongjian, L., Liying, Z., Xiaofei, Z., 2002, "The Study on Composition Changes of Heavy Oils during Steam Stimulation Processes", *Fuels*, 81 (13), 1733–1738.

Huc, A.-Y., Ed., 2011, *Heavy Crude Oils: From Geology to Upgrading: An Overview*, Technip, Paris, France, and references therein.

Hyne, J. B., 1986, "Aquathermolysis A Synopsis of Work on the Chemical Reaction between Water (Steam) and Heavy Oil Sands during Simulated Steam Stimulation", Synopsis Report No.: 50, AOSTRA Contracts No.: 11, 103, 103B/C April.

Hyne, J. B., Greidanus, J. W., Tyrer, J. D., Verona, D., Rizek, C., Clark, P. D., Clarke, R. A., Koo, J., 1982, "Aquathermolysis of Heavy Oils", Proc. of 2nd Int. Conf. Heavy Crude and Tar Sands, Caracas, Venezuela, 7–17 February, 404–411.

Jha, R. K., Kumar, M., Benson, I., Hanzlik, E., 2012, "New Insights into Steam-Solvent Co-Injection Process Mechanism", SPE No. 159277, presented at the SPE Annual Technical Conf. and Exhibition held in San Antonio, TX, 8–10 October.

Jha, R. K., Kumar, M., Benson, I., Hanzlik, E., 2013, "New Insights Into Steam/Solvent-Coinjection-Process Mechanism, *SPE J.*, 18(5), 867–877.

Jia, N., Zhao, H., Yang, T., Ibatullin, T., Gao, J., 2016, "Experimental Measurements of Bitumen–Water Aquathermolysis during a Steam-Injection Process", *Energy Fuels*, 30, 5291–5299.

Jiang, S., Liu, X., Liu, Y., Zhong, L., 2005, "In Situ Upgrading Heavy Oil by Aquathermolytic Treatment under Steam Injection Conditions", SPE No. 91973, presented at SPE International Symposium on Oilfield Chemistry, The Woodlands, TX, 2–4 February.

Kapadia, P. R., Wang, J., Kallos, M. S., Gates, I. D., 2012, "New Thermal-Reactive Reservoir Engineering Model Predicts Hydrogen Sulfide Generation in Steam Assisted Gravity Drainage", *J. Pet. Sci. Eng.*, 94–95, 100–111.

Kapadia, P. R., Wang, J., Gates, I. D., 2014, "On In Situ Hydrogen Sulfide Evolution and Catalytic Scavenging in Steam-Based Oil Sands Recovery Processes", *Energy*, 64, 1035–1043.

Kapadia, P. R., Kallos, M. S., Gates, I. D., 2015, "A Review of Pyrolysis, Aquathermolysis, and Oxidation of Athabasca Bitumen", *Fuel Proc. Technol.*, 131, 270–289.

Krumrine, P. H., Falcone, J. S., Lefenfeld, M., 2017, "Enhanced Crude Oil Recovery Using Metal Silicides", US Patent No. 9,657,549, 23 May.

Kuhlmann, E. J., Jung, D. Y., Guptill, R. P., Dyke, C. A., Zang, H. K., 1984, "Coal Liquefaction Using Hydrogenated Creosote Oil Solvent", *Fuel*, 64, 1552–1557.

Lamoureux-Var, V., Lorant, F., 2005, "H_2S Artificial Formation as a Result of Steam Injection for EOR: A Compositional Kinetic Approach", SPE No. 97810, presented at SPE International Thermal Operations and Heavy Oil Symposium, Calgary, Alberta, Canada, 1–3 November.

Lin, R., Song, D., Wang, X., Yang, D., 2016, "Experimental Determination of In Situ Hydrogen Sulfide Production during Thermal Recovery Processes", *Energy Fuels*, 30, 5323–5329.

Liu, Y., Fan, H., 2002, "The Effect of Hydrogen Donor Additive on the Viscosity of Heavy Oil during Steam Stimulation", *Energy Fuels*, 16, 842–846.

Maity, S. K., Ancheyta, J., Marroquin, G., 2010, "Catalytic Aquathermolysis Used for Viscosity Reduction of Heavy Crude Oils: A Review", *Energy Fuels*, 24, 2809–2816.

Mohammad, A. A. A., Mamora, D. D., 2008, "Insitu Upgrading of Heavy Oil under Steam Injection with Tetralin and Catalyst", SPE No. 117604, presented at International Thermal Operations and Heavy Oil Symposium, Calgary, Alberta, Canada, 20–23 October.

Monin, J. C., Audlbert, A., 1988, "Thermal Cracking of Heavy-Oil/ Mineral Matrix Systems", SPE No. 16269, *SPE Res. Eng.*, 3 (4), 1243–1250.

Montgomery, W., Court, R. W., Rees, A. C., Sephton, M. A., 2013, "High Temperature Reactions of Water with Heavy Oil and Bitumen: Insights into Aquathermolysis Chemistry during Steam-Assisted Recovery", *Fuel*, 113, 426–434.

Montgomery, W., Sephton, M., Watson, J., Zeng, H., 2014, "The Effects of Minerals on Heavy Oil and Bitumen Chemistry When Recovered Using Steam-Assisted Methods", SPE No. 170035, presented at SPE Heavy Oil Conference-Canada, Calgary, Alberta, Canada, 10–12 June.

Montgomery, W., Watson, J. S., Lewis, J. M. T., Zeng, H., Sephton, M. A., 2018, "Role of Minerals in Hydrogen Sulfide Generation during Steam-Assisted Recovery of Heavy Oil", *Energy Fuels*, 32, 4651–4654.

Muraza, O., Galadima, A., 2015, "Aquathermolysis of Heavy Oil: A Review and Perspective on Catalyst Development", *Fuel*, 157, 219–231.

Noguchi, T., Ed., 1991, *Heavy Oil Processing Handbook*, Research Association for Residual Oil Processing, Tokyo, Japan, pp 48–64.

Ovalles, C., Hamana, A., Rojas, I., Bolívar, R. A., 1995, "Upgrading of Extra-Heavy Crude Oil by Direct Use of Methane in the Presence of Water. Deuterium Labeled Experiments and Mechanistic Consideration", *Fuel*, 74, 1162.

Ovalles, C., Garcia, M. del C., Lujano, E., Aular, W., Bermudez, R., Cotte, E., 1998a, "Structure/Interfacial Activity Relationships and Thermal Stability Studies of Cerro Negro Crude Oil and Its Acid, Basic and Neutral Fractions", *Fuel*, 77 (3), 121–126.

Ovalles, C., Filgueiras, E., Rojas, I., Morales, A., de Jesus, J. C., Berrios, I., 1998b, "Use of Dispersed Molybdenum Catalyst and Mechanistic Studies for Upgrading Extra-Heavy Crude Oil Using Methane as Source of Hydrogen", *Energy Fuel*, 12, 379–385.

Ovalles, C., Vallejos, C., Vásquez, T., Martinis, J., Peréz-Peréz, A., Cotte, E., Castellanos, L., Rodríguez, H., 2001, "Crude Oil Downhole Upgrading Process Using Hydrogen Donors under Steam Injection Conditions", SPE No. 69692, presented at SPE International Thermal Operations and Heavy Oil Symposium, Porlamar, Margarita Island, Venezuela, 12–14 March.

Ovalles, C., Vallejos, C., Vasquez, T., Rojas, I., Ehrman, U., Benitez, J. L., Martinez, R., 2003a, "Downhole Upgrading of Extra-Heavy Crude Oil Using Hydrogen Donors and Methane under Steam Injection Conditions", *Pet. Sci. Technol.*, 21 (1–2), 255–274.

Ovalles, C., Rengel-Unda, P., Bruzual, J., Salazar, A., 2003b, "Upgrading of Extra-Heavy Crude Using Hydrogen Donor under Steam Injection Conditions. Characterization by Pyrolysis GC-MS of the Asphaltenes and Effects of a Radical Initiator". Preprints of *Symposia* – American Chemical Society, New Orleans, Division of Fuel Chemistry, 48 (1), 59–60.

Ovalles, C., Rodriguez, H., 2008, "Extra-Heavy Crude Oil Downhole Upgrading Using Hydrogen Donors under Cyclic Steam Injection Conditions. Physical and Numerical Simulation Studies", PETSOC-08-01-43, *J. Can. Pet. Technol.*, 47, 43–51.

Perez-Perez, A., Kamp, A.M., Soleimani, H., Darche, G., 2011, *Numerical Simulation of H_2S and CO_2 Generation during SAGD*, World Heavy Oil Congress, Edmonton, Alberta, Canada, March 14–17.

Phillips, C. R., Haidar, N. I., Poon, Y. C., 1985, "Kinetic Models for the Thermal Cracking of Athabasca Bitumen: The Effect of the Sand Matrix", *Fuel*, 64, 678–691.

Ramirez-Corredores, M. M., 2017, *The Science and Technology of Unconventional Oils: Finding Refining Opportunities*, 1st Ed., Elsevier, London, Chapter 2, p 2, and references therein.

Robinson, C. J., 1971, "Low Resolution Mass Spectrometric Determination of Aromatics and Saturates in Petroleum Fractions", *Anal. Chem.*, 43, 1425–1434.

Shen, J., Iion, M., 1994, "Heat Treatment of Coals in Hydrogen-Donating Solvents at Temperatures as Low as 175–300°C", *Energy Fuels*, 8, 978–983.

Sheng, Q., Wang, G., Duan, M., Ren, A., Yao, L., Hu, M., Gao, L., 2016, "Determination of the Hydrogen-Donating Ability of Industrial Distillate Narrow Fractions", *Energy Fuels*, 30 (12), 10314–10321.

Shijun, H., Qian, H., Hao, L., Linsong, C., Zifei, F., Zhao L., 2017, "A Modified Model for Aquathermolysis and Its Application in Numerical Simulation", *Fuel*, 207, 568–578.

Siskin, M., Brons, G. R., Katritzky, A., Balasubramanian, M., 1990, "Aqueous Organic Chemistry. 1. Aquathermolysis: Comparison with Thermolysis in the Reactivity of Aliphatic Compounds", *Energy Fuels*, 4, 475–482.

Speight, J. G. 2007, *The Chemistry and Technology of Petroleum*, 4th Ed., CRC Press, Boca Raton, FL, and references therein.

Stapp, P. R., 1989, "In Situ Hydrogenation", Report from Nat. Inst. of Pet. Ener. Res. NIPER-434, Bartlesville, OK, December, and references therein.

Steinberg, M., 1986, "The Direct Use of Natural Gas for Conversion of Carbonaceous Raw Materials to Fuels and Chemical Feedstocks", *Int. J. Hydrogen Energy*, 11 (11), 715–720.

Stine, L. O., 1984a, "Method for In Situ Conversion of Hydrocarbonaceous Oil", US Patent No. 4,444,257, April 24.

Stine, L. O., 1984b, "Method for In Situ Conversion of Hydrocarbonaceous Oil", US Patent No. 4,448,251, May 15.

Stine, L. O., Barger, P. T., 2012, "Oil Recovery by In-Situ Cracking and Hydrogenation", US Patent No. 8,230,921, July 31.

Strausz, O. P., Jha, K. N., Montgomery, D. S., 1977, "Chemical Composition of Gases in Athabasca Bitumen and in Low-Temperature Thermolysis of Oil Sand, Asphaltene and Maltene", *Fuel*, 56, 114–120.

Thimm, H. F., 2007, "Hydrogen Sulphide Measurements in SAGD Operations", presented at the Canadian International Petroleum Conference, Calgary, Alberta, Canada, 12–14 June.

Thimm, H. F., 2014, "Aquathermolysis and Sources of Produced Gases in SAGD", SPE No. 170058, presented at SPE Heavy Oil Conference-Canada, Calgary, Alberta, Canada, 10–12 June.

Vallejos, C., Vasquez, T., Ovalles, C., 1999, "Process for the Downhole Upgrading of Extra Heavy Crude Oil", US Patent No. 5,891,829, 6 April.

Viloria, A., Hernandez, J., Borges, L., 1987, "Efecto de la Atmosfera Autógena Producida Durante la Acuatermólisis de un Crudo Extrapesado de la FPO", *Rev. Tec. INTEVEP*, 7 (1), 75–80.

Viloria, A., Parisi, S., Borges, L., Rodriguez, R., Martinez, E., Ramos, A. D., Brito, J. D., Campos, R. G., 1985, "Simulación y Predicción de la Generación de H_2S en Yacimientos de Crudos Pesados Sometidos a la Inyección de Vapor", *Rev. Tec. INTEVEP*, 5 (2), 147–155.

Wang, Q., Guo, L., Wang, Z., Mu, B., Guo, A., Liu, H., 2012, "Hydrogen Donor Visbreaking of Venezuelan Vacuum Residue. *J. Fuel Chem. Technol.*, 40 (11), 1317–1322.

Ware, C. H., Rose, L. C., Allen, J. C., 1986, "Recovery of Oil by In Situ Hydrogenation", US Patent No. 4,597,441, 1 July.

Yan, T. Y., Espenscheid, W. F., 1983, "Liquefaction of Coal in a Petroleum Fraction under Mild Conditions", *Fuel. Proc. Technol.*, 7, 121–133.

Zeng, H., Court, R. W., Sephton, M. A., Rees, A., Montgomery, W., Watson, J. S., 2013, "Quantitative Laboratory Assessment Of Aquathermolysis Chemistry During Steam-Assisted Recovery of Heavy Oils And Bitumen, with a Focus on Sulfur", SPE No. 165404, presented at SPE Heavy Oil Conference-Canada, Calgary, Alberta, Canada, 11–13 June.

8 Catalytic Hydrogen Addition

In this route (Figure 1.9), catalysts and hydrogen-donating compounds are used in the presence an energy source (steam or other) for the subsurface upgrading (SSU) of heavy crude oils and bitumens (HO/B) [Weissman 1997, Maity *et al.* 2010, Muraza and Galadima 2015, Guo *et al.* 2016]. As with the thermal-only processes (Chapter 7), most of the hydrocarbon-bond cracking (carbon–carbon, sulfur–carbon, etc.) is due to the high temperature (> 230°C) and residence times (days), but there is the possibility of catalytic cracking if the catalyst contains an acidic functionality. By injecting a metal-containing catalyst downhole, the hydrogenation reaction is greatly enhanced, and lower viscosity and higher API, hydrodesulfurization, and residue conversion can be obtained in comparison with thermal-only processes. Also, the stability of the upgraded products is increased with the concomitant improvements in the economic benefits of the process [Guo *et al.* 2016].

However, these hydrogenation processes suffer several disadvantages such as poor mixing between the catalyst, hydrogen, and heated heavy crude oil, and lack of availability of inexpensive and abundant catalytic materials, and hydrogen and energy sources. Also, due to the lack of control, maintaining the activity and avoiding catalyst deactivation represent significant challenges. The chemistry and mechanistic pathways involved in the catalytic hydrogenation of hydrocarbons were discussed in Section 4.2 [Noguchi 1991, Speight 2007, Huc 2011, Gray 2015, Ramirez-Corredores 2017].

This chapter starts by discussing various issues (Section 8.1) encountered during the development of catalytic downhole processes and continues with the literature review of the different hydrogen addition concepts that use water (Section 8.2), hydrogen gas (Section 8.3), and hydrogen donor compounds (Section 8.4). The final part (Section 8.5) is devoted to the description of the SSU field tests reported in this area.

8.1 CATALYST ISSUES

From the general point of view, there are two main types of metal-containing catalysts used for hydrogen addition to HO/B, i.e., heterogeneous and homogeneous catalysts. The first type is generally composed for one or several metals supported on highly porous solids (e.g., alumina or silica), and the second are metallic-containing compounds which are squeezed downhole in a highly dispersed fashion (i.e., nanoparticles) using aqueous or oleic media. The most common metals are taken from the groups 6 to 11 of the Periodic Table (Mo, Mn, Fe, Co, Ni, Cu, etc.), and their concentrations varied from few hundreds of ppm to weight-percentage levels.

8.1.1 INITIAL CONSIDERATIONS

Several issues should be considered and addressed adequately for the successful operation of metal-containing catalysts for the subsurface upgrading of heavy oil and bitumens. The first issue is the transportation, handling, and downhole injection of these materials. Heterogeneous catalysts can be visualized as a gravel pack type of solids whereas homogenous catalysts are in liquid phase solutions at surface conditions. Thus, it can be envisioned that the same procedures used for handling oilfield chemicals (e.g., demulsifiers, antifoaming, etc.) and solid suspensions (e.g., drilling muds) in surface facilities at or near the wellhead can be employed with success. Another advantage of using homogeneous catalysts is that remote preparation of the catalysts may be possible [Weissman 1997]. However, new protocols and standard operating practices are to be developed to handle these catalytic materials that are new or not generally used in upstream operations.

Another critical issue to be considered is the catalyst cost. In catalytic SSU, heterogeneous or homogeneous catalysts are injected downhole and are not expected to be recovered. Thus, the need for using less expensive metals is of paramount importance to reduce the operating cost of the process. Metals such as iron, nickel, or copper have relatively low prices per pound whereas noble metals such as platinum, palladium, and rhodium, even though they have high hydrogenation activities, are not recommended. Other relatively expensive metals such as cobalt, molybdenum, and tungsten could be employed, but their concentration should be kept to a minimum. Also, the use of inorganic supports tends to increase the cost of the catalyst not only due to the price of the metallic element *per se* but also for the high manufacturing costs. The use of dispersed or nanocatalysts eliminates this issue as discussed in Section 8.2.3.

8.1.2 PLACEMENT OF THE CATALYST DOWNHOLE

The effective catalyst placement is another important consideration because it is necessary to provide the required contact between the catalyst and heated bitumen in the reservoir to make sure the upgrading reactions take place. Commonly practiced production engineering methodologies may be adaptable to catalyst injection or solid placement [Weissman *et al.* 1996, Weissman and Kessler 1996, Weissman 1997]. For example, using conventional techniques for cased or open-hole gravel pack completions or proppant injection, solid catalysts can be placed around a wellbore and into the oil-bearing formation [Coulter *et al.* 1987]. By this way, the crude oil could flow through the catalytic material during the oil production with the subsequent upgrading. Commercially available Hydroprocessing catalysts have particles sizes in the 12–40 mesh and crush strengths similar to gravel pack solids used in upstream operations [Weissman 1997]. Additionally, heterogeneous catalysts can be placed in a second slotted liner within the outer liner, forcing the hot fluids to flow through the catalyst. These two catalyst-placement strategies were used for the lab and field tests of the THAI-CAPRI process [Shah *et al.* 2010, Hart 2014]. The authors did not report plugging issues, but some reduction in the catalytic activity was observed.

For the case of homogeneous or fluid-phase metal catalysts, the downhole injection of the precursor materials into an oil-bearing formation may be practicable if the catalyst can be distributed throughout the reservoir or in an extended volume around the producing well. By spreading the catalyst to a greater extent than would be possible using solid catalysts, a much more significant amount can be available for processing. However, placing and utilizing a dispersed catalyst creates an additional set of concerns and procedural difficulties, the most notable being the need to ensure oil/catalyst contacting. Additionally, injected aqueous solutions may adversely interact with clays in the oil-bearing formation, leading to a loss of fluid permeability [Weissman 1997].

Pereira-Almao *et al.* reported that the retention of nanocatalyst particles in Athabasca sand beds was higher than in clean silica sand under similar conditions [Pereira-Almao 2012]. It seems that the surface roughness and the presence of very fine mineral particles in the Athabasca sand play a role in that increased catalyst retention. In the case where a catalyst is squeezed into the formation around the producer, it is expected that part of the injected nanocatalysts is retained by the sand media via a deep-bed-filtration process. This phenomenon varies with flow conditions such as sand permeability, flow velocity, catalyst particle size distribution, temperature, and presence or absence of fluids such as gas, connate water, and steam.

However, it is expected that some catalysts are going to be produced back over time diminishing the concentration retained around the producer. Therefore, large volumes of catalyst for achieving the required residence times would be necessary. If we assume a 1 BBL/day per foot length of horizontal producer wellbore (~60 BBL/d), the residence time vs. the radius catalyst zone can be estimated as shown in Figure 8.1. As seen, an area of ~2 m (6 ft) around the producer has to be "invaded" by the catalysts in order to achieve a residence time of ~ten days. These values do not seem out of reach, but a clear understanding of the physics of the countercurrent flows in the reservoir is needed to quantify the contact times better and maximize the success of the SSU process.

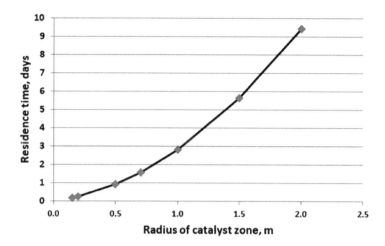

FIGURE 8.1   Residence times for a flow 1 BBL/day/ft in a horizontal wellbore (0.5 m³/d/m).

8.1.3   CATALYST CONTROL AND RESIDENCE TIMES

Some of the most critical issues and uncertainties for the catalytic subsurface upgrading of heavy oils are the control, dispersion, and activity of catalysts within the reservoir. As time progresses impurities, brine, and metals deposited from the crude [Weissman and Kessler 1996] may collect on the catalyst, resulting in higher pressure drops around the wellbore and possibly leading to decreased production rates and lowered catalyst activities. Thus, robust catalytic systems must be designed so they can operate unattended for extended periods of time and without operators' intervention.

The effect of water or steam on catalysts is another area of concern. It has been found that exposing conventional hydroprocessing catalysts to a high-pressure water environment did not affect the dispersion or sulfide state of the active metals; however, overall catalytic activity was reduced [Laurent and Delmon 1994]. Also, W and Mo nitrides and carbides, hydrated molybdenum oxides and zeolites, the hydrophobic zeolites such as Y, Beta, ZSM-5, and Mordenite could be good candidates for the subsurface upgrading of HO/B due to their excellent hydrothermal stability properties [Muraza and Galadima 2015].

Catalysts with larger pores are found to be more effective for heavier feeds and more resistant to deactivation by metal deposition. These small differences in activities between Ni-Mo and Co-Mo for heavy oil processing indicate that the most readily available catalyst is the most appropriate choice [Weissman 1997].

The residence time for the HO/B inside the reservoir is another critical consideration for *in-situ* catalytic upgrading. Estimated residence times for reservoir-mobilized bitumen are more than 100 h when SAGD is used as the recovery method. In effect, lab results showed that a trimetallic catalyst formulation (at 1,200 ppm) upgraded an 8°API bitumen to a 17°API oil using a batch reactor at 340°C with 2,480 kPa of hydrogen pressure and 72 h of residence time [Pereira-Almao 2012].

8.1.4   ENVIRONMENTAL CONCERNS

In the next decades, environmental requirements are expected to significantly increase in Europe, the United States, and around the world [Huc 2011]. The use of heterogeneous or homogeneous catalysts downhole could be formally considered as a release of hazardous chemicals to the environment unless detailed chemical qualifications studies were carried out that proved otherwise. Many of the heavy metals used as catalytic centers (Mo, W, Co, Ni, Fe, Cr, etc.) are very toxic if they find their way to the underwater aquifers or the surface where it can be ingested by humans and fauna. Heavy metal toxicity has proven to be a significant threat, and there are several health risks associated with it. In general, the presence of heavy metallic elements can lower the human energy levels and damage the functioning of the brain, liver, lungs, kidney, blood composition, and other vital organs [Jaishankar et al. 2014]. These metals sometimes act as a pseudo enzymatic active site while at certain times they may even interfere with metabolic processes. In general, metal toxicity depends upon the absorbed dose, the route of exposure, and duration of exposure, i.e., acute or chronic [Jaishankar et al. 2014].

Also, these metals have not been assessed for their long-term impacts on the environment and human and animal health. Long-term exposure can lead to gradually progressing physical, muscular, and neurological degenerative processes that imitate diseases such as multiple sclerosis, Parkinson's disease, Alzheimer's disease, and muscular dystrophy. Repeated long-term exposure of some metals and their compounds may even cause cancer. Hence thorough knowledge of heavy metals toxicity is essential to provide proper defensive measures against their excessive contact [Jaishankar et al. 2014].

On the other hand, it has been reported that metallic elements such as Co, Cu, Cr, Fe, Mg, Mn, Mo, and Ni are essential micronutrients that are needed for various biochemical and physiological functions [Tchounwou et al. 2012]. In small amounts, they are necessary for maintaining good health, but in more substantial quantities they can become toxic or dangerous. Therefore, qualifications of chemicals should consist of both testing and toxicological evaluations for each field application, as hazard assessments are site and process specific.

In conclusion, environmental concerns should be addressed early during the development of catalytic SSU processes, and adequate measures should be taken to avoid pollution and damages to the biosphere.

8.2 CATALYTIC AQUATHERMOLYSIS

As described in Section 7.2, Aquathermolysis is a hydrogen addition route (see Figure 1.9) that involves the cracking of heavy crude oils, bitumens, or oil-containing sands in the presence of water as a source of hydrogen to generate an upgraded crude oil and small amounts of hydrogen sulfide, carbon dioxide and monoxide, methane, and hydrogen (Equation 7.4) [Clark et al. 1983, Viloria et al. 1985, Brons and Siskin 1994, Maity et al. 2010, Kapadia et al. 2015, Muraza and Galadima 2015]. The use of homogeneous or heterogeneous catalysts to further enhance the properties of the upgraded crude oils was first reported by Hyne et al. in 1982 [Hyne et al. 1982]. Since then, several research groups around the world have studied a large variety of catalytic systems based on water-soluble (Section 8.2.1), oil-soluble (Section 8.2.2), amphiphilic (Section 8.2.3) catalysts, and dispersed or nanocatalysts (Section 8.2.4). Also, other types of catalytic systems have received attention such as industrial heterogeneous and amphiphilic materials (Section 8.2.5). The effect of the mineral formation on the Aquathermolysis reaction was previously discussed in Section 7.2.2. In all previous cases, water was used as the *only* source of hydrogen for Aquathermolysis. To further improve the properties of upgraded products, the use of additional H_2-donating compounds such as hydrogen gas or hydrogen donor solvents is described in Sections 8.3 and 8.4, respectively

It is important to mention that the area of subsurface upgrading via catalytic Aquathermolysis has been the subject of several review articles. The first review was by Weissman in 1997 [Weissman 1997] followed by Maity and coworkers in 2010 [Maity et al. 2010]. In the last few years, Muraza and Galadina [2015] and Guo et al. [2016] have reported recent developments and perspectives in this area.

8.2.1 WATER-SOLUBLE CATALYSTS

The use of aqueous-soluble catalysts for the Aquathermolysis of HO/B is well documented in the literature [Maity et al. 2010, Muraza and Galadima 2015, Guo et al. 2016]. The utilization of water solutions as catalyst carriers is a relatively simple process and easy to implement in the field. It is envisioned that the only equipment required is a tank and a pump and that the catalyst solution could be co-injected downhole in the same line as the one used for steam. As mentioned in Section 8.1, the catalysts are not expected to be recovered but could find their way to the surface in the produced water or as an emulsion within the produced oil. Thus, further treatment of the effluents may be necessary to comply with environmental regulations.

Table 8.1 shows several examples reported in the literature on the upgrading of HO/B via catalytic Aquathermolysis using aqueous-soluble catalysts and water as the only hydrogen source. The role of metal-containing catalysts was first studied by Hyne and coworkers [Hyne et al., 1982; Hyne and Clark 1985]. These authors treated homogenized Athabasca oil sands in a pressurized autoclave without agitation. The autoclave and its contents were thoroughly flushed with nitrogen before the reaction to ensure anaerobic conditions. Distilled water was added to the autoclave in an amount to obtain a ratio of between 0.2 and 0.6 with respect to the oil sand. The experiments were carried out at 240°C and ~0.7 MPa using 0.15–1 wt. % of V-, Fe-, Ni-, Mo-, Cu-, and Co-containing water-soluble salts as catalysts for 14 d [Hyne *et al.* 1982, Hyne and Clark 1985].

As seen in Table 8.1, the upgrading experiments showed small increments on the API gravity in the order of 0.4–1.3°API with respect to the thermal run experiment (no catalyst) and viscosity reductions in the 20–70% range vs. the Canadian feed. Figure 8.2 shows the viscosity values of the upgraded crude oils from the same experiments in function of the type of metal employed as catalyst. As seen, the thermal run gave a viscosity slightly higher than that found in the feed (~600 vs. 520 mPa s at 90°C) whereas the $FeSO_4$- and $(MoO_4)^{-2}$-containing runs showed viscosity as low as ~170 mPa s. Other metal catalysts ($CuSO_4$ and $CoSO_4$) also had some viscosity-reduction activity reaching values in the 55–57% range [Hyne 1986].

Hyne and coworkers also reported the percentages of desulfurization (%DS) during the aquathermolytic experiments. As shown in Table 8.1, the thermal-only test gave an increase in the sulfur content (~6%) whereas the Ni- and Co-containing runs showed %DS of 4% [Hyne *et al.* 1982, Hyne and Clark 1985]. These results indicated that there was a small but significant effect of the metal catalysts in properties of the upgraded crude oils.

During Hyne's autoclave experiments, hydrogen gas was generated and quantified. Figure 8.3 shows the production of H_2 gas from the Aquathermolysis of Venezuelan oil-sand using water-soluble metal catalysts at the same reaction condition (240°C, 28 days). As seen, hydrogen production was observed in the thermal and catalytic experiments ranging from 90 to 640 mL of H_2 per kg of sand. Similar results were obtained using the Athabasca oil sand [Hyne 1986].

The authors proposed that the hydrogen generation was due to the water–gas-shift reaction (WGSR) and that the metal salts were acting as water-soluble catalysts (see Equation 8.1). As discussed earlier (see Equation 7.4), carbon monoxide is a product of the Aquathermolysis reaction, so it is not surprising that CO reacted with the water present, in the presence of the metal catalysts, to generate carbon dioxide and hydrogen gas [Clark and Hyne 1984]. Also, Hynes and coworker carried out experiments in the presence of added carbon monoxide, and greater viscosity reductions were observed [Hyne 1986].

$$H_2O + CO \xrightarrow{\text{Catalyst}} CO_2 + H_2 \qquad (8.1)$$

Furthermore, the author proposed that the generated H_2 could be used to further enhance the properties of the upgraded crude oils via thermal or catalytic hydrogen addition [Hyne 1986]. This proposal can explain why lower amounts of hydrogen gas were measured using the V- and Ni-containing catalysts (see Figure 8.3) in comparison with the other runs. V and Ni are known hydrogenation catalysts, so this proposal is with merits. As seen, the $(MoO_4)^{-2}$-containing experiment was the most active for hydrogen generation, so it was used to study the composition of the upgraded crude oils by SARA analysis.

Figure 8.4 shows the SARA composition for the feed and the Aquathermolysis products of Canadian and Venezuelan oil sands under thermal and catalytic conditions using water-soluble

TABLE 8.1

Catalytic Aquathermolysis Using Selected Aqueous-Soluble Catalysts and Water as the Hydrogen Source

Catalyst	Crude Oil (Country)	°API	Viscosity (mPa s @ °C)[a]	Temp. (°C)	Press. (MPa)	Time	Incr. °API[b]	% Visc. Red.[c]	% DS[d]	% Asp. Conv.[e]	References
Thermal	Athabasca oil sand (Canada)	ND	520 (@90°C)	240	0.69	14 days	–	–11	–6	ND	Hyne et al. [1982], Hyne and Clark [1985]
VO(SO₄)							0.52[f]	22	1		
FeSO₄							0.75[f]	68	0		
NiSO₄							0.50[f]	38	4		
MoO₄⁻²							ND	62	ND		
CuSO₄							1.34[f]	53	–1		
CoSO₄							0.43[f]	57	4		
NiSO₄	Cerro Negro (Venezuela)	8.4	490 (@90°C)	240	ND	30 min	ND	97	3	ND	Rivas et al. [1988]
FeSO₄	Cold Lake (Canada)	ND	2,540 (@39°C)	415	18.6–20.7	3 h	ND	99	20	48	Clark and Kirk [1994]
RuCl₃	Peace River (Canada)		2,140 (@39°C)				ND	99	36	48	
V(O)SO₄ Fe₂(SO₄)₃ NiSO₄	Liaohe (China)	ND	88,500 (@50°C)	240	ND	24 h	ND	86[f]	50	28	Fan [2001]
K₃PMo₁₂O₄₀	G540 (China)	ND	812,000 (@50°C)	200	3.0	24 h	ND	80	ND	18	Chen et al. [2009]
Iron-Aromatic Sulfonate	DF32005 (China)	ND	91,000 (@50°C)	200	3.0	24 h	ND	96	ND	50	Wang et al. [2010]
Mo-Aromatic Sulfonate							ND	99	ND	66	
Cu-p-Toluene-sulfonate	Shengli (China)	ND	180,000 (@50°C)	200	3.0	24 h	ND	93	ND	30	Li [2013]
Fe-p-Toluene-sulfonate	F10223 (China)		85,000 (@50°C)					95	ND	ND	
Ni- Mannich base	Yumen (China)	25.5	12,050 (@50°C)	180	ND	24 h	ND	75	ND	ND	Chen et al. [2017b]
Fe (Lactate)₃	Tazhong (China)	ND	4,000 (@50°C)	180	ND	24 h	ND	85	ND	59	Chen et al. [2017c]

Feed Characteristics columns: Crude Oil (Country), °API, Viscosity (mPa s @ °C)[a]. *Reaction Conditions* columns: Temp. (°C), Press. (MPa), Time. *Results* columns: Incr. °API[b], % Visc. Red.[c], % DS[d], % Asp. Conv.[e]

ND = Not disclosed.

[a] Viscosity (μ) in mPa s measured at the given temperature.

[b] Increase in API gravity of the upgraded crude oil. Negative values indicate increase with respect to the feed.

[c] Percentage of decrease in viscosity of the upgraded crude oil vs. the feed calculated as $((\mu_{feed} - \mu_{product}) - \mu_{feed}) \times 100$. Negative values indicate increase with respect to the feed.

[d] Percentage of desulfurization in the upgraded crude oil vs. the feed calculated as $((\%S_{feed} - \%S_{product}) - \%S_{feed}) \times 100$. Negative values indicate increase with respect to the feed.

[e] Percentage of reduction of asphaltenes in the upgraded crude oil vs. the feed calculated as $((\%Asp_{feed} - \%Asp_{product}) - \%Asp_{feed}) \times 100$.

[f] With respect to the control run in the presence of only water but no catalyst (thermal experiment). Feed characteristics were not disclosed.

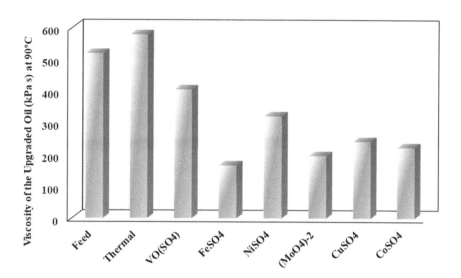

FIGURE 8.2   Viscosity of the upgraded crude oils from the Aquathermolysis of a *Canadian* oil sand using water-soluble metal catalysts (240°C, 28 days). Data taken from Hyne [1986].

molybdenum catalysts in the presence of added carbon monoxide [Hyne 1986]. As mentioned in Section 7.2.1, for both oils, the asphaltene content increased whereas the resin fraction decreased upon *thermal* Aquathermolysis. As shown in Chapter 2 and reported in the literature [Speight 2007], the asphaltene and resin fractions have the highest amount of sulfur in comparison with those found in the saturates and aromatics. Indeed, the organosulfur components of the oils are one of the most reactive species in the thermal Aquathermolysis reaction; it might be expected that a substantial change would occur in those fractions.

On the other hand, in the presence of the metal catalysts and for the Canadian oil sand, the amount of asphaltenes decreases and resin increased (see Figure 8.3). For the Venezuelan case, the asphaltene concentration was approximately the same, but the resin content decreased, and the aromatics increased. These results are consistent with a catalytic hydrocracking mechanism and indicate the complexity of the reactions occurring during Aquathermolysis. Based on the experimental

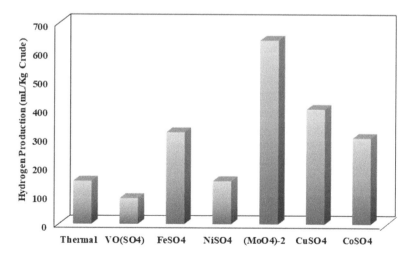

FIGURE 8.3   Production of hydrogen gas from the Aquathermolysis of *Venezuelan* oil-sand using water-soluble metal catalysts (240°C, 28 days). Data taken from Hyne [1986].

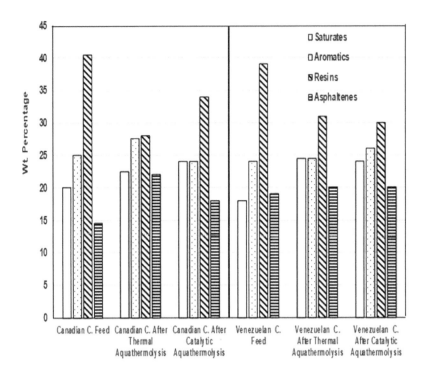

FIGURE 8.4 SARA composition for the feed and the Aquathermolysis products of Canadian and Venezuelan oil sands under thermal and catalytic conditions using water-soluble molybdenum catalysts in the presence of added carbon monoxide (240°C, 28 days). Data taken from Hyne [1986].

evidence, the authors concluded that the metal catalysts are involved in the initial cleavage of the organosulfur bonds, in the water–gas-shift reaction, and the hydrocracking and hydrodesulfurization steps [Hyne 1986].

Rivas *et al.* studied the effect of a water-soluble nickel catalyst on the upgrading of Cerro Negro crude oil (from the Orinoco Basin) at a temperature of 240°C under typical of Venezuelan steam injection conditions (see Table 8.1). The experimental plan consisted of both static and dynamic tests with $NiSO_4$ concentration up to 1.5 wt. % [Rivas *et al.* 1988]. Reconstituted and whole core samples from two Venezuelan bitumen reservoirs were used in their experiments [Rivas *et al.* 1988].

Figure 8.5 shows the viscosity of the original crude oil and the upgraded products under dynamic Aquathermolysis in function of time and wt. percentage of Ni. As seen, when the Ni catalysts were added, viscosity reductions were observed at shorter reaction times. At low Ni loadings (< 0.8 wt. %), the viscosity remained constant for the rest of the experience whereas, at an Ni concentration of 1.5 wt. %, this value decreased from almost 500 mPa s at 90°C to around 20 mPa s after a contact time of only 30 minutes. The author reported that these viscosity reductions were never observed in static tests. Thus, they concluded that the intimate contact of sand, heavy crude oil, and water led to more drastic results in shorter times than those obtained using batch cells [Rivas *et al.* 1988].

Clark and Kirk carried out thermal treatments (375–415°C) of Peace River and Cold Lake bitumens with and without water in the presence of aqueous solutions of iron or ruthenium chlorides as catalysts at 19–20 MPa for 3 h [Clark and Kirk 1994]. As expected for high-temperature Aquathermolysis experiments, the results showed significant gas production (CH_4, C_2–C_5, H_2S, CO_2, CO, and H_2) and viscosity reductions of almost two orders of magnitude at 415°C (see Table 8.1). Characterization of the upgraded products by Gel Permeation Chromatography (GPC) showed that significant molecular changes (cracking) had occurred in the presence of the catalysts both with and without water, which led to the formation of the low-viscosity upgraded products. They also

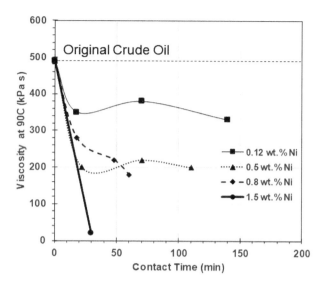

FIGURE 8.5 Viscosity of the original Cerro Negro crude oil and the upgraded products under dynamic Aquathermolysis in function of time and wt. percentage of a water-soluble nickel catalyst. Data taken from Rivas *et al.* [1988].

observed sulfur removal in the 20–36% range and reductions on the asphaltene content of almost 50 wt. %. The use of the iron catalyst ($FeSO_4$) minimized the formation of insolubles in experiments conducted above 400°C. However, treatment with a ruthenium-containing solution resulted in the increase in insolubles formation in comparison to all other conditions studied. The authors proposed that metal sulfides act as the active phase during the aquathermolytic experiments and that the upgrading reaction occurs primarily by C–S bond cleavage [Clark and Kirk 1994].

Fan and coworkers carried out laboratory Aquathermolysis reactions using Liaohe heavy oils in the presence of mineral formation (see Section 7.2.2.) and a $VO^{2+}/Ni^{2+}/Fe^{3+}$- containing catalyst at 240°C for 24 h [Fan *et al.* 2001]. As shown in Table 8.1, the authors found 86% of viscosity reduction vs. the thermal experiment and 50% of sulfur removal vs. the feed. Figure 8.6 shows the SARA composition for the feed and the Aquathermolysis products in the presence of the mineral formation, the water-soluble V-Fe-Ni-catalysts, and a combination of all. As found before for other

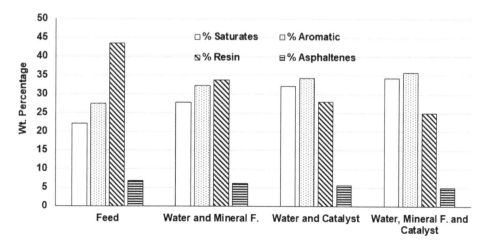

FIGURE 8.6 SARA composition for the feed and the Aquathermolysis products in the presence of mineral formation, water-soluble V-Fe-Ni-catalysts, and a combination of all. Data were taken from Fan [2001].

authors [Hyne and Clark 1985, Rivas *et al.* 1988, Clark and Kirk 1994, Hongfu *et al.* 2002], all Aquathermolysis experiments decreased the asphaltene contents in the 10–28% range with respect to the feed. Similarly, significant reductions in the resin fraction were found in the presence of the mineral formation (~34 wt. %) or the catalyst (~28 wt. %) in comparison with the original heavy crude oil (~44 wt. %). These reductions were accompanied by an increase in the saturates and aromatic fractions [Fan *et al.* 2001].

In the presence of water, mineral formation, and the metal-containing catalyst a further decreased in the resins were obtained (~25 wt. %) with the concomitant increase in the saturate (~34 wt. %) and aromatic (~36 wt. %) fractions (Figure 8.6.). Based on VPO, gas, and elemental analyses, a synergistic effect between the mineral and the catalyst was proposed that can potentially increase the chances of success of this subsurface upgrading concept [Fan *et al.* 2001].

The authors proposed that, when the catalyst solution is put in contact with the mineral formation, the metal ions, such as VO^{2+} and Ni^{2+}, can be adsorbed on the surface of clay via electrostatic forces. Under this circumstance, the minerals act as the catalyst support in the same fashion that alumina or silica are used in conventional refinery processes [Fan *et al.* 2001]. It is also proposed that these supported metal catalysts have a high activity to accelerate the decomposition of the sulfur compounds regardless of whether the sulfur was in an aromatic or aliphatic environment [Fan *et al.* 2001].

Chen and coworkers carried out the upgrading of Chinese heavy oils using a Nano-keggin-$K_3PMo_{12}O_{40}$ catalyst under Aquathermolysis conditions [Chen *et al.* 2009]. This hetero-poly acid salt has the capability of acid, redox, and liquid-phase transfer reactions [Maity *et al.* 2010]. The upgrading experiments were carried out using an oil-to-water weight ratio of 7:3 and 0.3 g of catalyst and a high-pressure reactor. The temperatures were in the 200–280°C range at 3 MPa of N_2 for 24 h (Table 8.1). The results showed a viscosity reduction by 92% for the heavy oil Zhen411 at 280°C, and by 80% and 90% for the G540 at 200°C and 280°C, with 9% and 12% conversions of heavy to light content, respectively (Table 8.1). Reductions down to 18% on the asphaltene contents were also observed [Chen *et al.* 2009].

By chemical and group-type characterization of the heavy oils before and after the reaction, the authors found significant changes in the concentration of oxygen-containing compounds during the catalytic Aquathermolysis [Chen *et al.* 2009]. Consistent with these results, by gas chromatography/mass spectrometry (GC/MS) analysis, they observed the appearance of new and small low-boiling-point molecules such as alcohols, phenols, olefins, alkanes, and ethers. They concluded that the Nano-keggin-catalyst promoted the pyrolysis of asphaltenes and resins to the lighter fractions by breaking the C–S and C–O bonds, with the concomitant viscosity reductions [Chen *et al.* 2009].

Wang *et al.* studied the activity and the mechanism of catalytic Aquathermolysis using water-soluble iron and molybdenum aromatic sulfonate catalysts [Wang *et al.* 2010]. At 200°C and 24 h, the authors found viscosity reduction in the 96–99% range as well as 50–66% of sulfur removal vs. the feed (Table 8.1). Figure 8.7 shows the SARA composition for the feed and the Aquathermolysis products. As found by Fan and coworkers [2001], the analysis showed that the iron catalyst caused reduction principally in the resin fraction, while the molybdenum-containing material led to decrease in the asphaltene content (66% for Mo and 50% for Fe).

Also, Wang *et al.* used thin-layer chromatography-flame ionization detection (TLC-FID), elemental analysis, ^{1}H-nuclear magnetic resonance (^{1}H NMR), and GC/MS to analyze the changes in heavy oil, reaction water, and produced gas during the catalytic Aquathermolysis experiments. The authors proposed that the mechanism HO/B upgrading via catalytic Aquathermolysis involves seven types of chemical reactions. Those are pyrolysis, hydrogenation, isomerization, ring opening, oxygenation, alcoholization, and esterification [Wang *et al.* 2010].

Li and coworkers [2013] utilized copper and iron-containing toluene sulfonates for the upgrading of Chinese heavy crude oils (Table 8.1). As found for other water-soluble catalysts [Wang *et al.* 2010], the results showed viscosity reduction in the 93–95% range and sulfur removal in the order

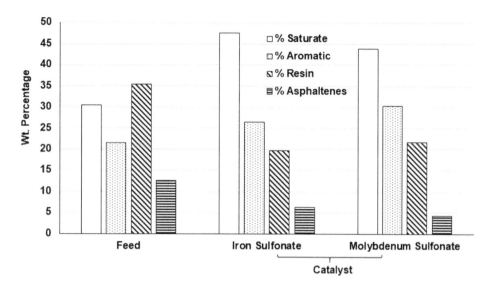

FIGURE 8.7 SARA composition for the feed and the Aquathermolysis products using water-soluble iron and molybdenum sulfonate catalysts. Data taken from Wang *et al.* [2010].

of 30%. Figure 8.8 shows the molecular weight determinations by GPC of the asphaltenes for the feed and the Aquathermolysis products of the thermal and catalytic runs. As shown, both metal-containing catalysts led to asphaltenes with lower molecular weights (~1,900–2,000 g/mol) than those found for the thermal reaction (~4,800 g/mol) and the feed (~7,000 g/mol). Between the two catalysts, the Cu-containing material gave the lowest molecular weight [Li *et al.* 2013].

Chen and coworkers synthesized a series of Mannich base-transition metal complexes (Fe, Co, Ni, Cu, and Zn) and evaluated them for the catalytic Aquathermolysis of a Chinese heavy oil [Chen *et al.* 2017b]. At a relatively low temperature of 180°C for 24 h, the Ni-containing complex was the most effective one with 75% of viscosity reduction (Table 8.1). As found by other authors [Hyne and Clark 1985, Rivas *et al.* 1988, Clark and Kirk 1994, Fan *et al.* 2001, Wang *et al.* 2010], the results

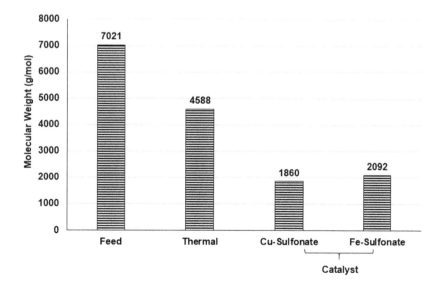

FIGURE 8.8 Molecular weight determinations by GPC of the asphaltenes for the feed and the Aquathermolysis products of the thermal and catalytic reactions. Data taken from Li [2013].

showed that the content of resin and asphaltenes decreased, while the content of saturate and aromatic increased after Aquathermolysis reaction [Chen *et al.* 2017b].

Chen and coworkers synthesized three iron (II) water-soluble catalysts by refluxing $FeCl_3$ with sodium citrate, sodium tartaric, or sodium lactate in water at the 1:3, 1:2, or 1:1 molar ratios for 2 h [Chen *et al.* 2017c]. The author carried out the upgraded reactions using a water/oil ratio of 0.3 in the presence of 0.5 wt. % of catalysts at 180°C for 24 h. The result showed that the iron lactate catalyst (Fe(Lactate)$_3$, see Table 8.1) was the most active for the viscosity reduction of Tazhong crude oil (85%) in comparison with the iron citrate (62%) and iron tartrate (58%) counterparts. As other authors have found, the SARA analysis showed reductions in the resins and asphaltenes from ~14 wt. % and ~25 wt. % to ~3 wt. % and 10 wt. %, respectively. Meanwhile, saturates and aromatics increased from 42 wt. % and ~20 wt. % to 60 wt. % and 27 wt. %, respectively [Chen *et al.* 2017c].

Based on the results presented in this section the role of the catalyst is to catalyze the WGSR to produce H_2 and to accelerate the breaking of C–S, C–O, and C–C bonds to further enhance the properties of the HO/B in contrast with thermal-only reactions. As discussed in Chapter 7, asphaltenes and resins were found to be the most reactive fraction of the crude oil in comparison with saturates and aromatics. Several chemical reactions have been proposed to occur during catalytic Aquathermolysis. Those are pyrolysis, cracking, hydrodesulfurization, decarboxylation, isomerization, and ring opening [Hyne *et al.* 1982, Hyne and Clark 1985, Clark and Kirk 1994, Fan *et al.* 2001, Hongfu *et al.* 2002, Chen *et al.* 2009, 2017b, Wang *et al.* 2010, Li *et al.* 2013].

8.2.2 OIL-SOLUBLE CATALYSTS

As with the water-soluble catalysts, the use of oil-soluble counterparts for the Aquathermolysis of HO/B has been reported in the literature [Maity *et al.* 2010, Muraza and Galadima 2015, Guo *et al.* 2016]. The objective is to increase the catalyst dispersion in the crude oil by using a catalyst precursor dissolved in an organic solvent. By this way, higher catalytic activity could be obtained with the concomitant improvement in the properties of the upgraded heavy oils and bitumens.

The utilization of organic solvents as catalyst carriers is also a relatively simple process and easy to implement in the field. As with the water-soluble catalysts, it is envisioned that the only equipment required is a tank and a pump and that the catalyst solution could be co-injected downhole with steam as a two-phase fluid. As mentioned in Section 8.1, the catalysts are not expected to be recovered but could find their way to the surface dissolved in the crude oil and may need to be removed by the desalter unit before further refining could be attempted.

Table 8.2 shows several examples reported in the literature on the upgrading of HO/B via catalytic Aquathermolysis using oil-soluble catalysts and water as the only hydrogen source. Wen and coworkers carried out laboratory experiments using Liaohe heavy crude oil in the presence of molybdenum oleate [Wen *et al.* 2007]. In this study, the molybdenum oleate organic catalyst was prepared by reacting MoO_3 with oleic acid to obtain an oil-soluble catalyst with 24.9 wt. % of Mo content in the organic phase. The tests were carried out in an autoclave reactor by taking 75 g of feed, 0.4 g of catalyst, and 25 g of water at 240°C for 24 h [Wen *et al.* 2007].

As seen in Table 8.2, the results showed an increase in the API gravity of the upgraded crude oil from 5.7°API to 17.5°API. At the same time, the authors reported up to 93% of viscosity reduction with 66% of sulfur removal and 30% of asphaltene conversion. During the upgrading runs, a significant amount of gases like CO_2, H_2S, and light hydrocarbons, mainly C_2–C_7, were generated. It was believed that these light hydrocarbons acted as a solvent and hence viscosity reduction of the upgraded crude was observed. Also, an increase in the hydrogen-to-carbon ratio was reported, and, at the same time, the amount of saturates and aromatics increased whereas the content of asphaltene and resin decreased [Wen *et al.* 2007]. This catalytic system was field tested in Qi-40 and Qi-108 blocks of Liaohe Oilfield, and the results are discussed in Section 8.5.

Yufeng and coworkers separated the resin and asphaltene fractions from a Liaohe heavy oil using pentane-to-crude mass ratio of 20 to 1 [Yufeng *et al.* 2009]. The catalytic Aquathermolysis of these

TABLE 8.2

Catalytic Aquathermolysis Using Selected *Oil-Soluble Catalysts* and *Water* as the Hydrogen Source

| Catalyst | Feed Characteristics | | | Reaction Conditions | | | Results | | | | References |
	Crude Oil (Country)	°API	Viscosity (mPa s @ °C)[a]	Temp. (°C)	Press. (MPa)	Time	Incr. °API[b]	% Visc. Red.[c]	% DS[d]	% Asp. Conv.[e]	
Mo-Oleate	Liaohe (China)	5.7	12,000 (@50°C)	240	ND	24 h	11.8	93.0	66	30	Wen et al. [2007]
Nickel Naphthenate	Asphaltenes and Resins from Liaohe (China)	9.9	ND	280	ND	48 h	ND	ND	ND	13	Yufeng et al. [2009]
Iron Naphthenate									21[f]	16	
Copper Aromatic Sulfonate	Shengli (China)	ND	178,000 (@80°C)	280	3	24 h	ND	95.5	ND	37	[Chao et al. 2012a] and [Chao 2012b]
Fe(acac)$_3$[g]	Ashal'cha (Tatarstan)	14.3	669 (@40°C)	250	3	6 h	–1.0	–22.0	–10	–8	Galukhin [2015]
Cu(MAC)(BA)[h]	Lukeqin (China)	ND	12,700 (@50°C)	240	2	24 h	ND	85.0	ND	3	Tang [2015]

ND = Not disclosed or not reported.

[a] Viscosity (μ) in mPa s measured at the given temperature.

[b] Increase in API gravity of the upgraded crude oil. Negative values indicate increase with respect to the feed.

[c] Percentage of decrease in viscosity of the upgraded crude oil vs. the feed calculated as $((\mu_{feed} - \mu_{product}) - \mu_{feed}) \times 100$. Negative values indicate increase with respect to the feed.

[d] Percentage of desulfurization in the upgraded crude oil vs. the feed calculated as $((\%S_{feed} - \%S_{product}) - \%S_{feed}) \times 100$. Negative values indicate increase with respect to the feed.

[e] Percentage of reduction of asphaltenes in the upgraded crude oil vs. the feed calculated as $((\%Asp_{feed} - \%Asp_{product}) - \%Asp_{feed}) \times 100$.

[f] Percentage of desulfurization in the asphaltene fraction only.

[g] acac = acetylacetonate ligand.

[h] Where MAC = Multicomponent copolymer, BA = Benzoate.

materials were investigated by using water-soluble ($NiSO_4$ and $FeSO_4$) and oil-soluble catalysts (nickel naphthenate and iron naphthenate) at 280°C for 48 h. As can be seen in Table 8.2, the authors reported up to 16% asphaltene conversion with 21% of sulfur removal. Figure 8.9 shows the effect of the type of catalyst on the percentage of conversion of the asphaltene fraction. In general, the oil-soluble catalysts showed higher activity than the water-soluble counterparts. The order of catalytic ability is as follows [Yufeng *et al.* 2009]:

$$\text{Fe-Naphthenate} > \text{Ni-Naphthenate} > FeSO_4 > NiSO_4 > \text{No catalyst} \tag{8.2}$$

In the presence of catalysts, the authors reported that the amount of H_2 and CO increased significantly, while the content of H_2S decreased [Yufeng *et al.* 2009]. From asphaltene Aquathermolysis, gas product, saturates, aromatics, resin, and toluene insoluble (coke) were obtained. Resins were also converted into gas product, saturates, aromatics, asphaltene, and toluene insolubles. For these two fractions, H/C ratio decreased, and the molecular weights increased after the upgrading reaction. The authors proposed that asphaltenes and resins were partially aggregated by Aquathermolysis. However, in the presence of the catalysts, the authors observed less aggregation than that without the catalyst [Yufeng *et al.* 2009].

Chao and coworkers evaluated several oil-soluble metal catalysts for the Aquathermolysis of Shengli crude oil using 0.2 wt. % catalyst and 25.0 wt. % water at 280°C for 24 h [Chao *et al.* 2012a,b]. The catalysts evaluated were iron-, nickel-, cobalt-, and copper-naphthenate, and iron-, and copper aromatic sulfonates. Their results showed that the latter material was the most active of the six catalysts evaluated obtaining ~96% of viscosity reduction and 37% of asphaltene conversion (see Table 8.2) [Chao *et al.* 2012 a, b].

Chao *et al.* characterized the resins and asphaltenes by elemental analysis and H-NMR spectroscopy before and after Aquathermolysis. The authors found lower sulfur, nitrogen, and oxygen contents in the upgraded products than those measured in the virgin fractions. By H-NMR, they reported that the aromaticity and aromatic condensation decreased in the treated resins and asphaltenes in comparison with the starting materials. These results were attributed to the cracking of carbon–heteroatom bonds and hydrogenation reactions. These two-chemical processes were accelerated in the presence of a highly active aromatic sulfonic copper catalyst [Chao *et al.* 2012].

Galukhin and coworkers studied the upgrading of Ashal'cha crude oil (Tatarstan) in the presence of iron(III) tris(acetylacetonate) catalyst ($Fe(acac)_3$) using cyclohexane as solvent

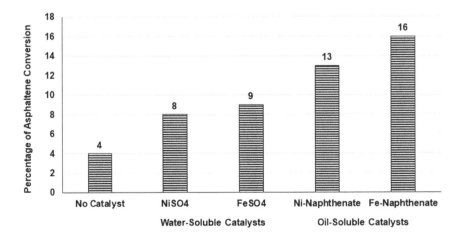

FIGURE 8.9 Percentage of conversion for the Aquathermolysis of asphaltenes in the presence of water- and oil-soluble metal catalysts. Data taken from Yufeng *et al.* [2009].

[Galukhin *et al.* 2015]. The experiments were carried out in an autoclave reactor using 70 g of oil samples and 30 g of water (oil/water ratio = 7:3) at 250°C, 3 MPa of nitrogen for 6 h. As seen in Table 8.2, the authors found lower API (an change of –1°API), higher viscosity (–22%), sulfur (–10% sulfur removal), and asphaltene content (–8%) after the upgrading tests. They attributed these results to the significant removal of light fractions during the dehydration step at the end of the Aquathermolysis experiments. However, the authors reported that the oil-soluble catalyst reduced the resin content from 28.8 to 18.6 wt. % and increased the aromatics from 43% to 51.3% [Galukhin *et al.* 2015]. In the presence of cyclohexane, the researchers found that $Fe(acac)_3$ forms magnetic nanoparticles (MNPs) during the Aquathermolysis runs without any addition of surfactants. The composition of MNPs was determined as a mixture of hematite, magnetite, and maghemite. The use of nanocatalyst for the subsurface upgrading of HO/B is discussed in the next section.

Tang and coworkers synthesized and evaluated an oil-soluble copper containing multicomponent acrylic copolymer (MAC) for the catalytic Aquathermolysis of Lukeqin heavy oil [Tang *et al.* 2015]. The catalyst was prepared by mixing MAC with benzoic acid (C_6H_5–COOH) in the presence of copper hydroxide at 90°C for 30 min. In this process, the following reaction occurred:

$$MAC - COOH + Cu(OH)_2 + C_6H_5 - COOH \rightarrow (MAC - COO^-)Cu(^-OOC - C_6H5) \qquad (8.3)$$

As shown in Table 8.2, the upgraded reactions were carried out at 240°C for 24h. Using the Cu(MAC)(BA) catalysts, the authors found 85% viscosity reduction and conversions of 10% of the resins and ~3% of the asphaltenes. After Aquathermolysis, the authors observed decreases in the long-chain saturated hydrocarbons and that the aromaticity and aromaticity condensation of the resins and asphaltenes decreased [Tang *et al.* 2015].

In general, the results presented in this section indicate that indeed oil-soluble catalysts have higher activity toward the Aquathermolysis reaction than their water-soluble counterparts. Both types of materials lead to lower viscosity in the upgraded products by catalyzing the cracking of C–S, C–O, and C–C bonds and the conversion of the asphaltene and resin fractions toward saturates and aromatics.

8.2.3 AMPHIPHILIC CATALYSTS

Water-soluble or oil-soluble catalysts only dissolve well in one phase, which indicates that the conversion and hydrogen transfer will not be efficient because most of the Aquathermolysis reactions take place at the oil–water interface [Guo *et al.* 2016]. Therefore, amphiphilic catalysts that combine the benefits of water-soluble and oil-soluble ones were proposed [Chen *et al.* 2008, Wang *et al.* 2014, Guo *et al.* 2016]. Furthermore, it is well known that resins and asphaltenes are active at the oil–water interface so an amphiphilic catalyst that is evenly distributed between the two liquid phases can maximize the conversion and hydrogen transfer to those two petroleum fractions with the concomitant improvement in the properties of the upgraded HO/B.

Table 8.3 shows several examples reported in the literature on the upgrading of HO/B via catalytic Aquathermolysis using amphiphilic catalysts and water as the hydrogen source. Chen and coworkers carried out the synthesis and spectroscopic characterization of an iron-aromatic sulfonate catalyst by reacting Fe or Fe_2O_3 with the sulfonic acid at 100°C for 2 h [Chen *et al.* 2008]. The upgrading of Henan heavy crude oil was performed in an autoclave at 200°C, 3 MPa for up to 72 h. As shown in Table 8.3, the evaluation of this iron-containing amphiphilic catalyst indicated that up to 91% of viscosity reduction was obtained with 15% conversion of heavy residue to light contents

Chen and coworkers also compared the iron-amphiphilic catalyst with other water- and oil-soluble analogs. Figure 8.10 shows the percentage of viscosity reduction for the Aquathermolysis Henan Crude oil in the presence of different types of catalysts at 200°C for 72 h [Chen *et al.* 2008].

TABLE 8.3

Catalytic Aquathermolysis Using Selected *Amphiphilic Catalysts and Water* as the Hydrogen Source

Catalyst	Feed Characteristics			Reaction Conditions			Results				References
	Crude Oil (Country)	°API	Viscosity (mPa s @ °C)[a]	Temp. (°C)	Press. (MPa)	Time	Incr. °API[b]	% Visc. Red.[c]	% DS[d]	% Asp. Conv.[e]	
Fe-Aromatic Sulfonate	Henan (China)	ND	28,867 (@50°C)	200	3	72 h	ND	91	ND	23	Chen et al. [2008]
Fe-Gemini- Sulfonate	Karamay (China)	ND	80,000 (@50°C)	170	3	24 h	ND	99	ND	55	Chen et al. [2010]
Fe(DBS)$_3$[f]	Heavy Crude oil (Egypt)	20.6	28.6 (@70°C)	250	ND	24 h	4.1	79	15.5	27	Desouky [2013]
Ni-(DBS)$_2$[f]				300	ND	24 h	3.9	73	18.5	25	
Fe(DBS)$_3$[f]	Shengli (China)	ND	8,957 (@80°C)	250	3	24 h	ND	45	ND	2	Wang et al. [2014]
Ni(DBS)$_2$[f]							ND	49	ND	11	
CoMo + NiMo)-PTS[g]	Heavy Oil (Oman)		1,490 (@70°C)	300	3	24 h	ND	95	15.6	17	Yusuf et al. [2016a]

ND = Not disclosed or not reported.

a Viscosity (μ) in mPa s measured at the given temperature.

b Increase in API gravity of the upgraded crude oil at 60°F. Negative values indicate increase with respect to the feed.

c Percentage of decrease in viscosity of the upgraded crude oil vs. the feed calculated as $((\mu_{feed} - \mu_{product}) - \mu_{feed}) \times 100$.

d Percentage of desulfurization in the upgraded crude oil vs. the feed calculated as $((\%S_{feed} - \%S_{product}) - \%S_{feed}) \times 100$.

e Percentage of reduction of asphaltenes in the upgraded crude oil vs. the feed calculated as $((\%Asp_{feed} - \%Asp_{product}) - \%Asp_{feed}) \times 100$.

f DBS = Dodecylbenzene sulfonate.

g P-toluene sulfonic acid.

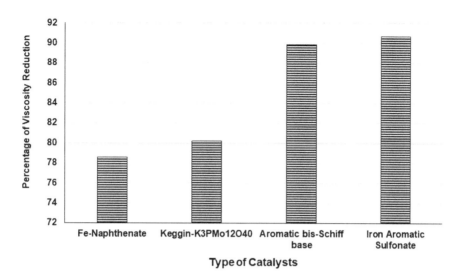

FIGURE 8.10 Percentage of viscosity reduction for the Aquathermolysis Henan crude oil in the presence of different types of catalyst at 200°C for 72 h. Data taken from Chen *et al.* [2008].

As seen, the amphiphilic catalyst led to the highest viscosity reduction of the ones evaluated. The order of reactivity was the following:

$$\text{Fe-Aromatic Sulfonate} > \text{Aromatic Schiff base} > \text{Keggin-K}_3\text{PMo}_{12}\text{O}_{40} > \text{Fe-Naphthenate} \qquad (8.4)$$

Unfortunately, the authors did not study the interfacial activity nor the critical micellar concentration to understand the higher activity of these amphiphilic catalysts in comparison with the water- and oil-soluble counterparts.

Chen and coworkers carried out the characterization of the upgraded crude oils by SARA analysis [Chen *et al.* 2008]. The results showed that, for the blank experiment without a catalyst, the resin contents decreased by ~2 wt. % and the asphaltenes by ~0.1 wt. %, while the saturate and aromatic fractions increased by ~0.9 wt. % and ~1 wt. %, respectively. In contrast, for the catalytic Aquathermolysis reaction, the resin content decreased by ~10 wt. % and asphaltenes by ~4 wt. % (23% reduction vs. feed, Table 8.3), while the saturate and aromatic fractions increased by ~8 wt. % and ~6 wt. %, respectively. As found for other water- and oil-soluble catalysts, these results indicate that the conversion and viscosity reduction occurred principally by decreasing the resin and asphaltene contents [Chen *et al.* 2008]. This amphiphilic catalytic system was field tested in the G61012, G6606 wells in the Henan Oilfield. The results are discussed in Section 8.5.

From the same research group, Chen *et al.* carried out the synthesis and spectroscopic characterization of an amphiphilic iron catalyst by reacting Fe_2O_3 with a Gemini sulfonate surfactant (GS) at 120°C for 0.5 h [Chen *et al.* 2010]. A GS consists of two conventional surfactant molecules chemically bonded together by a spacer. It has been reported that a GS can self-assemble at much lower concentrations and have higher in-surface activity as compared to conventional surfactants [Menger and Littau 1991, Menger and Keiper 2000].

As seen in Table 8.3, the evaluation of the Fe-Gemini sulfonate catalyst for the upgrading of Karamay crude oil led to a viscosity reduction of 99% at 170°C in 24 h [Chen *et al.* 2010]. This temperature is at least 30°C lower than those used for other iron-containing catalysts (see Tables 8.1., 8.2, and 8.3). Furthermore, the authors compared the Fe-Gemini catalyst with other oil- and water-soluble analogs, and they found the following order of reactivity (numbers in brackets are the percentages of viscosity reduction):

$$\text{Fe-Gemini catalyst} \left(99\%\right) > \text{Fe-naphthenate} \left(84\%\right) > \text{Fe}_2\left(\text{SO}_4\right)_3\left(69\%\right) > \text{FeCl}_3\left(66\%\right) \qquad (8.5)$$

As before, the authors did not carry out interfacial activity nor critical micellar concentration studies, so the understanding of the highest activity of the Gemini catalysts was not reported.

Also, Chen and coworkers evaluated the activity of the amphiphilic catalyst for four types of heavy oils from different regions in China (Du 813-39-36, G540, Zheng 411, and DF32025). The authors found viscosity reductions in the 92–99% range [Chen *et al.* 2010]. Finally, they carried out spectroscopic studies (FTIR and NMR, and elemental analysis) of the SARA fractions separated from the upgraded crude oils. They found that the aromaticity and aromatic condensation of asphaltenes and resins decreased after Aquathermolysis. The decrease of the former was attributed to the hydrogenation of unsaturated groups while the reduction in the last parameter was caused by the ring opening and reconstruction of the aromatic system by the amphiphilic catalysis [Chen *et al.* 2010].

In 2013, Desouky and coworkers synthesized two amphiphilic catalysts by reacting dodecylbenzene sulfonic acid (DBSH) with $FeCl_3$ or $NiCl_2$ at 85°C for 5 h [Desouky *et al.* 2013]. The compounds, iron (III) dodecylbenzene sulfonate ($Fe(DBS)_3$) and nickel (II) dodecylbenzene sulfonate ($Ni(DBS)_2$), were characterized by FTIR and UV spectroscopies and were evaluated for the catalytic Aquathermolysis of an Egyptian heavy crude oil in the 250–300°C temperature range for 24h [Desouky *et al.* 2013]. As shown in Table 8.3, the authors found increases the API gravity of the upgraded products of ~4°API, viscosity reductions of 73% (Ni) and 79% (Fe), percentages of hydrodesulfurization of ~15% (Fe) and ~19% (Ni), and reductions of the asphaltene content of 25% (Ni) and 27% (Fe) [Desouky *et al.* 2013].

The chemical and physical properties of heavy oils both before and after the reaction were investigated by FT-IR, dynamic viscosity, molecular weight, and SARA analysis. The results indicated that the contents of resins and asphaltenes, and average molecular weights of heavy oil were reduced after the catalytic Aquathermolysis. These results are consistent with the finding reported by other authors [Maity *et al.* 2010, Guo *et al.* 2016] and were attributed to the cracking of C–C and C–S bonds during the upgrading reactions [Desouky *et al.* 2013].

Desouky and coworkers determined the interfacial tensions of the dodecylbenzene sulfonate acid surfactant (starting material) and the amphiphilic catalysts $Fe(DBS)_3$ and $Ni(DBS)_2$. The authors found a value of 5 mN/m for the DBSH, 4 mN/m for the $Fe(DBS)_3$, and 3 mN/m for the $Fe(DBS)_3$, which indicate their high efficiency at the interface [Desouky *et al.* 2013]. Unfortunately, no comparisons with other water- or oil-soluble catalysts were carried out.

The authors also reported the critical micelle concentrations (CMC) of the amphiphilic catalysts and found that the nickel-containing material has a lower value (1.99×10^{-3} M) in comparison with iron analog (2.57×10^{-3} M) and DBSH (1×10^{-2} M). In general, a low CMC value indicates a higher surface activity of the compound in the bulk of the solution. The authors concluded that the $Ni(DBS)_2$ was the most surface-active compound at the interface. However, the $Fe(DBS)_3$ catalyst showed moderate surface activity [Desouky *et al.* 2013]. These results were attributed to the presence of two and three sulfonic molecules in the $Ni(DBS)_2$ and $Fe(DBS)_3$ amphiphilic catalysts which decreased the interaction between the metal-surfactant and the aqueous phase [Desouky *et al.* 2013].

Wang and coworkers evaluated the same two amphiphilic catalysts, $Fe(DBS)_3$ and $Ni(DBS)_2$, for the Aquathermolysis of a Shengli (China) heavy crude oil at 250°C for 24 h [Wang *et al.* 2014]. As shown in Table 8.3, the authors reported viscosity reductions of 45% (Fe) and 49% (Ni) with asphaltene conversions of 2% (Fe) and 11% (Ni). Consistent with the results disclosed by Chen *et al.* [2008, 2010], the amphiphilic catalyst was the most active of the ones studied [Wang *et al.* 2014]. The order of reactivity for the reduction of viscosity was the following (numbers in brackets indicate the percentage of viscosity reduction):

$$Ni(DBS)_2 \; (49\%) > Ni\text{-Naphthenate} \; (36\%) > NiSO_4 \; (29\%) > \text{No Catalyst} \; (25\%)$$

Amphiphilic Catalyst	Oil-Soluble Catalyst	Water-Soluble Catalyst	(8.6)

Wang and coworkers studied the water–Shengli heavy oil interfacial tension in the presence of the amphiphilic catalysts. At the concentration of 0.2 wt. %, Ni(DBS)$_2$ reduction of the interfacial tension from about 20 mN/m to 1.2 mN/m and Fe(DBS)$_3$ to 7.0 mN/m were observed. Unfortunately, no comparison with the oil- and water-soluble catalysts were made, nor were CMC studies carried out.

After the Aquathermolysis reaction, Wang *et al.* observed asphaltene conversions of 2–11% (Table 8.3) and reductions in the resin contents of 23–28% [Wang *et al.* 2014]. By FTIR, NMR, and elemental analysis, they found lower sulfur content, shorter length of alkyl chain, and aromaticity in the upgraded fractions than those observed in the virgin material. Based on their results, the authors concluded that the amphiphilic catalyst Fe(DBS)$_3$ was suitable for heavy oils with high sulfur and rather low asphaltene contents, whereas Ni(DBS)$_2$ was more active for naphthene-based crudes [Wang *et al.* 2014]. In other words, the matching between the catalyst type and composition of the heavy oil is crucial to the subsurface application of catalytic Aquathermolysis.

Yusuf and coworkers conducted the upgrading of Omani heavy oil under aquathermolytic conditions in the presence of the amphiphilic catalysts NiMo-PTS, CoMo-PTS (where PTS = p-toluene sulfonic acid), and their mixture (1:1 wt. ratio) at 300°C, 5 wt. % water concentrations for 24 h [Yusuf *et al.* 2016a]. The catalyst efficiency for viscosity reduction follows the trend (CoMo-PTS + NiMo-PTS) > NiMo-PTS > CoMo-PTS. As shown in Table 4.3, the authors reported a maximum of 95% using the CoMo + NiMo mixture. The upgraded-product distributions showed an increase in < C14 components and a reduction in > C21 fraction, which suggested cracking of longer hydrocarbon chains. By X-ray fluorescence, the researchers found a decrease in the sulfur content of up to 15.6% and confirmed the importance of C–S bond cleavage and hydrogenation in the reduction of heavy oil viscosity [Yusuf *et al.* 2016a].

Kayukova *et al.* carried out the physical simulation of catalytic Aquathermolysis of heavy crude oils from the Tatarstan Republic, Russia [Kayukova *et al.* 2018]. The catalytic processes were conducted at a temperature of 300–350°C in the presence of kaolin (to simulate the mineral formation) and catalysts composed of amphiphilic transition metal (Fe, Co, and Cu) carboxylates, and CO$_2$ and water [Kayukova *et al.* 2018]. Unfortunately, the physical properties (API gravity and viscosity) and elemental composition (H, C, and S contents) of the upgraded crude oils were not reported. The author concentrated on the detailed characterization of the SARA fraction by FTIR and GC/MS. As other authors have found [Chen *et al.* 2008, 2010, Desouky *et al.* 2013, Wang *et al.* 2014], the results showed increases of saturated fractions content and a decrease of resins and asphaltenes content in the products of experiments [Kayukova *et al.* 2018].

Based on the results presented in this section, it can be concluded that amphiphilic metal catalysts are more active for viscosity reduction, sulfur removal, and asphaltene conversions than the water- and oil-soluble counterparts. This enhanced catalytic activity toward Aquathermolysis was attributed to the higher tendency to distribute at the water-to-oil interface. In general, all types of metal catalysts lead to cracking of C–S, C–O, and C–C bonds and the conversion of the asphaltene and resin fractions toward saturates and aromatics.

8.2.4 Dispersed and Nanocatalysts

As described in Section 4.2.4, there has been considerable interest in the use of highly dispersed metals as nanocatalysts for the upgrading of heavy crude oils and bitumens (HO/B). According to ASTM, nanoparticles are a sub-classification of an ultrafine particle with lengths in two or three dimensions higher than 1 nm and smaller than about 100 nm [ASTM 2006]. Generally, metal-containing nanoparticles do not dissolve in water or oil but instead they disperse in either liquid phase. As the particle size drops into the nanoscale, the properties of the nanoparticles are different than the bulk material and are more affected by the behavior of the few surface atoms present. Also, nanosized catalysts offer the fundamental advantage of having a significantly superior surface-to-volume ratio in the absence of a pore structure. This feature increases the mass transfer of the substrate to the catalytic-active center considerably and simultaneously reduces the catalyst-deactivation issues [Pereira-Almao 2012]. Furthermore, nanoparticles are significantly smaller than the pores in the reservoirs which may benefit future applications for the subsurface upgrading of HO/B by making their injectivity less problematic [Hamedi Shokrlu and Babadagli 2010]. Thus, these nanostructured materials have the potential to generate vastly superior catalytic systems and at the same time, reduce the amount of metal used with the concomitant decrease in operating costs.

Table 8.4 shows several examples using highly dispersed metals as nanocatalysts for the Aquathermolysis of HO/B in the presence of water as the hydrogen source. Li and coworkers synthesized nickel nanoparticles using a methylcyclohexane-water-n-octanol microemulsion system and nona-ethylene glycol mono-dodecyl ether as a non-ionic surfactant at 30°C. By Transmission Electron Microscopy (TEM) the authors found that the particles were in spheroidal form, and the mean particle size was estimated as 6.3 nm [Li et al. 2007].

The Ni-nanoparticles were used for the viscosity reduction process of Liaohe extra-heavy oil by the Aquathermolysis reaction. The upgrading experiments were carried out in a stainless-steel batch reactor in the absence of a porous medium using 100 g extra-heavy oil, 10 mL of nickel microemulsion, and 50 g water under 6.4 MPa of nitrogen at 280°C for 24 h [Li et al. 2007]. As shown in Table 8.4, the authors reported that the viscosity of upgraded sample decreased from 139,800 mPa s to 2,400 mPa s at 50°C, i.e., ~99% viscosity reduction. Furthermore, they found that compared with the original crude oil sample, the average molecular weight of the upgraded products decreased, the content of sulfur changed from 0.45 wt. % to 0.23 wt. % (49% sulfur removal), and resin and asphaltene contents reduced 15.8% and 15.3%, respectively [Li et al. 2007]. Again, these findings are consistent with the breakage of C–S and C–C bonds as reported in the previous sections. However, no comparisons with other water- or oil-soluble metal catalysts were presented so the relative catalytic activity of the nanoparticles could not be determined.

By GC/MS, Li and coworkers found toluene in the light ends of the upgraded products. Thus, they proposed that the oil phase used to prepare the metal nanoparticles (methylcyclohexane) could be dehydrogenated by the Ni catalyst (Equation 8.7), which in turn could be used to upgrade the crude oil (Equation 8.8). This proposal means that the $C_6H_{10}CH_3$ solvent can act as a hydrogen donor [Li et al. 2007]. This topic is discussed in Section 8.2.3

$$(8.7)$$

$$\text{Crude Oil} \xrightarrow[\text{Ni}]{+[H]} \text{Upgraded Crude Oil} \qquad (8.8)$$

Wu et al. also studied the catalytic activity of nickel nanoparticles for the Aquathermolysis of a Chinese heavy oil (San56-13-19). The nanocatalyst was prepared by the microemulsion technique and characterized by TEM. The results showed that the size distribution of the nickel nanoparticles was homogeneous and that the average particle size was about 4.2 nm [Wu et al. 2013]. As shown in Table 8.4, using the Ni-nanocatalyst, the authors found a viscosity reduction of 90% at 200°C. Again, after the upgrading reaction, they reported that the resins and asphaltenes decreased by 6.5% and 5.3% (35% reduction), respectively, while the saturate and aromatic hydrocarbons increased by 6.7% and 5.5%.

Furthermore, it was found that the molecular weight of the asphaltenes decreases from 3,322 g/mol to 2,390 g/mol. The authors concluded that the conversion of asphaltenes played a crucial role in reducing the viscosity of the heavy oil [Wu et al. 2013]. Unfortunately, as with Li et al. [2007], no comparisons with other metal catalysts were studied, so the relative catalytic activity of the nanoparticles was not determined.

TABLE 8.4

Catalytic Aquathermolysis Using Highly Dispersed Metals as *Nanocatalysts and Water* as the Hydrogen Source

Catalyst (Part. Size)[a]	Feed			Reaction Conditions			Results				References
	Crude Oil (Country)	°API	Viscosity (mPa s @ °C)[b]	Temp. (°C)	Press. (MPa)	Time	Incr. °API[c]	% Visc. Red.[d]	%DS[e]	% Asp. Conv.[f]	
Ni (6.3 nm)	Liaohe (China)	9.4	139,800 (@50°C)	280	6.4	24 h	ND	99	49	15.8	Li [2007]
Ni (4.2 nm)	San56-13-19 (China)	ND	ND	200	ND	ND	ND	90	ND	35	Wu *et al.* [2013]
Ni(100 nm)	ND	15.1	8,500 (@25°C)	300	3	24 h	ND	30[g]	ND	5[g]	Hamedi Shokrlu and Babadagli [2013]
Ni-W-Mo Carbide (10 nm)	Heavy Oil (Mexican)	ND	1,130 (ND)	200	3	24 h	ND	97	ND	ND	Olvera *et al.* [2014]
CuO (< 50 nm)	Kohe Mond (Iranian)	13	16,000 (@7°C)	Steam (ND)	ND	12 h	ND	ND	ND	ND	Tajmiri and Ehsani [2016]
NiO/PdO/SiO₂(1.3–2.2 nm)	EHO (Colombian)	7.2	37,200 (@25°C)	Steam (ND)	ND	ND	4.9	59	ND	40	Franco *et al.* [2016]
Fe₂O₃(200 nm)	Ashal'cha (Tatarstan)	16.5	ND	375	20	3.5 h	ND	52[g]	36	53	Lakhova *et al.* [2017]
γ-Al₂O₃(40 nm)				350	11	ND	ND	−22[g]	29	58	
Ni(40–70 nm)	Heavy Oil (Mexican)	10.8	45,590 (@25°C)	220	ND	~7 h	ND	50[g]	ND	6	Yi *et al.* [2018]

ND = Not disclosed or not reported.

a   Catalyst composition and particle size (in nm) by Transmission or Scanning Electron Microscopy.

b   Viscosity in mPa s measured at the given temperature.

c   Increase in API gravity of the upgraded crude oil. Negative values indicate increase with respect to the feed.

d   Percentage of decrease in viscosity (μ) of the upgraded crude oil *vs.* the feed calculated as $((\mu_{feed} - \mu_{product}) - \mu_{feed}) \times 100$. Negative values indicate increase with respect to the feed.

e   Percentage of desulfurization (%DS) in the upgraded crude oil *vs.* the feed calculated as $((\%S_{feed} - \%S_{product}) - \%S_{feed}) \times 100$. Negative values indicate increase with respect to the feed.

f   Percentage of reduction of asphaltenes in the upgraded crude oil *vs.* the feed calculated as $((\%Asp_{feed} - \%Asp_{product}) - \%Asp_{feed}) \times 100$.

g   With respect to the control run in the presence of only water but *no catalyst* (thermal experiment).

Hamedi Shokrlu and Babadagli did compare the catalytic activity of nickel nanoparticles with a conventional Raney nickel catalyst (85 wt. % Ni, 2 wt. % Al, and 9 wt. % water) for the upgrading of heavy crude oil (15°API) under Aquathermolysis conditions [Hamedi Shokrlu and Babadagli 2013]. These authors purchased the nickel nanoparticles from a commercial vendor and characterized them by scanning electron microscopy (SEM). The authors reported that the nanomaterial was agglomerated with a particle size of 100 nm. Thus, a water-suspension of nickel nanoparticles was prepared using Xanthan gum polymer as a stabilizing agent and employed as a catalyst for Aquathermolysis. The upgrading experiments were carried out in a batch reactor in the absence of a porous medium with a crude oil-to-water ratio of 1:2 w/w at 300°C under 3 MPa of N_2 for 24 h. As shown in Table 8.4, using 500 ppm of Ni-nanocatalyst, the authors found a viscosity reduction of 30% in comparison with the experiment without catalyst (thermal test). Conversely, in the presence of the conventional Raney nickel catalyst, the authors reported a viscosity reduction of 22%. Furthermore, the authors found that the Ni-nanoparticles reduced the asphaltene content by 5% whereas the Raney nickel decreased by only 1.8% with respect to the thermal experiment. Hamedi Shokrlu and Babadagli concluded that nickel nanomaterial was a better catalyst than the conventional Raney nickel under batch conditions [Hamedi Shokrlu and Babadagli 2013].

Next, these authors studied the catalytic effect of nickel nanoparticles in the presence of sand pack as a porous medium (6 D) at 300°C under 6 MPa of N_2 for different time spans (6, 12, 24, and 36 h) [Hamedi Shokrlu and Babadagli 2013]. The catalyst was dispersed in deionized water using Xanthan gum polymer, mixed with the sand, and dried. Next, the sand pack was saturated with heavy oil, and water was added to obtain a water/oil ratio of two. The system was kept at the constant temperature for the duration of the test. In the end, the oil and water were separated and sent for analysis. A control experiment was carried out in the absence of the nickel nanoparticles. As can be seen in Figure 8.11, the authors found that a viscosity reduction of ~44% in the presence of catalysts vs. the control experiment at 6 h of reaction time. Similarly, after 36 h, the viscosity reduction was 39%. Consistent with those results the H_2S generation (1213 mL/Kg of crude oil), light hydrocarbons (565 mL/Kg of crude oil), and the recovery factor (67%) were higher in the presence of the nanocatalyst than those measured in the thermal-only experiment (767 mL/Kg of oil, 30 mL/Kg of crude oil, and 62%, respectively) [Hamedi Shokrlu and Babadagli 2013].

The same authors carried out a kinetic study using oil, water, and sandstone in the presence and absence of the nickel nanoparticles [Hamedi Shokrlu and Babadagli 2014a]. Eighteen experiments were conducted at three different temperatures (240, 270, and 300°C) and lengths of time (1, 3, and 5 days). The temperature at which the difference between the rate of the reaction of the

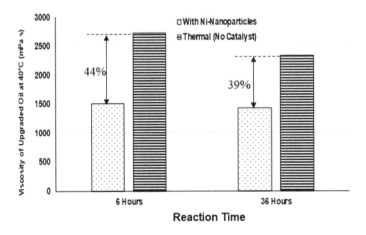

FIGURE 8.11   Viscosity of the upgraded crude oil (at 40°C) after the Aquathermolysis experiments in the presence and absence of nickel nanoparticles. Data taken from Hamedi Shokrlu and Babadagli [2013].

catalyzed and the thermal runs was a maximum was 270°C. The kinetic analysis showed that nickel nanoparticles reduced the activation energy of the Aquathermolysis reactions from 69 to 38 kJ/mol, confirming the catalytic effect of the nanomaterial for the upgrading of HO/B [Hamedi Shokrlu and Babadagli 2014a].

Using a displacement cell, three injections/soaking/production cycles were carried out at ~4 MPa of steam pressure (~250°C) with a sand pack having a permeability of 4.2 D and porosity of 32% [Hamedi Shokrlu and Babadagli 2013]. The results showed that the nickel nanoparticles increased the recovery factor by approximately 22% when the nanomaterial was injected with a cationic surfactant (cetyl-trimethylammonium bromide) and xanthan gum polymer. This increase in recovery was about 7% higher than that of the experiment conducted with the surfactant and polymer only. Based on their results, the authors concluded that the nickel nanoparticles are active catalysts for C–S and C–C bond breakage reactions, generating CO, H_2S, and lighter hydrocarbons (Equation 8.9), which in turn led to increasing the recovery factor [Hamedi Shokrlu and Babadagli 2013, 2014a]. As reported by Hyne [Hyne 1986], the produced CO reacts with water via water–gas-shift reaction (Equation 8.1) to generate hydrogen, which in turn can be used to upgrade the heavy oil.

$$\text{Heavy Oil} + H_2O \xrightarrow{\text{Ni}} CO + H_2S + \text{Lighter Hydrocarbons} \qquad (8.9)$$

Hamedi Shokrlu and Babadagli also studied the injectivity and transport of the micron- and nano-sized nickel particles during travel through a porous media [Hamedi Shokrlu and Babadagli 2013]. The injected nickel-containing suspensions were stabilized by use of xanthan gum polymer and ultrasonication. The particle stability was determined using zeta-potential measurements at different pH values. The results of the injectivity and transport tests confirmed that the nanoparticles had lower retention and better stability than the micron-sized counterparts. Also, the authors found that at the range of metal particles concentration required for catalytic Aquathermolysis (0.1–0.5 wt. %), only a small change in thermal conductivity was observed with micron-sized particles. However, the use of nanoparticles resulted in ~25% faster heat transfer to the oil [Hamedi Shokrlu and Babadagli 2014b].

Olvera et al. carried out the catalytic Aquathermolysis of Mexican heavy oil (Table 8.4) in the presence of nickel-tungsten-molybdenum carbides using water as a source of hydrogen at 200°C for 24 h [Olvera et al. 2014]. The catalysts were mixed and mechanically processed at room temperature for different grinding times: 0, 40, 80, 120, 160, 200, and 240 h. The authors found that, as the milling time increased, the percentage of the viscosity reduction of the heavy oil also increased (from 80% to 97%). X-ray diffraction showed that the nanostructured metallic carbide phases were synthesized and high-resolution TEM indicated that the carbides had crystallite size ranging from 10.1 to 125.6 nm. Again, the authors reported that the viscosity reductions are consistent with changes in the contents of resins and asphaltenes during the catalytic Aquathermolysis reaction [Olvera et al. 2014].

Tajmiri and coworkers carried out two-dimensional SAGD displacement tests of Iranian heavy oil cores (Table 8.4) in the presence and absence of copper oxide nanoparticles (< 50 nm) at steam temperatures for 12 h [Tajmiri and Ehsani 2016]. The experimental results showed that by adding nanomaterial, the ultimate oil recovery increased from ~52% to 80% vs. OOIP, and the steam-to-oil ratio decreased by 58% in comparison with the steam-only test. Unfortunately, the properties of upgraded crude oils and the characterization of the catalyst before and after the Aquathermolysis reaction were not disclosed [Tajmiri and Ehsani 2016].

Franco et al. also carried out 1-D displacement tests of a Colombian heavy oil in the absence and presence of 2 wt. % of NiO and PdO supported on nanostructured silica (1.3–2.2 nm) at 300°C [Franco et al. 2016]. The authors observed that the oil recovery increased up to 46% for the system assisted by nanoparticles in comparison with the experiment without the nanocatalyst. The API gravity of upgraded crude oil increased from 7.2 to 12.1°API with 59% viscosity reduction

(Table 8.4). Similarly, the asphaltene content decreased by 5.2 wt. % (~40% reduction) whereas the residue content (620°C+) reduced 47% after vapor injection in the presence of nanoparticles in comparison with the virgin EHO crude oil. The authors attributed the increase in oil recovery during the nanoparticle injection to the following three reasons: i) wettability alteration of the porous media, ii) viscosity reduction due to the reduction of the asphaltene content, and iii) increase in the amount of distillable material [Franco et al. 2016].

As shown in Table 8.4, Lakhova and coworkers used 6 wt. % of Fe_2O_3 (Hematite with a particle size of 200 nm) as a catalyst for the Aquathermolysis of a Tatarstan heavy oil (16.5°API) at 375°C and 20 MPA for 3.5 h [Lakhova et al. 2017]. The authors reported a 52% viscosity reduction of the upgraded oil in comparison with the experiment carried out in the absence of a catalyst. Also, they found 36% of desulfurization, and the asphaltene content decreased from 7.7 to 3.6 wt. %, and the low boiling fraction and saturated hydrocarbons increased by 30% and 10.5%, respectively [Lakhova et al. 2017].

Using a 6 wt. % of Al_2O_3 (particle size of 40 nm), Lakhova et al. observed the most significant reduction in asphaltene content (58%) and reduction of sulfur-containing compounds of 29%. However, due to the loss of volatiles during the upgrading reaction, an increase of 22% in the viscosity of the upgraded product was observed in comparison with the experiment without catalyst [Lakhova et al. 2017].

Yi and coworkers conducted cyclic steam stimulations displacement tests using a sand pack saturated with Mexican heavy oil (10.8°API) in the presence of nickel nanoparticles (40–70 nm) at temperatures of 220°C for ~7 h [Yi et al. 2018]. Experimental results showed that the best concentration of nickel nanomaterial that gives the highest ultimate oil recovery factor (38% vs. OOIP) was 0.20 wt. %. At a lower temperature (150°C) a much lower recovery factor (14%) was obtained in comparison to the results at 220°C. The authors attributed this finding to a lower level of Aquathermolysis reactions at 150°C [Yi et al. 2018].

By GC and SARA analyses of produced oil and gas samples, Yi et al. confirmed that, during the Aquathermolysis reaction, the principal reaction mechanism was the breakage of the C–S bond [Yi et al. 2018]. The nickel nanoparticles acted as a catalyst for the upgrading process, but the catalytic effect became less remarkable from cycle to cycle. The displacement test aimed to evaluate the impact of particle-penetration depth and revealed that the nickel nanomaterial mainly distributed near the injection port, which in turn significantly contributed to increasing the ultimate recovery factor [Yi et al. 2018].

From the results discussed in this section, it can be concluded that highly dispersed metals and metal oxides are active nanocatalysts for the subsurface upgrading of heavy oils and bitumens via the Aquathermolysis reaction. This research area is a relatively new topic and has good potential for commercial applications. As with the other types of catalysts discussed earlier (i.e., water- and oil-soluble and amphiphilic), the reaction mechanism involves the cracking of C–S, C–O, and C–C bonds and the conversion of the asphaltenes and resins to saturates and aromatics. However, more work is needed for understanding the relative activity of nanocatalysts vs. micron-sized or soluble counterparts. Also, the traveling and absorption of nanoparticles through different porous media under downhole conditions still have some uncertainties that need to be addressed.

8.3 CATALYTIC USE OF HYDROGEN GAS

In this section, hydrogen gas is used as the hydrogen source for the subsurface upgrading of heavy crude oils and bitumens in the presence of catalysts and an energy source such as steam or other [Guo et al. 2016]. The chemistry and mechanisms of the two main pathways involved in this route are Hydrotreatment and Hydroprocessing (see Section 4.2.3). These processes are aimed at increasing the hydrogen-to-carbon molar ratio of the upgraded products by carrying out the catalytic hydrogenation (Section 4.2.2.2), hydrodesulfurization (Section 4.2.3.1), hydrodemetallization (Section 4.2.3.3), and hydrodeoxygenation (Section 4.2.3.4) reactions. In this way, lower viscosity products with higher value and stability are obtained. Also, in catalytic hydrogen addition, no low-cost and difficult to dispose of byproducts (e.g., coke, pitch, etc.) are generated, which in turn decreases the environmental impact of this SSU route.

On the other hand, the use of catalytic systems downhole for the activation of hydrogen gas suffers several disadvantages. As mentioned in Section 8.1, the downhole mixing of the catalysts with the HO/B and the difficulties in maintaining the activity of the catalytic solids represent significant challenges. Secondly, low solubility and poor mixing between $H_2(g)$ and the heated heavy crude oil are expected in the presence of the porous media under subsurface conditions. Thirdly, there is a need to generate hydrogen gas on site, which increases the CAPEX and OPEX of the process [Speight 2007, Chapter 22, Gray 2015, Chapter 12]. Finally, the recovery of $H_2(g)$ from the produced gases to enable recycling of this valuable material adds additional complexity and cost to the SSU process.

In general, it was reported that hydrogen gas is relatively inactive even in the presence of catalysts [Gray 2015]. In refining, the operating pressures of Hydrotreatment and Hydroprocessing are in the 14–20 MPa range (2,000–3,000 psi) to increase the rate of hydrogen addition significantly. However, HO/B are typically found in relatively shallow reservoirs at pressures between 0.8 and ~10 MPa (see Table 2.1 and 2.2). Thus, an SSU process that uses $H_2(g)$ as the source of hydrogen should operate under those conditions and for environmental and safety considerations cannot exceed the reservoir fracture pressure. In Canada, the typical depth at which oil-sands deposits are found allows operation up to ~5 MPa maximum [Pereira-Almao 2012]. If steam were used to introduce heat in the reservoir, the temperature limit for steam injection would be 280°C. Deeper bitumen or extra-heavy oil reservoirs would allow operating at even higher pressure, which considerably favors the Hydroprocessing reactions [Pereira-Almao 2012].

In the next section, the description of the early experiments of using H_2 as a hydrogen source for the SSU of HO/B is presented, followed by the discussion of the process developed at the University Of Calgary. This section finishes with a literature review of lab experiments carried out by other research groups.

8.3.1 EARLY EXPERIMENTS

In 1989, Stapp performed batch experiments for the upgrading of Cat Canyon crude oil (10.7°API) in the absence and presence of commercial heterogenous and water-soluble catalysts at 343°C under 13.7 MPa of hydrogen (2,000 psi) for three days [Stapp 1989]. All experiments were carried out with 10 wt. % of brine. Even though Stapp's temperature was much higher than those potentially achievable in HO/B reservoirs, the results represent a proof-of-concept of using hydrogen gas for the subsurface upgrading of heavy crude oil and bitumens.

As seen in Table 8.5, Stapp reported increases in the API gravity from 3.3°API (control reaction) to ~12°API by using a commercial hydrogenation catalyst (Shell 324). Also, the percentages of viscosity reduction and hydrodesulfurization were higher than those found in the absence of the solid (99.7% vs. 97% and 45%, vs. 21%, respectively). The use of a water-soluble molybdenum catalyst achieved lower degrees of upgrading with only 6°API increase, 98.9% of viscosity reduction, and 36% of HDS [Stapp 1989].

Stapp characterized the upgraded crude oils by elemental and ^{13}C-NMR analyses. The results indicated that the Shell 324 catalyst led to an increase of the H/C molar ratio from 1.508 (feed) to 1.679 [Stapp 1989]. Under the same conditions, the reaction carried out in the absence of the catalysts (control) gave an H/C molar ratio of 1.583. Similarly, the author found that the percentage of aromatic carbon by ^{13}C-NMR decreased from 24.9% (feed) to 24%. These results indicated that hydrogenation reactions indeed occurred for the run with Shell 324 catalyst. On the other hand, the water-soluble molybdenum catalyst gave only a slightly increased H/C molar ratio (1.589) which is consistent with the lower activity of this compound in comparison with the commercial counterpart [Stapp 1989].

TABLE 8.5

Catalytic Upgrading of Heavy Crude Oil and Bitumens Using *Hydrogen Gas* as the Hydrogen Source

	Feed Characteristics			Reaction Conditions			Results				
Catalyst	Crude Oil (Country)	°API	Viscosity (mPa s @ °C)[a]	Temp. (°C)	H$_2$ Press. (MPa)	Time	Incr. °API[b]	% Visc. Red.[c]	% HDS[d]	% Asp. Conv.[e]	References
-(Control)	Cat Canyon (USA)	10.7	2,955 (@54°C)	343	13.7	3 d	3.3	97.0	21	ND	Stapp [1989]
Commercial Catalyst							11.7	99.7	45	ND	
Ammonium Molybdate							6.0	98.9	36	ND	
-(Control)	Cold Lake (Canada)	ND	2,540 (@39°C)	400	18.6–20.7	1 h	ND	84.0	11	24	Clark and Kirk [1994]
RuCl$_3$								87.0	20	42	
-(Control)	Middle Eastern	14.4	ND	300	10.3	2 d	-0.8	ND	30	3	Weissman and Kessler [1996]
Co-Mo over Al$_2$O$_3$						1 d	3.0		59	47	
Ammonium Molybdate							0.4		28	23	

ND = Not disclosed or not reported.

[a] Viscosity (μ) in mPa s measured at the given temperature.

[b] Increase in API gravity of the upgraded crude oil at 60°F.

[c] Percentage of decrease in viscosity of the upgraded crude oil *vs.* the feed calculated as $((\mu_{feed} - \mu_{product}) - \mu_{feed}) \times 100$.

[d] Percentage of hydrodesulfurization in the upgraded crude oil *vs.* the feed calculated as $((\%S_{feed} - \%S_{product}) - \%S_{feed}) \times 100$.

[e] Percentage of reduction of asphaltenes in the upgraded crude oil *vs.* the feed calculated as $((\%Asp_{feed} - \%Asp_{product}) - \%Asp_{feed}) \times 100$.

Furthermore, Figure 8.12 shows the effect of the temperature on the API gravity and the percentage of desulfurization of Cat Canyon crude using the commercial catalyst under 13.7 MPa of hydrogen after ten days (288°C), three days (343°C), and one day (371°C) [Stapp 1989]. As expected, the °API and %HDS increased with the temperature reaching values of 25.6°API and 52% HDS at 371°C. At 288°C, the upgraded product showed similar °API values and percentages of aromatic carbon as those found in the feed. Thus, the author concluded that at this temperature, there was not a significant hydrogenation reaction [Stapp 1989].

Clark *et al.* conducted the upgrading of Cold Lake and Peace River crude oils (crude/water = 4:1 w/w) under batch conditions using an aqueous ruthenium catalyst at 400°C, 18.6–20.7 MPa of H_2 for 1 h [Clark and Kirk 1994]. As seen in Table 8.5 and for the Cold River crude, the authors found higher viscosity reduction (87% vs. 84%) and percentages of HDS (20% vs. 11%) and asphaltene conversion (42% vs. 24%) in the presence of the catalyst than those measured for the control experiment (no catalyst). Similar results were obtained for the Peace River feed [Clark and Kirk 1994]. The authors rationalized these observations on the basis that the ruthenium catalyst promoted C–S bond cleavage and that those free radicals were capped by reaction with gas phase hydrogen instead of undergoing retrogressive coupling processes. Additionally, ruthenium sulfides, produced *in-situ* by reaction of H_2S with aqueous ruthenium species, would directly catalyze desirable hydrogenation reactions [Clark and Kirk 1994].

In 1996, Weissman and coworkers studied the Hydroprocessing of heavy crude oil using solid and water- and oil-soluble catalysts under batch conditions at 300°C, 10.3 MPa of H_2 for one to two days [Weissman and Kessler 1996]. All experiments contained a brine solution and 20/40 mesh quartz sand to simulate downhole conditions. As shown in Table 8.5, the run carried out in the presence of a Co-Mo/Al_2O_3 heterogeneous catalyst using a Middle Eastern crude led to an increase of 3.0°API with 59% HDS and 47% asphaltene conversion. Similarly, the authors found 48% residue conversion and 32% of nitrogen removal. Those results were higher than those observed for the control run [Weissman and Kessler 1996].

On the other hand, none of the oil- or water-soluble additives (Fe, Co, Mo, or combinations of additives tested) were effective in increasing the API gravity of upgraded products or removing

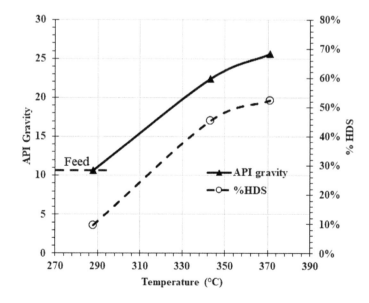

FIGURE 8.12 Effect of the temperature on the API gravity and the percentage of desulfurization for the upgrading of Cat Canyon feed (10.7°API). Experiments carried out using a commercial catalyst under 13.7 MPa of hydrogen (2,000 psi) after ten days (288°C), three days (343°C), and one day (371°C). Data taken from Stapp [1989].

sulfur from the crude beyond that observed using thermal processing conditions. As an example, Table 8.5 shows the result only for the water-soluble ammonium molybdate catalyst [Weissman and Kessler 1996]. However, these catalytic systems were active for asphaltene conversion as reported in Sections 8.2.1 and 8.2.2.

Weissman and coworkers performed several experiments to determine the kinetics of heavy oil upgrading using a conventional Hydroprocessing pilot plant, which was modified to allow water co-injection [Weissman and Kessler 1996]. Five runs were completed at reaction temperatures from 250 to 325°C, an oil-to-water ratio of 1:1 w/w, 1,500 psi of H_2, and a linearly hourly space velocity (LHSV) from 0.6 to 2.0 h^{-1}. The authors found that the hydrodesulfurization reaction proceeded by pseudo-first-order kinetics and that the density reduction increased with increasing reaction temperature. As expected, optimum reactivity occurred at the highest hydrogen partial pressures, although upgrading is still observed at lower H_2 pressures [Weissman and Kessler 1996].

Besides the Middle Eastern crude oil, Weissman and coworkers evaluated on- and off-shore Californian (12°API and 21°API, respectively), Alberta (12.5°API), and Venezuelan (10°API) crudes. The authors observed wide variability in oil reactivity with some oils being almost non-reactive while others showed significant thermal and catalytic upgrading. They could not find any correlation between the oil properties and the amounts of upgrading obtained. The authors concluded that each crude oil needs to be evaluated in the laboratory to show the application of subsurface Hydroprocessing to a particular heavy oil reservoir [Weissman and Kessler 1996].

8.3.2 University of Calgary Process

Since 2005, the University of Calgary has been particularly active in the area of subsurface upgrading of heavy crude oils and bitumens using metal dispersed nanocatalysts. They worked with a consortium of several Canadian and international oil companies to develop a process called *In-Situ Upgrading Technology* (ISUT), or Residue Assisted *In-Situ* Catalytic Upgrading (RAISCUP) [Pereira 2011, Pereira-Almao 2012, Pereira-Almao *et al.* 2015a, 2015b, Scheele-Ferreira *et al.* 2017]. The University of Calgary process has not been field tested, but at the moment of writing this book, several pilot tests are being planned. Those were in the Aguacate field (fracture carbonate reservoir) and Samaria oil-sands in Mexico, a Colombian wet oil-sands, and a Canadian thin pay reservoir.

Firstly, Pereira-Almao and coworkers carried out thermodynamic calculations on the relative reactivity of SARA fractions toward hydrogen production, hydrogen transfer (Section 4.2.1.2 and 4.4.2), hydrocracking (Section 4.2.2), and hydrotreating (Section 4.2.3) [Pereira-Almao 2012]. The principal finding was that the hydrogenation of aromatics, resins, and asphaltenes could be achievable at a relatively low temperature (< 350°C). Also, they reported that Hydrocracking is feasible for all heavy molecules within the 200–300°C range, but the large aromatics would need to be first hydrogenated before being cracked to produce distillates. From the thermodynamic study, the author concluded that a feasible option for subsurface upgrading would be one in which hydrogen could be effectively inserted to the aromatic molecules, followed by molecular cracking in the presence of hydrogen [Pereira-Almao 2012].

For surface upgraders, temperatures and pressures in the order of < 350°C and 4 MPa (580 psi) are not practical because the vessel size would become significantly large, resulting in dramatic cost escalations and operational problems. Thus, the Calgary group focused on increasing residence time to compensate for lower temperature and pressure conditions [Galarraga and Pereira-Almao 2010a]. At reservoir level, large chamber volumes and long residence times are naturally available. For example, estimated residence times for steam in SAGD are in the order of 48 hours, whereas for injected liquids they are more than 100 hours [Pereira-Almao *et al.* 2010].

8.3.2.1 Description of the Process

Figure 8.13 shows the general scheme of the ISUT process as patented by University of Calgary [Pereira-Almao *et al.* 2015a]. As seen, the atmospheric or vacuum residues are separated at the

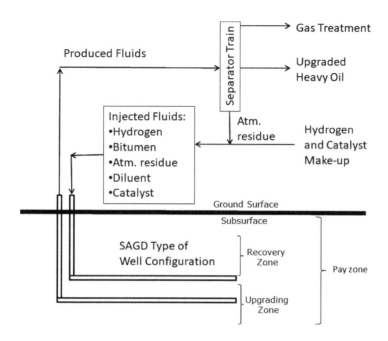

FIGURE 8.13   General scheme of the *In-Situ* Upgrading Technology (ISUT) patented by University of Calgary.

surface and used as a heat carrier and for the downhole injection of the ultradispersed nanocatalyst into the reservoir, along with hydrogen, makeup catalyst, and heated diluent. The authors stated that pre-mixing of hydrogen with injected residue before the injection is necessary to protect it against premature coking. Furthermore, pre-mixing hydrogen improves the effectiveness of the catalytic processes because mass transfer issues would not limit kinetics considerations. Hydrogen would have the time to saturate the residue during transportation toward the injection well [Pereira-Almao *et al.* 2010, 2015a].

The preferred configuration of the ISUT process is a two-well SAGD array. The injected fluids are pumped down through the top well to heat the reservoir to 350 ± 20°C. These hot fluids create a chamber that propagates upwards as well as laterally. The authors stated that the exothermic upgrading reactions also provide thermal energy to the upgrading zone, thereby minimizing energy consumption and natural gas usage on the surface [Pereira *et al.* 2010]. By injecting the residue fraction, *in-situ* thermal cracking/upgrading reactions are occurring within the formation and affect the overall efficiency of the process as the heavy oil fractions are most reactive [Pereira-Almao *et al.* 2015a]. The authors proposed that the nanocatalyst is contained within the reservoir and remains active for months to a few years (see Section 8.3.2.2). Under steady state conditions, the residence time for the injected residue may vary between approximately 24–2,400 h (estimated upper limit about 500 h), depending on the chamber volume and the porosity and permeability of the porous media. For these reasons, the top zone of the chamber is called the recovery zone whereas the bottom half is the upgrading zone [Pereira-Almao *et al.* 2010, 2015a].

As in SAGD, the produced fluids are recovered through the bottom well (see Figure 8.13) and contain the upgraded crude oil, excess hydrogen, unconverted bitumen, and atmospheric residue, diluent (if used), other gases, and unretained catalyst. The authors claimed that the upgrading at these moderate conditions produces oil with < 300 cP of viscosity (at 30°C) and 14–15 °API as well as sulfur removal and reduction of the micro-carbon residue value [Galarraga and Pereira-Almao 2010a]. The results of the lab experiments and numerical simulations are presented in Section 8.3.2.3.

The authors claimed that RAISCUP has the advantage that the recovered oil contains much less water than that produced with conventional steam injection processes [Pereira-Almao *et al.* 2015a].

Accordingly, injecting hot residue could eliminate the downhole injection of water, such that the only water in the reservoir is connate water. As a result, water treatment and/or water disposal costs are eliminated or substantially reduced. However, during start-up, steam could be injected into the injection well to begin growing the chamber during the start-up phases, in which case the steam is then progressively replaced with hot residue over time. Thus, during start-up, water treatment and recovery may be required. The authors suggested that the selection of either steam and/or heated oil to achieve connectivity between the top and bottom wells is based on the specifics and the economics of each project [Pereira-Almao et al. 2015a].

Finally, the authors proposed that hydrogen gas could be produced on site by small steam-methane reforming plants and distributed via pipeline to the producing pads [Pereira-Almao et al. 2010]. Alternatively, small-scale alkaline electrolysis units could be employed to produce hydrogen gas from water. However, hydrogen generation could represent a significant CAPEX and OPEX so the economics of the project should be carefully considered. This point is discussed in Section 8.3.2.4.

8.3.2.2 Catalyst and Flow Through the Porous Media

The general use, preparation, and typical residue conversion of slurry or highly dispersed catalysts and processes were discussed in Section 4.2.4. Even though catalytic subsurface upgrading has been under investigation since the pioneering work by Stapp in 1989 [Stapp 1989], the proper introduction of the catalyst into the reservoir has been a significant issue for any commercial application. The dispersion of the catalyst in a hydrocarbon feed is one of the most promising methods of introducing catalytic materials into reservoirs. The preparation of the metal nanocatalyst used in the ISUT process has been disclosed in four US patents [Pereira-Almao et al. 2011, 2012a, 2012b, 2012c] and several publications [Galarraga and Pereira-Almao 2010a, 2010b, Rodriguez-DeVecchis et al. 2015, Scott et al. 2015, Scheele-Ferreira et al. 2017]. As shown in Figure 8.14, the general procedure for the preparation of the metal nanocatalyst consists in the emulsification of Athabasca bitumen with aqueous solutions of Fe, Ni, Mo, and W salts under high-shear mixing. Ammonium sulfide and thiourea were used as sulfiding agents and ammonium hydroxide for activation of natural surfactants present in the bitumen [Scheele-Ferreira et al. 2017]. Then, the oil-in-water emulsion was thermally decomposed at 300°C, and the catalysts were characterized by elemental analysis, X-ray photoelectron spectroscopy, and particle size determinations [Rodriguez-DeVecchis et al. 2015].

The results showed the particle sizes were no bigger than 100 nm, and ammonium sulfide gave the smallest metal particles whereas thiourea was a more efficient sulfiding agent. Interactions between Ni and Mo in the obtained solids were confirmed through the presence of the NiMoS phase on the catalyst surface. Conversely, tungsten sulfiding was not obtained for any of the preparations [Scheele-Ferreira et al. 2017].

Loria et al., also from University of Calgary, developed a 3-D convective/dispersive mass transfer model to simulate the transport and settling of nanocatalysts through pipelines and vessels using

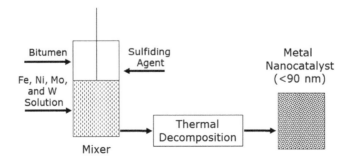

FIGURE 8.14 General procedure for the preparation of the metal nanocatalysts used in the ISUT process from the University of Calgary.

a deep-bed-filtration process [Loria *et al.* 2009a, 2009b]. This model enabled the prediction of particle concentration profiles and the determination of conditions under which sedimentation would occur. The model was experimentally validated and showed that particles less than 200 nm in diameter did not significantly deposit within a pipeline [Loria *et al.* 2009a, 2009b].

In the following work, Zamani *et al.* conducted lab experiments to examine the propagation of sub-micrometer-sized catalyst particles suspended in oil into a sand pack [Zamani *et al.* 2010]. The results showed that it was possible to propagate the catalyst suspension through the porous media. However, a fraction of the catalyst particles was retained by the sand (~14–18%), and much higher retention occurred in the entrance region of the bed. Particles appeared to be deposited on sand surfaces by an attachment mechanism, but larger particles were strained by mechanical trapping near the inlet face. The deposition of particles was found to be almost irreversible and could not be remobilized from the sand by reverse flow of the suspending medium [Zamani *et al.* 2010].

Rodriguez-DeVecchis and coworkers from the University of Calgary studied the effect of temperature and residence time on the deposition of nanoparticles of Ni, Mo, and W dispersed in an oil media over a sand pack [Rodriguez-DeVecchis *et al.* 2018]. High particle retention (over 95%) was obtained in the range of temperature (300°C) and residence time (24 h) representative of the ISUT process. The morphology of deposited particles showed significant differences based on the deposition conditions. At low temperatures (150–200°C) the metal particles were found in relatively large agglomerates (1,000–2,000 nm) throughout the full length of the sand pack. For high temperatures (300°C) it was evidenced that near the reactor entrance particles deposited as individual entities (~140 nm). The depths of the well-dispersed deposited nanoparticles were strongly related to the residence time, i.e., as residence time increased, the depth of the dispersed nanoparticles also increased from 0 cm at 3 h up to 10 cm at 24 h [Rodriguez-DeVecchis *et al.* 2018].

8.3.2.3   Physical and Numerical Simulations

In the last ten years, the University of Calgary has carried out a significant amount of laboratory and numerical simulation work to demonstrate the technical feasibility of the ISUT process. A summary of the principal findings is presented in this section. Further reading of the original articles is recommended to understand the state of this technology fully.

Table 8.6 shows several examples of the physical simulation of the *In-Situ* Upgrading Technology (ISUT) Process (Figure 8.13). As seen, Hashemi *et al.* conducted experiments using a Ni-Mo-W nanocatalyst (72, 240, and 408 ppm, respectively) for the Athabasca bitumen upgrading in a packed-bed flow reactor (1-D) at 3.5 MPa of H_2 pressure, temperatures from 320 to 340°C, and up to 144 h [Hashemi *et al.* 2014a, 2014b, Carbognani-Ortega *et al.* 2016]. For the thermal runs, API increments of 7.5°API with 93% viscosity reductions, 53% HDS, and 21% residue conversion were reported. In the presence of the nanocatalyst, further improvement on the upgraded crude oil was observed, i.e., 8.5 API increase, 95–97% viscosity reduction, 56% HDS, and 28–38 residue conversion [Hashemi *et al.* 2014a].

The content of micro-carbon residue (MCR) is a useful technique that is considered as a measure for potential coke formation. The authors observed that, in the thermal experiments, the MCR content increased with the temperature. On the contrary, in the presence of the nanocatalyst, raising the temperature favored the reduction in the MCR content. Thus, the authors suggested that the trimetallic nanocatalysts effectively favored the hydrogenation reactions and inhibited the coke formation [Hashemi *et al.* 2014a, 2014b].

Hovsepian and coworkers carried out the two-dimensional bench scale plant for the experimental simulation of production and upgrading of an Athabasca oil sand via the ISUT process, also called Dense Hot Fluid Injection (DHFI) processing [Hovsepian *et al.* 2016]. In these experiments, vacuum residue (VR) was injected at 350°C and 3.5 MPa of H2 (500 psi) at different residence times (24, 72, and 288 h) to study its conversion levels. As expected, API gravity (from 2.4°API to 7.7, 10.4, and 16.4°API) and the VR conversion (22, 49, and 60%) increased as a function of residence time whereas viscosity (from 40.475 Pa s to 523, 51, and 8 mPa s at 80°C) and sulfur content (from 4.5%

TABLE 8.6
Physical Simulations of the *In-Situ* Upgrading Technology (ISUT) Process from University of Calgary

Conditions[a]	Feed, Rock (Country)	Feed		Reaction Conditions			Results				References
		°API	Viscosity (mPa s @ °C)[b]	Temp. (°C)	Press. (MPa)	Time	Incr. °API[c]	% Visc. Red.[d]	%HDS[e]	% Res. Conv.[f]	
No Catalyst 1-Dimension	Athabasca, Sand (Canada)	9.5	7,550 (@40°C)	340	3.5	144 h	7.5	93	53	21	Hashemi *et al.* [2014a]
Ni, Mo, W1-Dimension				320			8.5	95	56	28	
				340			8.5	97	56	38	
Ni, Mo, W 2-Dimension	VR[g], Sand (Canada)	2.4	40,475 (@80°C)	350	3.5	288 h	14.0	99.9	49	60[h]	Hovsepian [2016]
No Catalyst 1-Dimension	VR[g], Carbonates (Mexico)	−0.1	7,025 (@150°C)	300	6.7	48 h	2.6	93	ND	4	Suarez *et al.* [2016]
Ni, Mo, W1-Dimension							5.9	96		18	
Ni, Mo, W1-Dimension	Heavy Oil Carbonates (Mexico)	8.0	4,000 (@40°C)	320	10.3	168 h	8.0	94	ND	47[i]	Chavez-Morales [2016]
Ni, Mo, W1-Dimension	VR[g], Carbonates (Mexico)	−2.8	ND	350		336 h	15.5	ND	ND	72[i]	

ND = Not disclosed or not reported.

[a] Lab scale experiments carried using 1-dimension or 2-dimension sand pack in the absence and presence of the trimetallic nanocatalyst [Galarraga 2010b] and H_2 gas as hydrogen source. See original references for details.

[b] Viscosity in mPa s measured at the given temperature.

[c] Increase in API gravity of the upgraded crude oil with respect to the feed.

[d] Percentage of decrease in viscosity (μ) of the upgraded crude oil vs. the feed calculated as $((\mu_{feed} - \mu_{product}) - \mu_{feed}) \times 100$.

[e] Percentage of hydrodesulfurization (%HDS) in the upgraded crude oil vs. the feed calculated as $((\%S_{feed} - \%S_{product}) - \%S_{feed}) \times 100$.

[f] Percentage of conversion of vacuum residue (VR defined as *545°C+ fraction*) in the upgraded crude oil vs. the feed calculated as $((\%VR_{feed} - \%VR_{product}) - \%VR_{product}) - \%VR_{feed}) \times 100$.

[g] Vacuum residue.

[h] Percentage of conversion of *500°C+ fraction*.

[i] Average 545°C+ residue conversion throughout the length of the experiment.

to 3.9, 3.2, and 2.3%) decreased. For comparison purposes, the best results are reported in Table 8.6. Also, the authors observed that high residence times led to produced oil instability, which suggested a threshold of permissible severities that should avoid high reservoir conversion levels (above 55% conversion of the 500°C$^+$ fraction) [Hovsepian et al. 2016].

The post-mortem mapping analysis of the sand pack showed the presence of extremely light material in the bottom horizontal branch of the 2-D cell [Hovsepian et al. 2016]. The almost complete depletion could be indicative of a solvent extraction mechanism due to the presence of light components. Overall, the most depleted sections were the high permeable zones (horizontal branches), while the vertical section presented the lowest production among the three arms of the 2-D setup. The author observed a reduction of the oil content in the post-mortem sands from 15.6 wt. % to 3.2 wt. % [Hovsepian et al. 2016].

As a first step toward the numerical simulation of the ISUT process, Loria et al. from the University of Calgary, studied a kinetic model for ultradispersed catalytic Hydroprocessing of Athabasca bitumen [Loria et al. 2011]. Kinetic parameters were estimated from experimental results obtained in a 1-D reactor in the temperature range 320–380°C under 2.76 MPa of H$_2$ (400 psi) and residence times of 9–51 h. A five-lump kinetic model was employed (see Equation 8.10) with seven chemical reactions (k$_1$–k$_6$, and k$_8$) to simulate the upgrading process. The pseudo-components were naphtha (IBP-216°C), distillates (216–343°C), VGO (343–550°C), and residue (> 550°C). The predicted product compositions were in good agreement with experimental values with average absolute errors of less than 7% [Loria et al. 2011].

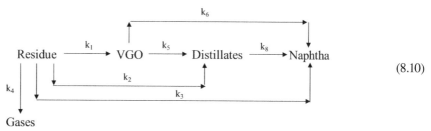

(8.10)

Additionally, the authors found that the viscosity of the experimental liquid products and residue conversions followed an exponential correlation. This correlation, in combination with the results from the kinetic modeling, was employed to calculate product distributions from bitumen Hydroprocessing and provided reaction conditions to achieve a desired product viscosity [Loria et al. 2011].

Nguyen et al. carried out the numerical simulation to evaluate the In-Situ Upgrading Technology (ISUT) using an SAGD well pad [Nguyen et al. 2017]. The commercially available software CMG-STARS (thermal and compositional) was used to build a half section of two horizontal SAGD wells. The five-lump kinetic model was employed (Equation 8.10) with seven chemical reactions (k$_1$–k$_6$, and k$_8$) to simulate the upgrading process [Loria et al. 2011]. The numerical model was run for one year in conventional steam injection mode to preheat the reservoir and then followed by injection of steam plus the trimetallic nanocatalyst, hydrogen, and vacuum residue (a process called ST-ISUT) until the end of the simulation. Several different injection strategies including continuous injection and alternating injection between steam and ST-ISUT were tested to compare oil recovery and to find the best injection method [Nguyen et al. 2017].

Based on the numerical simulations, the authors found that the ST-ISUT fluid did promote reactions of converting heavy oil components into lighter oil products in a heavy oil reservoir. The steam reduced the viscosity of bitumen and expanded the heated chamber leading to the production of more mobile oil toward the producer. The numerical simulations also showed that ST-ISUT method increased the oil recovery factor by 36% and lowered the requirement of steam by 50% in comparison with the conventional steam injection method (SAGD). Additionally, the produced oil had better quality as measured in terms of density and viscosity [Nguyen et al. 2017].

Suarez and coworkers carried out the ISUT physical simulation of Mexican heavy oil recovery from fractured Silurian dolomite cores using a cylindrical holder set-up (1-D) [Suarez *et al.* 2016]. The cores were fully saturated with heavy oil, and the fractures were filled with hydrogen and VR containing trimetallic nanocatalyst (240 ppm of Ni, 200 ppm of Mo, and 380 ppm of W) at 300 °C, 6.7 MPa of H_2 (1,000 psig) for 48 h (see Table 8.6). The produced oils were collected, and the recovery factor for each experiment was determined. The authors reported 62% of oil recovery vs. the OOIP which was almost twice as much as those obtained under N_2 and CO_2 [Suarez *et al.* 2015]. Furthermore, as shown in Table 8.6, in the presence of H_2 and the nanocatalyst, the API of the produced oil increased ~6°API whereas the control reaction (no catalyst) only led to 2.6°API [Suarez *et al.* 2016]. Similarly, the viscosity reduction and the residue conversion were higher in the reaction with the nanocatalyst than those obtained in the control experiments (96% vs. 93% and 18% vs. 4%, respectively). Based on these results the author concluded that the upgrading of heavy crude oil has a good potential for applications in carbonate reservoirs [Suarez *et al.* 2016].

Chavez-Morales *et al.* conducted laboratory experiments to simulate the Dense Hot Fluid Injection (DHFI) process in a reservoir located in the Gulf of Mexico at 10.3 MPa of H_2 (1,500 psi) and 320°C (heavy oil as feed) and 350°C (VR as feed) for 168 h and 336 h, respectively [Chavez-Morales and Pereira-Almao 2016]. As seen in Table 8.6, the API of the produced oils increased 8 and 15°API with average viscosity reductions of 94% and average residue conversions of 47% and 72% for the two temperatures studied. Using the commercial software CMG-STARS, the numerical modeling was carried out using a similar reaction sequence as shown in Equation 8.10. The authors obtained a good match between the experimental and calculated results. Based on the physical and numerical simulations the author proposed that, because of the temperature increase during the ISUT, the matrix rock expanded and expelled the oil. When the temperatures decreased, the pores in the matrix contracted, generating additional oil expulsion. Due to this phenomenon of expansion–contraction of the reservoir, oil production was increased [Chavez-Morales and Pereira-Almao 2016].

Finally, and as mentioned earlier, the process developed by the University of Calgary for the subsurface upgrading of heavy crude oil and bitumens has not been field tested yet. At the moment of writing this book, several pilot tests have been considered in Mexico, Colombia, and Canada.

8.3.3 OTHER LAB EXPERIMENTS

Besides the University of Calgary, other research groups have studied the upgrading of HO/B in the presence of nanocatalysts using H_2 as a source of hydrogen at subsurface conditions. Specifically, Shuwa *et al.* conducted the physical simulation of the enhanced recovery and subsurface upgrading of Omani heavy crude oil (14.3°API and viscosity of 3,951 mPa s at 30°C) using a 1-D reactor filled with Berea stone as porous media in a steam simulation process [Shuwa *et al.* 2016]. The experiments were carried out in the presence of a Ni-Co-Mo dispersed catalyst (2,000 ppm) at 280–300°C, 2 MPa of H_2 or N_2 for 24 h [Shuwa *et al.* 2016]. The results from the recovery tests showed a higher oil recovery (15% vs. OOIP) in the presence of the catalyst than that obtained with steam injection only (using N_2 to keep the pressure constant). The authors observed an increase in API gravity of ~1.4°API and viscosity reduction of 25% and 26% of HDS. By analysis of solid and gaseous products recovered from the experiments with catalyst, the authors concluded that thermal expansion and viscosity reduction due to catalytic hydrocracking were the dominant mechanisms for oil recovery [Shuwa *et al.* 2016].

Guo and coworkers synthesized different nickel nanoparticles (~12 nm) supported on three carbon nanomaterials, i.e., ketjenblack carbon, carbon nanotubes, and graphene nanoplatelets, and used them for the upgrading of heavy crude oil (9.3°API) in a batch reactor at 200, 250, 300, and 350°C using 11 MPa of H_2 gas as hydrogen source for 2 h [Guo *et al.* 2017]. The results showed viscosity reduction in the 85–95% range, and the Ni supported on ketjenblack carbon was the most active of the solids studied, indicating a possible synergistic effect between the nickel nanoparticles and the support. Reactions conducted with and without catalysts in N_2 and H_2, respectively, showed that the viscosity reduction is higher for the runs carried out under hydrogen (85%) than that in nitrogen (70%) [Guo *et al.* 2017].

Furthermore, from the molecular characterization obtained by Fourier Transform Ion Cyclotron Resonance Mass Spectrometry, the authors attributed the viscosity reduction and catalytic upgrading to the conversion of large molecular weight carboxylic acid compounds to derivatives with smaller carbon numbers and higher saturation. Thus, this work showed the potential application of *carbon*-based nanocatalysts for the *in-situ* upgrading and recovery of heavy crude oils [Guo *et al.* 2017].

8.4   CATALYTIC USE OF HYDROGEN DONORS

As discussed in Section 4.2.1.2, hydrogen donors are chemical compounds that can easily transfer hydrogen to the HO/B during subsurface upgrading processes. 1,2,3,4-tetrahydronaphthalene, or tetralin, and decahydronaphthalene, or decalin (cis or trans), are the most commonly used hydrogen donors (see Equation 4.16). Ancheyta *et al.* [Alemán-Vázquez *et al.* 2016] carried out a literature review on the use of H-donors for the upgrading of heavy crude oils and reported that partially hydrogenated polynuclear aromatics of pyrene, anthracene, phenanthrene, fluoranthene, and basic nitrogen compounds such as quinoline and benzoquinolines can function as effective hydrogen transfer agents and with better performance than tetralin or decalin. Additionally, oxygen-containing compounds such as alcohols, cyclic ethers, and occasionally formic and ascorbic acids have been found to be effective hydrogen donors [Alemán-Vázquez *et al.* 2016]. In general, the choice of H-donor is based on the reservoir properties, its availability, and solubility in the reaction medium.

As discussed in Section 7.3, under thermal conditions, the mechanism of hydrogen transfer from the H-donor to the accepting hydrocarbons follows a pathway known as the capping of free radicals (Equations 4.20–4.21) and hydrogen transfer and disproportionation (Equations 4.23–4.26). As mentioned in Section 4.2.1.3, by reacting very actively with olefins, H-donors can reduce the polymerization and coke formation (Equation 4.27) during SSU processes. Also, in the presence of hydrogen donor and steam, the Aquathermolysis reactions are affected due to the higher rate of hydrogen transfer of the H-donor in comparison with water. Finally, catalysts have been used in the presence of hydrogen donors to enhance the SSU of HO/B further.

8.4.1   PHYSICAL SIMULATIONS

Table 8.7 shows the results of the lab or physical simulations of the catalytic upgrading of heavy oils and bitumens using hydrogen donors under steam injection conditions. Liu and Fan studied the effect of tetralin as a hydrogen donor on the viscosity of Liaohe heavy oils during steam stimulation [Liu and Fan 2002]. The experiments were carried out in an autoclave reactor with 0.2 wt. % of a $V(O)SO_4$, $Fe_2(SO4)_3$, $NiSO_4$ catalyst, 20 wt. % of water at 240°C for 24–96 h. The authors observed that, in the absence of tetralin, the viscosity of the upgraded products increased after 20 days. This result indicated that, after Aquathermolysis, there are active free radical chains in the treated oil that can react with each other leading to a viscosity increase. As shown in Table 8.7, in the presence of 0.8 wt. % of tetralin, the authors reported up to 83% of viscosity reduction with 24% of asphaltene conversion [Liu and Fan 2002].

Mohammad and Mamora from Texas A&M University conducted upgrading experiments in a vertical injection cell using a mixture of sand, water, and Jobo oil (see Table 8.7) [Mohammad and Mamora 2008]. Three cases were considered: Pure steam injection; steam injection with tetralin; and steam injection with tetralin and catalyst (organic-soluble iron acetylacetonate, $Fe(acac)_3$ at 750 ppm). The result showed that the addition of 5 wt. % of tetralin increased oil recovery by 15% more than that measured with pure steam injection. The pre-mixing the tetralin-catalyst solution with the sand yielded 20% higher oil recovery than the pure steam injection. When the tetralin-catalyst

TABLE 8.7

Catalytic Upgrading of Heavy Oils and Bitumen Using Hydrogen Donors Under Steam Injection Conditions

Catalyst-H-Donor[a]	Feed			Reaction Conditions			Results				References
	Crude Oil (Country)	°API	Viscosity (mPa s@°C)[b]	Temp. (°C)	Press. (MPa)	Time	Incr. °API[c]	% Visc. Red.[d]	%HDS[e]	% Asp. Conv.[f]	
V(O)SO$_4$ -Fe$_2$(SO$_4$)$_3$- NiSO$_4$ Tetralin	Liaohe (China)	ND	88,500 (@50°C)	240	ND	96 h	ND	83	ND	24	Liu [2002]
Fe (acac)$_3$[g]Tetralin	Jobo (Venezuela)	12.4	1,400 (@50°C)	273	3.4	4 h	1.3	87	ND	ND	Mohammad and Mamora [2008]
Fe(acac)$_3$[g]Decalin	Jobo (Venezuela)	11.4	1,108 (@50°C)	300	ND	48 h	1.3	73	ND	ND	Zhang et al. [2012]
Ni (~50 nm)Decalin	Athabasca (Canada)	8.2	6,209 (@60°C)	240	ND	12 h	ND	81	ND	ND	Hendraningrat et al. [2014]
Fe$_2$O$_3$ (~20 nm)Tetralin	Hamaca (Venezuela)	9.1	18,000 (@60°C)	315	6.2	24 h	8.0	89	26	27	Ovalles et al. [2015]
Cu-DMBSDMBS[h]	Shengli (China)	11.2	178,000 (@70°C)	240	3.0	24 h	14.2	82	ND	20	Ren et al. [2015]
NiMo-oleateGlycerol	Heavy Oil (Oman)	ND	1,490 (@70°C)	277	3.0	30 h	ND	69	ND	ND	Yusuf et al. [2016b]
Cobalt-CitrateEthanol	Tazhong (China)	16.4	4,080 (@50°C)	180	ND	24 h	ND	93	23	ND	Chen et al. [2017a]

ND = Not disclosed or not reported.

a   Hydrogen donor used during the upgrading experiments.

b   Viscosity in mPa s measured at the given temperature.

c   Increase in API gravity of the upgraded crude oil.

d   Percentage of decrease in viscosity (μ) of the upgraded crude oil vs. the feed calculated as $((\mu_{feed} - \mu_{product}) - \mu_{feed}) \times 100$. Negative values indicate increase with respect to the feed.

e   Percentage of hydrodesulfurization (%HDS) in the upgraded crude oil vs. the feed calculated as $((\%S_{feed} - \%S_{product}) - \%S_{feed}) \times 100$. Negative values indicate increase with respect to the feed.

f   Percentage of reduction of asphaltenes in the upgraded crude oil vs. the feed calculated as $((\%Asp_{feed} - \%Asp_{product}) - \%Asp_{feed}) \times 100$.

g   acac = acetylacetonate ligand.

h   Dimethylbenzenesulfonic.

solution was injected as a slug, oil recovery was similar to that found with the addition of tetralin. Acceleration in oil production was observed for all the H-donor-containing runs but was more pronounced in the presence of the catalyst. As shown in Table 8.7, a small increase in the API (1.3°API) and 87% viscosity reduction was found in the presence of tetralin and iron acetylacetonate additive [Mohammad and Mamora 2008].

Zhang and coworkers, also from Texas A&M University, studied the *in-situ* upgrading of Jobo crude oil in an autoclave reactor using steam, tetralin or decalin, and a catalyst ($Fe(acac)_3$) at temperatures of 250°C, 275°C, and 300°C for 24 hours, 48 hours, and 72 hours [Zhang 2012]. The authors found that the use of H-donor and steam led to viscosity reductions of 44% (tetralin) and 39% (decalin). However, in the presence of the Fe-containing catalyst, lower viscosity was observed, and decalin was a more active H-donor than tetralin (73% vs. 72%, respectively). As shown in Table 8.7, the reaction with decalin led to an increase of 1.3°API of the upgraded products [Zhang 2012].

Hendraningrat *et al.* studied various metal nanoparticles (Cu, Zn, Ni, and Fe,) as catalysts for the upgrading of Athabasca bitumen in the presence of steam and decalin as a hydrogen donor at 240°C for 12 h [Hendraningrat *et al.* 2014]. As shown in Table 8.7, the authors reported up to 81% viscosity reduction using nickel nanoparticles of 50 nm of average particle size [Hendraningrat *et al.* 2014].

Ovalles and coworkers carried out batch experiments for the upgrading of Hamaca crude oil (Venezuela) under steam injection conditions (280–315°C) using tetralin as H-donor in the presence of three silica supported Fe_2O_3 nanocatalysts (20nm, 60nm, and 90nm) at 6.2MPa of CH_4 and residence time of 24 h [Ovalles *et al.* 2015]. Figure 8.15 shows the API gravity of the upgraded products vs. the reaction temperature for the iron nanoparticles. As seen, all nanocatalysts increased the API gravity of the treated crudes. The best results were obtained (8°API increase) with the Fe_2O_3 with a particle size of 20 nm at 315°C. Also, as shown in Table 8.7, a viscosity reduction of 89% and an asphaltene conversion of 27% were found [Ovalles *et al.* 2015].

Furthermore, Figure 8.16 shows the percentage of hydrodesulfurization for the upgrading of Hamaca crude oil vs. the reaction temperature for the iron-containing catalysts [Ovalles *et al.* 2015]. As seen, the % HDS increased in going from 280°C to 315°C. Consistent with the previous results, the most active catalyst was the Fe_2O_3 with a particle size of 20 nm.

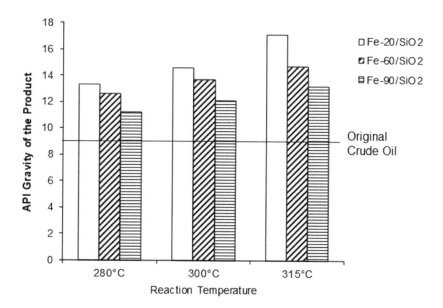

FIGURE 8.15 API gravity of the upgraded products vs. reaction temperature using iron nanoparticles supported on silica as catalysts using tetralin as hydrogen source and residence time of 24 h [Ovalles *et al.* 2015].

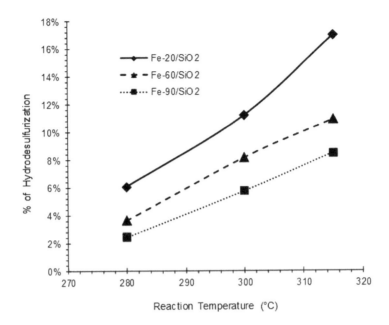

FIGURE 8.16   Percentage of hydrodesulfurization for the upgrading of Hamaca crude oil vs. reaction temperature using iron nanoparticles supported on silica as catalysts using tetralin as hydrogen source and residence time of 24 h [Ovalles *et al.* 2015].

The iron nanocatalysts were characterized by different spectroscopy and microscopy techniques before and after the upgrading reactions [Ovalles *et al.* 2015]. The results indicated the presence of hematite (Fe$_2$O$_3$) as the predominant iron phase and that the catalysts were deactivating by particle sintering (~20% increase in particle size) and carbon deposition.

The mechanism of catalytic methane activation was discussed in Section 4.2.2.4. For iron-supported nanocatalysts, the authors carried out ^{1}H-NMR analyses on the upgraded crude oil [Ovalles *et al.* 2015]. They found that the percentages of α-hydrogen bonded to aromatic rings and aromatic protons increased in the reaction products of the three iron-supported nanocatalysts increases. The authors attributed these findings to the metal-catalyzed incorporation of the methyl groups to the crude oil molecules followed by thermal or catalytic aromatization of the naphthenic compounds, as shown in Equation 8.11 [Ovalles *et al.* 2015]:

$$R \Big\langle \bigcirc \Big\rangle \; + \; CH_4 \; \xrightarrow{\text{Cat.}} \; R \Big\langle \bigcirc \Big\rangle^{CH_3} \; \xrightarrow{-6H} \; R \Big\langle \bigcirc \Big\rangle^{CH_3} \tag{8.11}$$

As discussed in Section 4.2.2.4, the use of methane as a hydrogen source for extra-heavy crude oil upgrading has been reported previously at a higher temperature (380–420°C) under thermal [Ovalles *et al.* 1995] and catalytic conditions (Mo-containing catalyst) [Ovalles *et al.* 1998]. The incorporation of methyl groups, coming from methane, into the crude oil molecules was confirmed by isotopic carbon distribution measurements (^{13}C/^{12}C) using ^{13}CH$_4$ as a source of hydrogen [Ovalles *et al.* 1998]. The higher percentages of HDS obtained for three iron-supported nanocatalysts (Table 8.7 and Figure 8.16) can also be explained due to the generation of hydrogen due to aromatization of the naphthenic compounds, as shown in Equation 8.11 [Ovalles *et al.* 2015].

Ren *et al.* [2015] developed a new type of catalyst, dimethylbenzenesulfonic copper (Cu-DMBS), which is composed of a hydrogen-donating ligand (DMBS) and a Cu^{2+} center [Ren *et al.* 2015]. They used this catalytic formulation (0.3 wt. %) for Aquathermolysis of Shengli heavy oil (China) under batch conditions at 240°C for 24 h [Ren *et al.* 2015]. As seen in Table 8.7, the authors reported an increase in the API of 14.2°API, 82% viscosity reduction, and 20% asphaltene conversion. The authors attributed the activity of the catalyst to the hydrogen-donating ability of the DMBS ligand and the dehydrogenation of naphthene-aromatic compounds occurring during the Aquathermolysis experiments [Ren *et al.* 2015].

In a continuation work with the same Cu-DMBS catalyst, Li and coworkers characterized the organosulfur compounds present in the feed and the Aquathermolysis products by using $PdCl_2$/silica gel ligand exchange chromatography, followed by GC/MS [Li *et al.* 2016]. The results showed that the non-conjugated C–S bonds of high-molecular-weight compounds were cleaved whereas the conjugated C–S bonds were stable under the catalytic Aquathermolysis condition [Li *et al.* 2016].

Yusuf *et al.* conducted the Aquathermolysis of Omani heavy oil in the presence of a NiMo-oleate soluble catalyst using glycerol as a hydrogen donor at 277°C for 30 h [Yusuf *et al.* 2016b]. As seen in Table 8.7, the authors reported a maximum of 69% of viscosity reduction. They attributed the heavy oil upgrading to the hydrogen generation by aqueous phase reforming of glycerol, as shown in Equation 8.12 [Yusuf *et al.* 2016b]:

$$2C_3H_5(OH)_3 + 2H_2O \rightarrow 4CO + 2CO_2 + H_2 \tag{8.12}$$

Chen and coworkers carried out the Aquathermolysis of Tazhong crude oil (China) using a cobalt citrate catalyst and ethanol as hydrogen donor at the relatively low temperature of 180°C [Chen *et al.* 2017a]. The authors studied the effects of the water, catalyst, and ethanol concentrations. As seen in Table 8.7, they reported up to 93% of viscosity reduction with 35% HDS using 0.5 wt. % of catalyst and 10 wt. % of ethanol [Chen *et al.* 2017a].

Based on the results presented in this section, the subsurface upgrading of HO/B can be effectively carried out in the presence of hydrogen donors, and the properties of the upgraded products can be enhanced further by using metal-containing catalysts. In Section 8.5, several field tests have been carried out using some of the previous catalytic systems. But first, two fundamental studies on the mechanism of catalytic HDS in the presence of H-donors are presented.

8.4.2 HDS MECHANISTIC STUDIES

As discussed in Section 2.3.1, heavy crude oils and bitumens contain significant quantities of sulfur-containing compounds. Thus, the sulfur removal via the hydrodesulfurization (Section 4.2.3.1) is of paramount importance during the subsurface upgrading via catalytic hydrogen addition. In this section, the mechanism of HDS is discussed to gain knowledge of chemical pathways and to help design more efficient catalysts for this process.

Khalil and coworkers studied the desulfurization of thiophene to investigate the catalytic activity of hematite (Fe_2O_3) nanoparticles in the Aquathermolysis of heavy oil [Khalil *et al.* 2015]. The experiments were carried out in a Teflon-lined autoclave reactor with thiophene-to-water ratios from 7:3 to 3:7, 10% v/v of a hydrogen donor (tetralin) in the temperature range of 120–180°C, and reaction times of 12–72 h. The results showed that thiophene conversion increased with reaction time, temperature, and catalyst concentration in the 10–30% range, but decreased with thiophene-to-water ratio, particle size, and with the presence of hydrogen donors. The authors attributed the latter phenomenon to the blocking of the catalytic sites on the catalyst surface by the absorption of the hydrogen donor [Khalil *et al.* 2015].

By FTIR and XRD analyses the authors proposed that thiophene underwent oxidative desulfurization to produce maleic acid, SO_2, and CO_2 (see Equation 8.13), whereas some areas of the hematite surface were transformed into magnetite (Fe_3O_4). In the next step (Equation 8.14), magnetite was re-oxidized back into hematite in the presence of water as the source of active oxygen [Khalil *et al.* 2015].

$$\text{(thiophene)} \xrightarrow[\text{Fe}_3\text{O}_4]{\text{Fe}_2\text{O}_3} \text{(maleic acid)} + SO_2 + CO_2 \qquad (8.13)$$

$$2Fe_3O_4 + H_2O \rightarrow 3Fe_2O_3 + H_2 \qquad (8.14)$$

Guo *et al.* synthesized nickel and cobalt nanoparticles with different sizes via the thermal decomposition of organometallic precursors [Guo *et al.* 2018]. Characterization using transmission electron microscopy confirmed that the Ni-nanoparticles had average sizes of 9 and 27 nm, whereas the Co analogs had an average size of 6 nm with narrow size distribution. Similar to Khalil and coworkers [2015], Guo *et al.* studied different parameters including particle size, catalyst dosage, hydrogen donor ratio, temperature, and reaction duration to optimize the catalytic HDS performance. The results showed that the 9 nm Ni-nanocatalyst exhibited the best HDS activity and stability compared with other catalysts. As found by Ovalles *et al.* for Fe-containing catalysts [Ovalles *et al.* 2015], Guo and coworkers concluded that well-dispersed Ni-nanoparticles are promising candidates for the subsurface upgrading of heavy crude oil and bitumens [Guo *et al.* 2018].

8.5 FIELD TESTS

All the field tests for the subsurface upgrading of HO/B via catalytic hydrogen addition have been carried out in Chinese oilfields (Liaohe, Henan, and Xinjiang) under steam stimulation process (i.e., Aquathermolysis conditions). For the first field, two types of catalysts have been evaluated, i.e., water-soluble iron salt and hydrogen donor [Zhong *et al.* 2003, Jiang *et al.* 2005] and oil-soluble molybdenum catalysts [Wen *et al.* 2007]. For the Henan field, an amphiphilic-aromatic sulfonic iron complex was used [Chen *et al.* 2008] whereas for the Xinjiang oilfield an alkyl ester sulfonate copper catalyst was employed [Chao *et al.* 2012]. The key learnings from these field tests are discussed next.

8.5.1 LIAOHE OILFIELD

The Liaohe Oilfield is located in the Liaoning Province, China. It was discovered in 1958 and began production in 1970. The total proven reserves of the Liaohe Oilfield are around 6.87 billion barrels. Currently, it is operated by the China National Petroleum Corporation and produces an estimated 200–250 MBD. Liaohe Oilfield has become one of the largest heavy oil productions utilizing steam injection processes [Zhang 1997, Höök *et al.* 2010].

8.5.1.1 Use of Water-Soluble Catalyst

In 2003, Zhong and coworkers studied the upgrading of Liaohe heavy oil under lab and field conditions [Zhong *et al.* 2003]. Firstly, laboratory experiments were carried out using an autoclave containing heavy oil, water, catalysts, and tetralin as a hydrogen donor at different temperature (160–260°C) and reaction times (24~240 h). Different metals salts (Fe^{2+}, Co^{2+}, Mo^{2+}, Zn^{2+}, Ni^{2+}, Al^{3+}, Mn^{2+}, and Cu^{2+}) were evaluated with a catalyst concentration of 0.02 mol/L at 240°C for 72 h. The most active metal was iron and led to 60% decrease in viscosity with respect to the untreated oil. The authors found further viscosity reduction (87%) in the upgraded crude oil by adding tetralin as a hydrogen donor. As shown in Table 8.8, the author also reported 24% of sulfur removal and 46% of asphaltene conversion. Average molecular weight, element, and SARA analyses showed the same synergetic effect between the catalyst and tetralin [Zhong *et al.* 2003].

TABLE 8.8
Laboratory and Field Tests for the Catalytic Upgrading of Heavy Oils Under Steam Injection Conditions (Aquathermolysis)

	Feed			Experiments			Average Results[a]				
Catalyst (Type)	Oilfield	Viscosity (mPa s@°C)[b]	Conditions (H-Donor)	Temp. (°C)	Time (h)	% Oil Rate Inc.[c]	% Visc. Red.[d]	%DS[e]	% Asp. Conv.[f]	References	
Fe Salt (Water-Soluble.)	Liaohe	220,000 (@80°C)	Laboratory (Tetralin)	240	72	–	87	24	46	Zhong et al. [2003]	
		322,000[a] (@80°C)	Field (Coal-Derived)	ND	ND	Yes[g]	73	ND	ND	Jiang et al. [2005]	
		660 (@50°C)	Field (Coal-Derived)	ND	ND	Yes[g]	80[h]	86[h]	28[h]	Wen et al. [2007]	
Mo-Oleate (Oil-Soluble)	Liaohe	12,000 (@50°C)	Laboratory (Water)	240	24	–	93	66	30		
		12,400 (@50°C)	Field (Water)	ND	ND	21	78[h]	65[h]	19[h]		
Fe-Aromatic Sulfonate (Amphiphilic)	Henan	28,867 (@50°C)	Laboratory (Water)	200	72	–	91	ND	23	Chen et al. [2008]	
		84,600 (@50°C)	Field (Water)	ND	ND	51[i]	81[i]	ND	ND		
Cu-Alkyl Ester Sulfonate (Amphiphilic)	Xinjiang	181,000 (@70°C)	Laboratory (Water)	240	24	–	91	ND	26	Chao et al. [2012]	
		85,000 (@70°C)	Field (Water)	ND	ND	Yes[g]	85[h]	ND	46[h]		

ND = Not disclosed or not reported.

a Average results obtained during the lab or field upgrading tests.

b Viscosity in mPa s measured at the given temperature.

c Percentage of increase in oil rate (BBL) after the addition of catalyst calculated as ((oil rate $after$ test) – (oil rate $before$ test)) / (oil rate $before$ test) × 100.

d Percentage of decrease in viscosity (μ) of the upgraded crude oil $vs.$ the feed calculated as (($\mu_{feed} - \mu_{product}$) – μ_{feed}) × 100.

e Percentage of desulfurization (%DS) in the upgraded crude oil $vs.$ the feed calculated as (($\%S_{feed} - \%S_{product}$) – $\%S_{feed}$) × 100.

f Percentage of reduction of asphaltenes in the upgraded crude oil $vs.$ the feed calculated as (($\%Asp_{feed} - \%Asp_{product}$) – $\%Asp_{feed}$) × 100.

g Yes = Incremental cumulative oil reported (tons) but no comparison before adding the catalyst was reported.

h Results reported for only one well.

i Results reported for only two wells.

Based on the lab results, Zhong *et al.* selected five wells in the Liaohe Oilfield for the field trial using the iron-based catalyst and a coal-derived hydrogen donor [Zhong *et al.* 2003]. All oil wells have previously been produced for several cycles (two to four) of cyclic steam stimulation (CSS, see Section 3.1.2.1). The wells were pre-heated by injecting steam alone followed by catalysts injection at 0.2 Kmol per ton of steam and hydrogen donor at 0.1 m^3 per ton of steam. The Fe-catalyst and hydrogen donor were co-injected or pumped down separately. As seen in Table 8.8, the results showed an average of 73% of viscosity reduction with an increase in cumulative oil production. However, no comparisons with the cases before the catalysts and hydrogen donor were injected, so no data on the percentage of the rate increase were reported [Zhong *et al.* 2003].

In 2005, the same research group from China reported a field trial in 20 oil wells from the Liaohe Oilfield using the Fe-catalysts and hydrogen donor [Jiang *et al.* 2005]. As before, the wells were previously produced by CSS, and increases in the cumulative oil were observed for all the cases, but no comparisons without catalysts were reported. One well was sampled, and the upgraded crude was analyzed. As shown in Table 8.8, 80% viscosity reduction was found with 86% sulfur removal and 28% asphaltene conversion [Jiang *et al.* 2005].

As pointed out by Maity and coworkers [2010], the percentages of viscosity reduction were always higher in the lab than those found in the field (87% vs. 73–80%, respectively). As presented in Section 8.1, there are several issues with the use of catalyst downhole that could explain this finding. Most probably, the poor mixing between the heated crude oil, steam, catalyst, and hydrogen donor is a significant challenge, as reported by Ovalles and Rodriguez in their lab and numerical simulation studies [Ovalles and Rodriguez 2008].

8.5.1.2   Use of Oil-Soluble Mo-Catalyst

Wen and coworkers evaluated an oil-soluble molybdenum oleate catalyst for the upgrading of Liaohe heavy crude oil using water as a hydrogen source (Aquathermolysis) at 240°C for 24 h [Wen *et al.* 2007]. As shown in Table 8.8, the authors found a 93% viscosity reduction with 66% sulfur removal and 30% asphaltene conversion in lab experiments. This Mo-catalyst was field test in two oil wells from the Qi-40 and Qi-108 blocks of the Liaohe Oilfield. These wells have been previously produced by CSS, but their production was declining. The authors disclosed that the reservoir was composed of sand, clay, and sandstone with 25–30% porosity and permeabilities between 1.5 and 2 D. As before, the wells were pre-heated with steam, and the Mo-catalyst was injected downhole followed by additional steam. After a soaking period of seven to ten days, the wells were put into production [Wen *et al.* 2007]. No hydrogen donor was used in this test.

The result showed an average of 21% increase in the oil rate in comparison with the case before catalyst injection (Table 8.8). Also, the authors reported that the characterization of the produced crude oil showed a viscosity reduction of 80%, 78% of desulfurization, and 28% of asphaltene conversion. Similarly, the authors found that the oxygen and nitrogen contents were lower in the upgraded crude than those in the untreated oil [Wen *et al.* 2007].

The results reported by Wen *et al.* indicated that the Aquathermolysis reaction was occurring downhole, and it is in agreement with the findings discussed in the previous section [Zhong *et al.* 2003, Jiang *et al.* 2005]. As before (Table 8.8), the viscosity reduction found in the field test was lower than that measured in the lab experiments (78% vs. 93%, respectively). As discussed, this finding could be attributed to poor downhole mixing so further studies must be carried out to address this and other challenges of using catalysts and H-donors downhole.

8.5.2   Henan Oilfield Using Sulfonic Fe-Catalyst

Chen and coworkers carried out lab experiments and a two-well field test of the subsurface upgrading of heavy oil from the Henan Oilfield [Chen *et al.* 2008]. This field is located in the Nanyang region of the Henan Province in China. It was discovered in the 1970s and has proven oil reserves of ~2,200 million of barrels. Currently, it is operated by Sinopec Henan Oilfield Company.

An amphiphilic iron-aromatic sulfonate was synthesized and characterized by spectroscopic methods and then used for catalytic Aquathermolysis [Chen *et al.* 2008]. The lab experiments were carried out in a batch reactor at 200°C for 72 h. As seen in Table 8.8, the author found 91% viscosity reductions and 23% of asphaltene conversion. The field test was carried out in two wells (G61012 and G6606) by the co-injection of the catalyst and steam (H-donor). In one case, nitrogen was used to increase the pressure of the downhole reservoir to achieve the temperature necessary for the upgrading reaction. After a soaking period of 14 d, the wells were produced, and samples were drawn for analysis [Chen *et al.* 2008].

The results showed an average of 51% increase in the oil rate in comparison with the case before the catalyst injection [Chen *et al.* 2008]. Similarly, the viscosities of the produced oils from well G61012 and G6606 were reduced by 80% and 82%, respectively. All the results indicated that the Aquathermolysis upgrading reactions were occurring at downhole conditions. As before (Table 8.8), lower viscosity reductions were found in the field than those determined in the lab (85% vs. 91%, respectively). Thus, better knowledge of downhole conditions is needed to optimize this subsurface upgrading process.

8.5.3 XINJIANG OILFIELD USING SULFONIC CU-CATALYST

The Xinjiang Oilfield was discovered in 1951, and it is one of the giant petroleum reservoirs of China [Höök *et al.* 2010]. Chao *et al.* used an amphiphilic Cu-alkyl ester sulfonate for the catalytic Aquathermolysis of Xinjiang heavy oil in the laboratory and a one-well field test [Chao *et al.* 2012]. As seen in Table 8.8, the batch laboratory experiment showed that the viscosity of upgraded heavy oil was reduced by 91% using 0.3 wt. % catalyst at 240°C for 24 h. Similarly, the authors found a 26% asphaltene reduction and 10% conversion of heavy content to light content and gas. Furthermore, reductions in the sulfur, nitrogen, and oxygen contents of the upgraded resins and asphaltenes were reported [Chao *et al.* 2012].

The field test for the Cu-containing catalytic system was conducted in the F10223 oil well, whose average petrophysical properties of its oil-bearing formation are as follows: Porosity of 30%, oil saturation of 73.3%, permeability of 1,052 mD, reservoir depth ranges from 159.5 m to 189.5 m, and pay zone thickness of 10 m [Chao *et al.* 2012].

Before the field test, the F10223 well had already been produced for two CSS periods, and as expected, its production conditions had declined [Chao *et al.* 2012]. To postpone the declining trend, the catalyst was added in the third CSS cycle. During the first month (31 days), the authors found that the oil production increased by 136 tons, and the viscosity of the produced oil was reduced by an average of 85% after the catalytic Aquathermolysis reaction. Also, they reported that the resin and asphaltene contents decreased by 16 wt. % and 6 wt. %, while the saturates and aromatic increased by 18 wt. % and 4 wt. %. These results gave evidence that the catalytic Aquathermolysis reactions were occurring [Chao *et al.* 2012].

Based on the results reported in this section for the field tests, it can be concluded that the use of downhole injection of water- or oil-soluble catalysts under CSS conditions has excellent potential for subsurface upgrading and the economical production of HO/B. Iron, molybdenum, and copper catalysts have been used with success. However, more knowledge is needed to address the issues associated with the injection of the catalytic system downhole regarding the placement, control, and environmental concerns.

REFERENCES

Alemán-Vázquez, L. O., Torres-Mancera, P., Ancheyta, J., Ramírez-Salgado, J., 2016, "Use of Hydrogen Donors for Partial Upgrading of Heavy Petroleum", *Energy Fuels*, 30, 9050–9060.
ASTM E-2456-06, 2006, "Standard Terminology Relating to Nanotechnology", ASTM International, West Conshohocken, PA.

Brons, G., Siskin, M., 1994, "Bitumen Chemical Changes during Aquathermolytic Treatments of Cold Lake Tar Sands", *Fuel*, 73 (2), 183–191.

Carbognani-Ortega, L., Hovsepian, C., Scott, C. E., Pereira-Almao, P., Moore, R. G., Mehta, S. A., Ursenbach, M. G., 2016, "Athabasca Bitumen In Situ Upgrading Reaction Monitoring: Operational Parameters versus Distillates Nature and Permanent Upgrading Achievable", *Energy Fuels*, 30, 5459–5469.

Chao, K., Chen, Y., Li, J., Zhang, X., Dong, B., 2012a, "Upgrading and Visbreaking of super-heavy oil by catalytic Aquathermolysis with aromatic sulfonic copper", *Fuel Proc. Technol.*, 104, 174–180.

Chao, K., Chen, Y., Liu, H., Zhang, X., Li, J., 2012b, "Laboratory Experiments and Field Test of a Difunctional Catalyst for Catalytic Aquathermolysis of Heavy Oil", *Energy Fuels*, 26, 1152–1159.

Chavez-Morales, S., Pereira-Almao, P., 2016, "Experimental and Numerical Simulation of Combined Enhanced Oil Recovery with In Situ Upgrading in a Naturally Fractured Reservoir", SPE No. 181207, presented at SPE Latin America and Caribbean Heavy and Extra Heavy Oil Conference, Lima, Peru, 19–20 October.

Chen, Y., Wang, Y., Wu, C., Xia, F., 2008, "Laboratory Experiments and Field Tests of an Amphiphilic Metallic Chelate for Catalytic Aquathermolysis of Heavy Oil", *Energy Fuels*, 22, 1502–1508.

Chen, Y., Wang, Y., Lu, J., Wua, C., 2009, "The Viscosity Reduction of Nano-keggin-K3PMo12O40 in Catalytic Aquathermolysis of Heavy Oil", *Fuel*, 88, 1426–1434.

Chen, Y., Wang, Y., Lu, J., Wua, C., 2010, "Gemini Catalyst for Catalytic Aquathermolysis of Heavy Oil", *J. Anal. Appl. Pyrol.*, 89, 159–165.

Chen, G., Yuan, W., Bai, Y., Zhao, W., Gu, X., Zhang, J., Jeje, A., 2017a, "Ethanol Enhanced Aquathermolysis of Heavy Oil Catalyzed by a Simple Co(II) Complex at Low Temperature", *Pet. Chem.*, 57 (5), 389–394.

Chen, G., Yuan, W., Wu, Y., Zhang, J., Song, H., Jeje, A., Song, S., Qu, C., 2017b, "Catalytic Aquathermolysis of Heavy Oil by Coordination Complex at Relatively Low Temperature", *Pet. Chem.*, 57 (10), 881–884.

Chen, G., Yan, J., Bai, Y., Gu, X., Zhang, J., Li, Y., Jeje, A., 2017c, "Clean Aquathermolysis of Heavy Oil Catalyzed by Fe(III) Complex at Relatively Low Temperature", *Pet. Sci. Technol.*, 35 (2), 113–119.

Clark, P. D., Hyne, J. B., Tyrer, J. D., 1983, "Chemistry of Organosulphur Compound Types Occurring in Heavy Oil Sands", *Fuel*, 62, 959.

Clark, P. D., Hyne, J. B., 1984, "Steam-Oil Chemical Reactions: Mechanisms for the Aquathermolysis of Heavy Oils", *AOSTRA J. Res.*, 1, 15–20.

Clark, P. D., Kirk, M. J., 1994, "Studies on the Upgrading of Bituminous Oils with Water and Transition Metal Catalysts", *Energy Fuels*, 8 (2), 380–387.

Coulter, A. W., Martinez, S. J., Fisher, K. F., 1987, "Remedial Cleanup, Sand Control, and Other Stimulation Treatments", In: *Petroleum Engineering Handbook*, H. B. Bradley, Ed., Society of Petroleum Engineers, Richardson, TX, Vol. 56, pp 56-1–56-9.

Desouky, S., Al Sabagh, A., Betiha, M., Badawi, A., Ghanem, A., Khalil, S., 2013, "Catalytic Aquathermolysis of Egyptian Heavy Crude Oil", *World Acad. Sci. Eng. Technol. Int. J. Chem. Molec. Eng.*, 7 (8) 638–643.

Fan, H.-F., Liu, Y.-J., Zhong, L.G., 2001, "Studies on the Synergetic Effects of Mineral and Steam on the Composition Changes of Heavy Oils", *Energy Fuels*, 15, 1475–1479.

Franco, C. A., Cardona, L., Lopera, S. H., Mejía, J. M., Cortés, F. B., 2016, "Heavy Oil Upgrading and Enhanced Recovery in a Continuous Steam Injection Process Assisted by Nanoparticulated Catalysts", SPE No. 179699 presented at SPE Improved Oil Recovery Conference, Tulsa, OK, 11–13 April.

Galarraga, C. E., Pereira-Almao, P., 2010a, "Athabasca Bitumen Upgrading with Ultradispersed Catalysts at Conditions Near to In-Reservoir Operation", *Prepr. Pap.-Am. Chem. Soc., Div. Petr. Chem.*, 55 (1), 32–34.

Galarraga, C. E., Pereira-Almao, P., 2010b, "Hydrocracking of Athabasca Bitumen Using Submicronic Multimetallic Catalysts at Near In-Reservoir Conditions", *Energy Fuels*, 24, 2383–2389.

Galukhin, G. V., Erokhin, A. A., Osin, Y. N., Nurgaliev, D. K., 2015, "Catalytic Aquathermolysis of Heavy Oil with Iron Tris(acetylacetonate): Changes of Heavy Oil Composition and *In Situ* Formation of Magnetic Nanoparticles", *Energy Fuels*, 29, 4768–4773.

Gray, M. R., 2015, *Upgrading Oilsands Bitumen and Heavy Oil*, The University of Alberta Press, Edmonton, and references therein.

Guo, K., Li, H., Yu, Z., 2016, "In-Situ Heavy and Extra-Heavy Oil Recovery: A Review", *Fuel*, 185, 886–902.

Guo, K., Zhang, Y., Shi, Q., Yu, Z., 2017, "The Effect of Carbon-Supported Nickel Nanoparticles in the Reduction of Carboxylic Acids for In Situ Upgrading of Heavy Crude Oil", *Energy Fuels*, 31, 6045–6055.

Guo, K., Hansen, V. F., Li, H., Yu, Z., 2018, "Monodispersed Nickel and Cobalt Nanoparticles in Desulfurization of Thiophene for In-Situ Upgrading of Heavy Crude Oil", *Fuel*, 211, 697–703.

Hamedi Shokrlu, Y., Babadagli, T., 2010, "Effects of Nano-Sized Metals on Viscosity Reduction of Heavy Oil/Bitumen during Thermal Applications", presented at Canadian Unconventional Resources and International Petroleum Conference, Calgary, Alberta, Canada, 19–21 October.

Hamedi Shokrlu, Y., Babadagli, T., 2013, "In-Situ Upgrading of Heavy Oil/Bitumen During Steam Injection by Use of Metal Nanoparticles: A Study on In-Situ Catalysis and Catalyst Transportation", *SPE Res. Eval. Eng.*, 16 (3), 333–344.

Hamedi Shokrlu, Y., Babadagli, T., 2014a, "Kinetics of the In-Situ Upgrading of Heavy Oil by Nickel Nanoparticle Catalysts and Its Effect on Cyclic-Steam-Stimulation Recovery Factor", *SPE Res. Eval. Eng.*, 17 (3), 355–364.

Hamedi Shokrlu, Y., Babadagli, T., 2014b, "Viscosity Reduction of Heavy Oil/Bitumen Using Micro-and Nano-Metal Particles during Aqueous and Non-Aqueous Thermal Applications", *J. Pet. Sci. Eng.*, 119, 210–220.

Hart, A., 2014, "The Novel THAI–CAPRI Technology and Its Comparison to Other Thermal Methods for Heavy Oil Recovery and Upgrading", *J. Pet. Exp. Prod. Technol.*, 4, 427–437.

Hashemi, R., Nassar, N. N., Pereira Almao, P., 2014a, "In Situ Upgrading of Athabasca Bitumen Using Multimetallic Ultradispersed Nanocatalysts in an Oil Sands Packed-Bed Column: Part 1. Produced Liquid Quality Enhancement", *Energy Fuels*, 28, 1338–1350.

Hashemi, R., Nassar, N. N., Pereira Almao, P., 2014b, "In Situ Upgrading of Athabasca Bitumen Using Multimetallic Ultradispersed Nanocatalysts in an Oil Sands Packed-Bed Column: Part 2. Solid Analysis and Gaseous Product Distribution", *Energy Fuels*, 28, 1351–1361.

Hendraningrat, L., Souraki, Y., Torsater, O., 2014, "Experimental Investigation of Decalin and Metal Nanoparticles-Assisted Bitumen Upgrading during Catalytic Aquathermolysis", SPE No. 167807 presented at SPE/EAGE European Unconventional Resources Conference and Exhibition, Vienna, Austria, 25–27 February.

Hongfu, F., Yongjian, L., Liying, Z., Xiaofei, Z., 2002, "The Study on Composition Changes of Heavy Oils during Steam Stimulation Processes", *Fuel*, 81 (13), 1733–1738.

Höök, M., Tang, X., Pang, X., Aleklett, K., 2010, "Development Journey and Outlook of Chinese Giant Oilfields", *Pet. Expl. Dev.*, 37 (2), 237–249.

Hovsepian, C. N., Carbognani-Ortega, L., Pereira-Almao, P., 2016, "Laboratory Two-Dimensional Experimental Simulation of Catalytic in Situ Upgrading", *Energy Fuels*, 30, 3652–3659.

Huc, A.-Y., Ed., 2011, *"Heavy Crude Oils: From Geology to Upgrading: An Overview"*, Technip, Paris, France, and references therein.

Hyne, J. B., Greidanus, J. W., Tyrer, J. D., Verona, D., Rizek, C., Clark, P. D., Clarke, R. A., Koo, J., 1982, "Aquathermolysis of Heavy Oils", Proc. of 2nd Int. Conf. Heavy Crude and Tar Sands, Caracas, Venezuela, 7–17 February, pp 404–411.

Hyne, J. B., Clark, P. D., 1985, "Additive for Inclusion in a Heavy Oil Reservoir Undergoing Steam Injection", US Patent No. 4,506,733, 26 March.

Hyne, J. B., 1986, "Aquathermolysis — A Synopsis of Work on the Chemical Reaction between Water (Steam) and Heavy Oil Sands during Simulated Steam Stimulation", Synopsis Report No.: 50, AOSTRA Contracts No.: 11, 103, 103B/C April.

Jaishankar, M., Tseten, T., Anbalagan, N., Mathew, B. B., Beeregowda, K. N., 2014, "Toxicity, Mechanism and Health Effects of Some Heavy Metals", *Interdiscip. Toxicol.*, 7 (2), 60–72.

Jiang, S., Liu, X., Liu, Y., Zhong, L., 2005, "In Situ Upgrading Heavy Oil by Aquathermolytic Treatment under Steam Injection Conditions", SPE No. 91973, presented at SPE International Symposium on Oilfield Chemistry, The Woodlands, TX, 2–4 February.

Kapadia, P. R., Kallos, M. S., Gates, I. D., 2015, "A Review of Pyrolysis, Aquathermolysis, and Oxidation of Athabasca Bitumen", *Fuel Proc. Technol.*, 131, 270–289.

Kayukova, G. P., Mikhailova, A. M., Kosachev, I. P., Feoktistov, D. A., Vakhin, A. V., 2018, "Conversion of Heavy Oil with Different Chemical Compositions under Catalytic Aquathermolysis with an Amphiphilic Fe-Co-Cu Catalyst and Kaolin", *Energy Fuels*, 32, 6488–6497.

Khalil, M., Lee, R. L., Liu, N., 2015, "Hematite Nanoparticles in Aquathermolysis: A Desulfurization Study of Thiophene", *Fuel*, 145, 214–220.

Lakhova, A., Petrov, S., Ibragimova, D., Kayukova, G., Safiulina, A., Shinkarev, Okekwe, R., 2017, "Aquathermolysis of Heavy Oil Using Nano Oxides of Metals", *J. Pet. Sci. Eng.*, 153, 385–390.

Laurent, E., Delmon, B., 1994, "Deactivation of a Sulfided NiMo/y-Al2O3 during the Hydrodeoxygenation of Bio-Oils: Influence of a High Water Pressure", In: *Catalyst Deactivation*, B. Delmon and G. F. Froment, Eds., Elsevier, Amsterdam, p 459.

Li, W., Zhu J.-H., Qi J.-H., 2007, "Application of Nano-Nickel Catalyst in the Viscosity Reduction of Liaohe Extra-Heavy Oil by Aqua-Thermolysis", *J. Fuel Chem. Technol.*, 35 (2), 176–180.

Li, J., Chen, Y., Liu, H., Wang, P., Liu, F., 2013, "Influences on the Aquathermolysis of Heavy Oil Catalyzed by Two Different Catalytic Ions: Cu^{2+} and Fe^{3+}", *Energy Fuels*, 27 (5), 2555–2562.

Li, G.-R., Chen, Y., An, Y., Chen, Y.-L., 2016, "Catalytic Aquathermolysis of Super-Heavy Oil: Cleavage of C\S Bonds and Separation of Light Organosulfurs", *Fuel Proc. Technol.*, 153, 94–100.

Liu, H., Fan, H., 2002, "The Effect of Hydrogen Donor Additive on the Viscosity of Heavy Oil during Steam Stimulation", *Energy Fuels*, 16, 842–846.

Loria, H., Pereira-Almao, P., Scott, C., 2009a, "A Model to Predict the Concentration of Sub-Micrometer Solid Particles in Viscous Media Confined Inside Horizontal Cylindrical Channels", *Ind. Chem. Eng. Res.*, 46, 4094–4100.

Loria, H., Pereira-Almao, P., Scott, C., 2009b, "A Model to Predict the Concentration of Dispersed Solid Particles in an Aqueous Medium Confined Inside Horizontal Cylindrical Channels", *Ind. Chem. Eng. Res.*, 48, 4088–4093.

Loria, H., Trujillo-Ferrer, G., Sosa-Stull, C., Pereira-Almao, P., 2011, "Kinetic Modeling of Bitumen Hydroprocessing at In-Reservoir Conditions Employing Ultradispersed Catalysts", *Energy Fuels*, 25, 1364–1372.

Maity, S. K., Ancheyta, J., Marroquin. G., 2010, "Catalytic Aquathermolysis Used for Viscosity Reduction of Heavy Crude Oils: A Review", *Energy Fuels*, 24, 2809–2816.

Menger, F. M., Littau, C. A., 1991, "Gemini-Surfactants: Synthesis and Properties", *J. Am. Chem. Soc.*, 113, 1451–1452.

Menger, F. M., Keiper, J. S., 2000, "Gemini Surfactants", *Angew. Chem. Int. Ed.*, 39, 1906–1920.

Mohammad, A. A. A., Mamora, D. D., 2008, "Insitu Upgrading of Heavy Oil Under Steam Injection with Tetralin and Catalyst", SPE No. 117604, presented at International Thermal Operations and Heavy Oil Symposium, Calgary, Alberta, Canada, 20–23 October.

Muraza, O., Galadima, A., 2015, "Aquathermolysis of Heavy Oil: A Review and Perspective on Catalyst Development", *Fuel*, 157, 219–231.

Noguchi, T., Ed., 1991, *Heavy Oil Processing Handbook*, Research Association for Residual Oil Processing, Tokyo, Japan, pp 48–64.

Nguyen, N., Chen, Z., Pereira Almao, P., Scott, C. E., Maini, B., 2017, "Reservoir Simulation and Production Optimization of Bitumen/Heavy Oil via Nanocatalytic in Situ Upgrading", *Ind. Eng. Chem. Res.*, 56, 14214–14230.

Olvera, J. N. R., Gutiérrez, G. J., Serrano, J. A. R., Ovando, A. M., Febles, V. G., Arceo, L. D. B., 2014, "Use of Unsupported, Mechanically Alloyed NiWMoC Nanocatalyst to Reduce the Viscosity of Aquathermolysis Reaction of Heavy Oil", *Catal. Commun.*, 43, 131–135.

Ovalles, C., Filgueiras, E., Morales, A., Rojas, I., de Jesus, J. C., Berrios, I., 1998, "Use of Dispersed Molybdenum Catalyst and Mechanistic Studies for Upgrading Extra-Heavy Crude Oil Using Methane as Source of Hydrogen", *Energy Fuels*, 12, 379–385.

Ovalles, C, Hamana, A, Rojas, I., Bolivar, R., 1995, "Upgrading of Extra-Heavy Crude Oil by Direct Use of Methane in the Presence of Water. Deuterium Labeled Experiments and Mechanistic Consideration", *Fuel*, 74, 1162–1168.

Ovalles, C., Rivero, V., Salazar, A., 2015, "Downhole Upgrading of Orinoco Basin Extra-Heavy Crude Oil Using Hydrogen Donors under Steam Injection Conditions. Effect of the Presence of Iron Nanocatalysts", *Catalysts*, 5, 286–297.

Ovalles, C., Rodriguez, H., 2008, "Extra-Heavy Crude Oil Downhole Upgrading Using Hydrogen Donors under Cyclic Steam Injection Conditions. Physical and Numerical Simulation Studies", PETSOC-08-01-43, *J. Can. Pet. Technol.*, 47, 43–51.

Pereira Almao, P., Sosa-Stull, C., Trujillo, G., Galarraga, C., Maini, B., Zamani, A., Chen, J., Li, J., 2010, "Residual Oil Assisted *In Situ* (In-Reservoir) Catalytic Upgrading", *Prepr. Pap. Am. Chem. Soc., Div. Petr. Chem.*, 55 (1), 40–43.

Pereira-Almao, P., Marcano, A., Vieman, A., Lopez-Linares, F., Vasquez, A., 2011, "Ultradispersed Catalyst Compositions and Methods of Preparation", US Patent 7,897,537, 1 March.

Pereira-Almao, P., 2012, "In Situ Upgrading of Bitumen and Heavy Oils Via Nanocatalysis", *Can. J. Chem. Eng.*, 90, 320–329.

Pereira-Almao, P., Marcano, A., Vieman, A., Lopez-Linares, F., Vasquez, A., 2012a, "Ultradispersed Catalyst Compositions and Methods of Preparation", US Patent 8,283,279, 9 October.

Pereira-Almao, P., Marcano, A., Vieman, A., Lopez-Linares, F., Vasquez, A., 2012b, "Ultradispersed Catalyst Compositions and Methods of Preparation", US Patent 8,298,982, 30 October.

Pereira-Almao, P., Marcano, A., Vieman, A., Lopez-Linares, F., Vasquez, A., 2012c, "Ultradispersed Catalyst Compositions and Methods of Preparation", US Patent 8,304,363, 6 November.

Pereira-Almao, P., Chen, Z., Maini, B., Scott-Algara, C., 2015a, "In Situ Upgrading Via Hot Fluid Injection", US Patent Appl. No. 20150114636, 30 April.

Pereira-Almao, P., Scott, C., Carbognani, L., Maini, B., Chen, J., 2015b, "Advancing the Unconventional Upgrading of Heavy Oils", World Heavy Oil Conf., Paper No. WHOC15 – 174, Edmonton, Alberta, Canada.

Ramirez-Corredores, M. M., 2017, *The Science and Technology of Unconventional Oils: Finding Refining Opportunities*, 1st Ed., Elsevier, London, Chapter 2, p 2, and references therein.

Rivas, O. R., Campos, R. E., Borges, L. G., 1988, "Experimental Evaluation of Transition Metals Salt Solutions as Additives in Steam Recovery Processes", SPE No. 18076, presented at SPE Annual Technical Conference and Exhibition, Houston, TX, 2–5 October.

Rodriguez-DeVecchis, V. M., Carbognani Ortega, L., Scott, C. E., Pereira-Almao, P., 2015, "Use of Nanoparticle Tracking Analysis for Particle Size Determination of Dispersed Catalyst in Bitumen and Heavy Oil Fractions", *Ind. Eng. Chem. Res.*, 54, 9877–9886.

Rodriguez-DeVecchis, V. M., Carbognani Ortega, L., Scott, C. E., Pereira-Almao, P., 2018, "Deposition of Dispersed Nanoparticles in Porous Media Similar to Oil Sands. Effect of Temperature and Residence Time", *Ind. Eng. Chem. Res.*, 57, 2385–2395.

Scheele-Ferreira, E., Scott, C. E., Perez-Zurita, M. J., Pereira-Almao, P., 2017, "Effects of the Preparation Variables on the Synthesis of Nanocatalyst for In Situ Upgrading Applications", *Ind. Eng. Chem. Res.*, 56, 7131–7140.

Ren, R., Liu, H., Chen, Y., Li, J., Chen, Y., 2015, "Improving the Aquathermolysis Efficiency of Aromatics in Extra-Heavy Oil by Introducing Hydrogen-Donating Ligands to Catalysts", *Energy Fuels*, 29, 7793–7799.

Scott, C. E., Perez-Zurita, M. J., Carbognani, L. A., Molero, H., Vitale, G., Guzmán, H. J., Pereira-Almao, P., 2015, "Preparation of NiMoS Nanoparticles for Hydrotreating", *Catal. Today*, 250, 21–27.

Shah, A., Fishwick, R. P., Leeke, G. A., Wood, J., 2010, "Experimental Optimization of Catalytic Process In-Situ for Heavy Oil and Bitumen Upgrading", SPE No. 136870, presented at the Can. Unconventional Resources & Int. Pet. Conf. held in Calgary, Alberta, Canada, 19–21 October.

Shuwa, S. M., Al-Hajri, R. S., Mohsenzadeh, A., Al-Waheibi, Y. M., Jibril, B. Y., 2016, "Heavy Crude Oil Recovery Enhancement and In-Situ Upgrading during Steam Injection Using Ni-Co-Mo Dispersed Catalyst", SPE No. 179766, presented at SPE EOR Conference at Oil and Gas West Asia, Muscat, Oman, 21–23 March.

Speight, J. G., 2007, *The Chemistry and Technology of Petroleum*, 4th Ed., CRC Press, Boca Raton, FL, and references therein.

Stapp, P. R., 1989, "In Situ Hydrogenation", Report from Nat. Inst. of Pet. Ener. Res. NIPER-434, Bartlesville, OK, December.

Suarez, R. G. S., Scott, C., Hejazi, S. H., Pereira-Almao, P., 2015, "Experimental Study of Heavy Oil In-Situ Upgrading Using High Temperature Gas-Oil Gravity Drainage in Naturally Fractured Reservoirs", SPE No. 177220, presented at SPE Latin American and Caribbean Petroleum Engineering Conference, Quito, Ecuador, 18–20 November.

Suarez, R. G. S., Scott, C. E., Pereira-Almao, P., Hejazi, S. H., 2016, "Experimental Study of Nanocatalytic In-Situ Upgrading for Heavy Oil Production from Naturally-Fractured-Carbonate Reservoirs", SPE No. 179619, presented at SPE Improved Oil Recovery Conference, Tulsa, OK, 11–13 April.

Tajmiri, M., Ehsani, M. R., 2016, "The Potential of CuO Nanoparticles to Reduce Viscosity and Alter Wettability at Oil- Wet and Water- Wet Rocks in Heavy Oil Reservoir", SPE No. 181298, presented at SPE Annual Technical Conference and Exhibition, Dubai, UAE, 26–28 September.

Tang, X.-D., Zhu, H., Li, J.-J., Wang, F., Qing, D.-Y., 2015, "Catalytic Aquathermolysis of Heavy Oil with Oil-Soluble Multicomponent Acrylic Copolymers Combined with Cu^{2+}", *Pet. Sci. Technol.*, 33, 1721–1727.

Tchounwou, P. B., Yedjou C. G., Patlolla A. K., Sutton D. J., 2012, "Heavy Metal Toxicity and the Environment", In: *Molecular, Clinical and Environmental Toxicology. Experientia Supplementum*, vol 101, A. Luch, Ed., Springer, Basel.

Viloria, A., Parisi, S., Borges, L., Rodriguez, R., Martinez, E., Ramos, A. D., Brito, J. D., Campos, R. G., 1985, "Simulación y predicción de la generación de H_2S en yacimientos de crudos pesados sometidos a la inyección de vapor", *Rev. Tec. INTEVEP*, 5 (2), 147–155.

Wang, J., Liu, L., Zhang, L., Li, Z., 2014, "Aquathermolysis of Heavy Crude Oil with Amphiphilic Nickel and Iron Catalysts", *Energy Fuels*, 28, 7440–7447.

Wang, W, Chen, Y., He, J., Li, P., Chao Yang, C., 2010, "Mechanism of Catalytic Aquathermolysis: Influences on Heavy Oil by Two Types of Efficient Catalytic Ions: $Fe3+$ and $Mo6+$", *Energy Fuels*, 24, 1502–1510.

Weissman, J. G., 1997, "Review of Processes for Downhole Catalytic Upgrading of Heavy Crude Oil", *Fuel Proc. Technol.*, 50, 199–213.

Weissman, J. G., Kessler, R. V., 1996, "Downhole Heavy Crude Oil Hydroprocessing", *Appl. Catal.*, 140, 1.

Weissman, J. G., Kessler, R. V., Sawicki, R. A., Belgrave, J. D. M., Laureshen, C. J., Metha, S. A., Moore, R. G., Ursenbach, M. G., 1996, "Down-Hole Catalytic Upgrading of Heavy Crude Oil", *Energy Fuels*, 10, 883.

Wen, S., Zhao, Y., Liu, Y., Hu, S., 2007, "A Study on Catalytic Aquathermolysis of Heavy Crude Oil During Steam Stimulation", SPE No. 106180, presented at International Symposium on Oilfield Chemistry, Houston, TX, 28 February–2 March.

Wu, C., Su, J., Zhang, R., Lei, G., Cao, Y., 2013, "The Use of a Nano-Nickel Catalyst for Upgrading Extra-Heavy Oil by an Aquathermolysis Treatment Under Steam Injection Conditions", *Pet. Sci. Technol.*, 31, 2211–2218.

Yi, S., Babadagli, T., Li, H. A., 2018, "Use of Nickel Nanoparticles for Promoting Aquathermolysis Reaction during Cyclic Steam Stimulation", *SPE J.*, 23 (1), 145–156.

Yusuf, A., Al-Hajri, R. S., Al-Waheibi, Y. M., Jibril, B. Y., 2016a, "Upgrading of Omani Heavy Oil with Bimetallic Amphiphilic Catalysts", *J. Taiwan Inst. Chem. Eng.*, 67, 45–53.

Yusuf, A., Al-Hajri, R. S., Al-Waheibi, Y. M., Jibril, B. Y., 2016b, "In-situ upgrading of Omani Heavy Oil with Catalyst and Hydrogen Donor", *J. Anal. Appl. Pyrol.*, 121, 102–112.

Yufeng, Y., Shuyuan, L., Fuchen, D., Hang, Y., 2009, "Change of Asphaltene and Resin Properties after CATALYTIC Aquathermolysis", *Pet. Sci.*, 6, 194–200.

Zamani, A., Maini, B., Pereira-Almao, P., 2010, "Experimental Study on Transport of Ultra-Dispersed Catalyst Particles in Porous Media", *Energy Fuels*, 24, 4980–4988.

Zhang, T., 1997, "Heavy Oil Recovery Techniques of Liaohe Oilfield", WPC-29200, presented at 15th World Petroleum Congress, Beijing, China, 12–17 October.

Zhang, Z., Barrufet, M. A., Lane, R. H., Mamora, D. D., 2012, "Experimental Study of In-Situ Upgrading for Heavy Oil Using Hydrogen Donors and Catalyst Under Steam Injection Condition", SPE No. 157981 presented at SPE Heavy Oil Conference Canada, Calgary, Alberta, Canada, 12–14 June.

Zhong, L. G., Liu, Y. J., Fan, H. F., Jiang, S. J., 2003, "Liaohe Extra-Heavy Crude Oil Underground Aquathermolytic Treatments Using Catalyst and Hydrogen Donors under Steam Injection Conditions", SPE No. 84863, presented at SPE International Improved Oil Recovery Conference in Asia Pacific, Kuala Lumpur, Malaysia, 20–21 October.

9 *In-Situ* Combustion

As mentioned in Section 3.1.5, *in-situ* combustion (ISC) or fireflood is the oldest of the thermal recovery methods to produce HO/B. In these processes, air or oxygen is injected downhole to create a combustion front that propagates from the injector to the producing well (Figure 3.14). At the same time, some of the heavy oil is burned, a residue is left behind, and the generated heat reduces the viscosity of the HO/B to increase the oil rate and total recovery. The inherent advantage is that no external energy sources are needed, and very high temperatures can be achieved for the conversion of the heavy crudes into lighter hydrocarbons. However, they suffer from significant drawbacks such as lack of control of the burning front, safety concerns, partial oxidation of the valuable distillable materials, generation of toxic emissions, and large volumes of flue gas, among others [Speight 2007, Huc 2011, Guo *et al.* 2016].

In-situ combustion processes have been studied for more than 60 years, and there is an impressive volume of technical literature in this area [Sarathi 1999]. Furthermore, more than 200 ISC field tests have been performed with different degrees of failures and successes [Turta *et al.* 2007, Manrique *et al.* 2010, Huc 2011]. Thus, the objective of this chapter is not to carry out a literature review of ISC-only processes and the oil-producing mechanisms but to discuss the subsurface upgrading concepts that use *in-situ* combustion to achieve their objectives. Most of the ISC lab and field tests have been aimed to enhance the oil rates and percentages of oil recovery and *not* to upgrade the HO/B to address the transportation and environmental issues or to increase the quality of the produced oil. However, during their ISC experimentation, several lab and field publications have reported improvements in the quality of the produced oil, as measured in terms of increases in the API, residue conversion, desulfurization, etc. [Marjerrison and Fassihi 1994, Weissman *et al.* 1996, Castanier and Brigham 2003]. These reports are discussed in Section 9.2.

Similarly, several articles have reported the use of water-soluble metal salts and heterogeneous catalysts for ISU lab experiments to improve further the quality of the produced oil in comparison with the thermally-only counterparts. During *in-situ* combustion, most of the hydrocarbon-bond cracking (carbon–carbon, sulfur–carbon, etc.) is due to the extremely high temperature achieved during combustion (> 500°C). The use of catalysts jointly with the thermally or catalytically generated hydrogen could be an intriguing alternative to upgrade and enhance further the stability of the produced oil [Hajdo *et al.* 1985, Weissman *et al.* 1996]. As discussed in Section 8.1.1, it has been proposed that heterogeneous catalysts could be placed as a gravel pack in the production wells after the combustion and heated zones [Weissman 1997]. Homogeneous and water-soluble catalysts could also be considered. However, there are several issues associated with the use of catalytic materials downhole during ISC experiments (see also Section 8.1). These reports are presented in Section 9.3.

This chapter finishes with the description of the Toe-to-Heel Air Injection Process (THAI) and its catalytic version CAPRI. A significant amount of work has been carried out in these concepts from lab scale to field test. Numerical simulations have been reported as well (Section 9.4).

9.1 GENERAL MECHANISM OF *IN-SITU* COMBUSTION

Before we start discussing the literature of SSU via *in-situ* combustion, we should present an overview of the chemical pathways that have been reported for ISC. Numerous bitumen combustion experiments conducted over the last decades have shown that the oxidation and upgrading mechanisms are very complex and depend not only on the origin and chemical properties of the HO/B but on the petrophysical and reservoir characteristics [Hascakir 2015, Kapadia *et al.* 2015].

For Athabasca bitumen, the chemical reactions are broadly classified into Low (LTO) and High (HTO) Temperature Oxidation. In LTO (Equation 9.1), water, carbon oxides, and oxygen-containing compounds such as carboxylic acids, ketones, aldehydes, alcohols, and hydroperoxides are formed in the temperature range from 150°C to 300°C. In general, LTO leads to an increase in the viscosity of the produced oil [Xu *et al.* 2001].

LTO Reaction

$$\begin{array}{c} \text{Heavy Oil} \\ \text{Or Bitumens} \end{array} + \begin{array}{c} \text{Air or} \\ \text{Oxygen} \end{array} \xrightarrow{150\ to\ 300°C} CO + CO_2 + H_2O + R\text{-}\overset{O}{\overset{\|}{C}}OH + R\text{-}\overset{O}{\overset{\|}{C}}\text{-}R$$
$$+$$
$$R\text{-}CHO + R\text{-}CH_2OH + R\text{-}CH_2OO\text{-}H \tag{9.1}$$

At temperatures higher than 300°C, the bitumen starts to pyrolyze (i.e., thermal cracking, see Section 4.1.2) to generate lighter hydrocarbons, heavy fractions, and coke (HTO, Equation 9.2). In turn, the last two materials get combusted at a higher temperature to produce carbon oxides and water. Additionally, smaller molecules formed during pyrolysis are oxidized, generating most of the heat to self-sustain the network of reactions during the *in-situ* combustion of HO/B [Burger and Sahuquet 1972, Belgrave *et al.* 1993, Kapadia *et al.* 2015, Klock and Hascakir 2015].

HTO Reaction

$$\begin{array}{c} \text{Heavy Oil} \\ \text{Or Bitumens} \end{array} + \begin{array}{c} \text{Air or} \\ \text{Oxygen} \end{array} \xrightarrow{>300°C} \begin{array}{c} \text{Lighter} \\ \text{Hydrocarbons} \end{array} + CO + CO_2 + H_2O$$
$$+$$
$$\text{Heavy Fractions} + \text{Coke} \tag{9.2}$$

In addition to the LTO and HTO reactions, other chemical pathways are occurring simultaneously such as Aquathermolysis (Equation 7.4, Section 7.2), water–gas shift (Equation 7.3) and dehydrogenation (Equation 4.10) to generate H_2, and coking-forming reactions (Section 4.1.3.1) such as dealkylation (Equation 4.9), cyclization (Equation 4.11), demethylation (Equation 4.12), aromatization (Equation 4.13), and aromatic condensation (Equation 4.14). Detailed discussions on the ISC mechanisms and kinetics have been reviewed by several authors [Cavanzo *et al.* 2014, Chen *et al.* 2014, Kapadia *et al.* 2015] and escape from the objectives of this book.

Gates *et al.* pointed out that, during ISC lab experiments, in addition to the physical properties of the bitumen, the measurements are aimed to determine oxygen uptake rates and the effect of temperature and partial pressure of oxygen on bitumen composition [Kapadia *et al.* 2015]. There have been several attempts to understand the underlying mechanisms of bitumen oxidation and upgrading in terms of generation or disappearance of hydrocarbon constituents and solubility fractions. For example, the reaction sequences of individual components of SARA group-type separation, i.e., Saturates, Aromatics, Resins, and Asphaltenes during the oxidation process and the kinetic behaviors in different oxidation stages have been reported [Kok and Karacan 2000, Freitag and Verkoczy 2005]. In particular, Hascakir studied the effect of initial oil saturation on the recovery and upgrading of a Canadian bitumen during ISC [Hascakir 2015]. This author used a reaction model based on SARA fractions and five combustion tube experiments to calibrate the model. The author found that the resin-to-aromatics ratio governed the viscosity reduction of the produced oil, and the molecular structure of the aromatic fraction seemed to be the leading factor for density and API gravity improvement [Hascakir 2015].

Finally, the effect of sulfur-containing compounds on thermal cracking, coking, and oxidation mechanisms remains unclear. Future studies should focus on the relationships between sulfur chemistry and hydrocarbon pyrolysis and oxidation [Kapadia *et al.* 2015].

9.2   UPGRADING DURING LAB AND FIELD TESTS

The effects of the *in-situ* upgrading reactions, i.e., LTO (Equation 9.1) and HTO (Equation 9.2) and Equations 4.9–4.13 and 7.3–7.4 on the produced crude oils by the ISC process have been widely documented in the literature [Prats 1978, Ramey *et al.* 1992, Xu *et al.* 2001, Zhao *et al.* 2015]. In general, Castanier and Brigham reported that numerous ISC lab and field observations had shown upgrading from 2° to 6°API for heavy oils undergoing combustion [Castanier and Brigham 2003]. In this section, we discuss some of those publications.

Marjerrison and coworkers from Amoco-Canada operated a pressure cycling fireflood at Morgan, in the Lloydminster area of Alberta, Canada, since the mid-1980s [Marjerrison and Fassihi 1994]. The reservoir contains a ~12°API heavy crude with a live oil viscosity of 6,800 kPa s at 70°F. The formation has an average thickness of 31 ft and has no permeability barriers present. The project has 43 wells on 10 acres/well spacing laid out in 30 acres, seven-spot patterns. The historical field performance was as follows. Firstly, after a brief primary production period, steam and air/steam cyclic stimulations were performed on individual production wells from 1981 to 1986 [Marjerrison and Fassihi 1994].

When the steam-to-oil ratios (SOR) and water-oil ratios (WOR) were below the acceptable values, the *in-situ* combustion project was started in 1986. Air was injected into a central well accompanied by production from its surrounding producers [Marjerrison and Fassihi 1994]. Once a predetermined pressure level was reached, air injection was stopped, and the producers were reactivated. The production was continued until the oil rate dropped below its economic level or when injection had to be restarted to maintain burning in the reservoir. The authors reported that the field performance was exceptional, with air-to-oil ratios of less than 2 MCF/STB (355 m^3/m^3) and oil rates of 30–150 B/D per well. The cumulative oil production as of December 1992 (end-of-project) was 23% of the OOIP or 15.2 MMBBLS [Marjerrison and Fassihi 1994].

Figure 9.1 shows the average values of the API gravity and the volume percentage of residue 650°F+ in the produced oil in August 1983 (original) and April 1989 (after ISC) [Marjrrison 1994]. As seen, an increase in the API was observed (from 11° to 22°API), and the percentage of residue conversion was 32% (from 81% to 55%). The authors mentioned that air injection had been stopped since June 1988 indicating a lag time of six months to the time the upgrading was observed. The

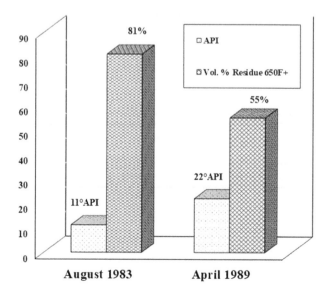

FIGURE 9.1   Average values of the API gravity and the volume percentage of residue 650°F+ in the produced oil on August 1983 (original) and April 1989 (After ISC). Data taken from Marjrrison and Fassihi [1994].

authors suggested that a standard frontal displacement combustion process only showed upgraded oil production once a significant portion of a pattern was burned and the light oil bank reached the production wells [Marjrrison 1994].

Furthermore, Marjerrison and coworkers carried out the numerical simulation and history match of the field results [Marjerrison and Fassihi 1994]. Their model incorporated the following production mechanisms: (1) A high-speed channel from the injector to the producer, (2) combustion gas entrainment by the oil phase, and (3) heavy oil upgrading. The simulations predicted the cumulative oil production, the producing gas-to-oil ratios (GOR), and the reservoir pressures reasonably well. The production timing of the upgraded crude which resulted from the pressure cycling process could not be matched closely. The authors reported that field limitations might have hindered an early detection of light oil bank [Marjrrison 1994]. Unfortunately, no details were given on how the upgrading reactions were included in the numerical simulations.

Xu *et al.* from the University of Calgary developed a two-stage LTO process, in which first the crude oil was contacted with air at mild conditions (80–120°C) followed by a period at higher temperatures (200–220°C) [Xu *et al.* 2001]. The authors proposed that the first step incorporated oxygen into some of the hydrocarbons, yielding labile bonds that were broken in the second stage. In a field situation, this process would be analogous to first injecting air into a formation at a low temperature, then starting a steam soak or steamflooding [Xu *et al.* 2001].

Xu and coworkers carried out laboratory batch experiments using Athabasca bitumen and examined the effects of oxygen partial pressure, temperature, reaction time, and the presence of rock and brine [Xu *et al.* 2001]. The upgrading of the bitumen was followed only by viscosity measurements at 70°C. The authors found viscosity reductions in the 30–40% range using bitumen/distilled water mixtures and shorter duration (six days) and lower temperature (80°C) first stages, followed by longer (nine days) and higher temperature (220°C) second periods [Xu *et al.* 2001]. In general, experiments carried out using brine gave a lower viscosity of the produced oil than those using distilled water. A run conducted with 71 wt. % core, 8 wt. % brine, and 20 wt. % bitumen gave the highest viscosity reduction (~50%). The authors concluded that Athabasca core has a definite "catalytic effect" on free radical initiation and cracking reactions and may catalyze the LTO reactions in the heavy oil [Xu *et al.* 2001].

Rahema and coworkers from Texas A&M University carried out physical and numerical simulations of a concept called Combustion Assisted Gravity Drainage (CAGD) [Rahnema *et al.* 2012a, 2012b]. Figure 9.2 shows the diagram for this process. As seen, it consists of two horizontal wells, one on top of the other, but drilled in the opposite direction. Air injection was carried out using the top well whereas oil production was performed with the lower one. The authors reported ISC experiments using a rectangular 3-D combustion cell with dimensions of 0.62 m, 0.41 m, and 0.15 m. Enriched air (50% O_2) was injected to create and sustain the combustion front of the model. Experimental results showed that oil displacement occurred mainly by gravity drainage. The authors observed vigorous combustion at the early stages near the heel of the injection well, where peak temperatures were in the of 530–690°C range. Analysis of the spent sand pack confirmed that coke deposition resulted in the formation of a gas seal barrier between injector and producer which enhanced the circulation of the injected air of the combustion chamber. The authors reported an ultimate oil recovery estimated at 72% OOIP [Rahnema *et al.* 2012a, 2012b].

Furthermore, Rahema *et al.* observed upgrading of the produced oils as measured by the API gravity [Rahnema *et al.* 2012a, 2012b]. Figure 9.3 shows the API of the produced oil from the lab experiments of the CAGD process using Peace River and Athabasca bitumens. At the beginning of the run, the authors reported that the produced oils were downgraded from 9 to 8°API for Peace River and from 8 to 7°API for Athabasca. They attributed this phenomenon to vaporization of the lighter components of the crude oil. After the combustion front became stable, they observed increases in the gravity of the produced oil of ~3°API for both bitumens [Rahnema *et al.* 2012a, 2012b].

Finally, Rahema and coworkers carried out numerical simulations and history matching of the Athabasca laboratory data [Rahnema *et al.* 2012a]. They used the kinetic scheme reported by

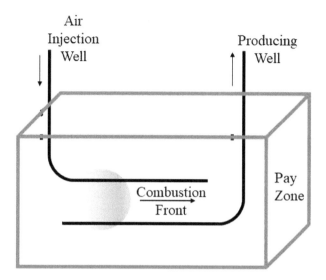

FIGURE 9.2 Concept diagram for the Combustion Assisted Gravity Drainage.

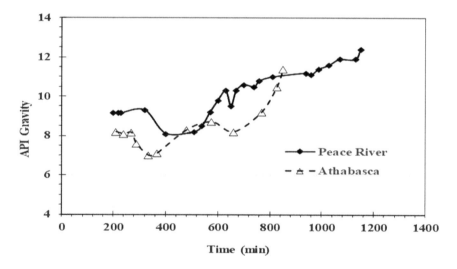

FIGURE 9.3 API of the produced oil from the lab experiments of the Combustion Assisted Gravity Drainage using Peace River and Athabasca bitumens. Data taken from Rahnema *et al.* [2012a, 2012b].

Belgrave *et al.* of two pseudo-components (maltenes and asphaltenes) and three chemical reactions, i.e., thermal cracking, LTO (Equation 9.1), and HTO (Equation 9.2) [Belgrave *et al.* 1993]. Simulation results showed agreement between numerical and experimental data in terms of fluid production rate, combustion temperature, and produced-gas composition. Unfortunately, no modeling of the bitumen upgrading reactions was performed.

Zhao and coworkers carried out an *in-situ* combustion field test in the Qian block of the Jiang oilfield, Junggar basin, China [Zhao *et al.* 2015]. The reservoir characteristics are the following. The depth of the reservoir ranged from 272 to 735 m, 580 m on average; the thickness was approximately 16–66 m, 46 m on average. This block covers an area of about 1.24 km^2 with 1.44×10^6 t of OOIP. After ten years of Huff and Puff (1997–2007) and one year of steamflooding, this block began the ISC field test in December 2009, with an estimated 40% of the OOIP still left in the reservoir. No details were given on the oil rates, cumulative oil production, and a recovery factor of the field test. However, the authors reported that "a favorable result" was obtained [Zhao *et al.* 2015].

Zhao *et al.* concentrated on studying the changes in crude oil properties during the combustion process using rheology tests, SARA, FTIR, GC/MS, and acid number measurements [Zhao *et al.* 2015]. During fireflood, the authors reported that the viscosity of the produced crude oil decreased significantly (one order of magnitude in 42 months) as the combustion front moved toward the producing well. This reduction varied according to different sampling wells with varying dip angles and distances [Zhao *et al.* 2015].

During the early stage of ISC, they observed an increase of the saturates accompanied by the decrease of aromatics content with the concomitant upgrading in the crude oil properties [Zhao *et al.* 2015]. For the resin fractions and using FTIR, polar functional groups were identified such as carboxylic acids, ketones, and alcohols. By GC/MS the authors detected significant quantiles of n-alkanes, normal or branched long-chain fatty acids, alkylphenol, and naphthalenes. Consistent with those results, the acid number of produced oils increased remarkably during the ISC process [Zhao *et al.* 2015]. All these findings were consistent with the general mechanisms discussed in Section 9.1.

A shown in the previous paragraphs, it can be concluded that several lab publications and field tests have found upgrading of the produced oil during their ISC experimentation as measured in terms of increases in the API, residue conversion, desulfurization, etc. [Marjerrison and Fassihi 1994, Castanier and Brigham 2003, Rahnema 2012a, 2012b]. The upgrading or the HO/B can be attributed to several chemical reactions occurring at the same time such as LTO (Equation 9.1) and HTO (Equation 9.2) reactions and Equations 4.9–4.13 and 7.3–7.4 as well as Aquathermolysis (Equation 7.4, Section 7.2), water–gas shift (Equation 7.3) and dehydrogenation (Equation 4.10) to generate H_2, and coking-forming reactions (Section 4.1.3.1) such as dealkylation (Equation 4.9), cyclization (Equation 4.11), demethylation (Equation 4.12), aromatization (Equation 4.13), and aromatic condensation (Equation 4.14).

9.3 USE OF METAL-CONTAINING HETEROGENEOUS CATALYSTS

Broadly, two types of metal-containing catalysts have been studied in the literature to enhance further the oil rates, cumulative production, and oil quality by *in-situ* combustion processes. Those materials are soluble metal compounds (water and organic) and heterogeneous catalysts.

9.3.1 ADDITION OF SOLUBLE METAL COMPOUNDS

Wichert *et al.* from the University of Calgary reported that sodium hydroxide inhibited LTO reactions as evidenced by a reduction in the coke formation during combustion experiments [Wichert *et al.* 1995]. These authors used a high-pressure three-phase batch reactor to evaluate the effect of an aqueous solution of NaOH (5 wt. %) at a relatively low temperature (80–120°C) for up to 24 days. They found that the produced oil had a minimum of one order of magnitude lower viscosity than that measured for the Athabasca bitumen feed. They showed that the bitumen upgrading occurred at oxygen uptakes in the range of 0.004 to 0.015 g O_2/g oil. Uptakes below this range were not sufficient to initiate the cracking process, while absorptions above this range promoted undesirable oxidation reactions. In some cases, the asphaltene content of each test sample increased. These results suggested that, for the instances where a reduction in viscosity was observed, cracking occurred in the lighter maltenes fraction of the oil. Thus, the authors concluded the presence of NaOH solutions catalyzed oxidation, increased the overall oxygen uptakes, inhibited the conversion of the asphaltenes to coke, and decreased the viscosity of the produced oil [Wichert *et al.* 1995].

Shallcross and coworkers from Stanford University studied the effects of various additives on the oxidation kinetics of Californian and Venezuelan crude oils [Shallcross *et al.* 1991]. Aqueous solutions of ten metallic salts ($FeCl_2$, $SnCl_2$, $CuSO_4$, $ZnCl_2$, $MgCl_2$, $K_2Cr_2O_7$, Al_2Cl_3, $MnCl_2$, $Ni(NO_3)_2$, and $CdSO_4$) were mixed with sand and Huntington Beach and Hamaca crude oils. The mixtures were subjected to a constant flow of air and a linear heating schedule while the effluent gases were analyzed for composition. Iron and tin salts were found to enhance fuel formation, while copper, nickel, and cadmium compounds had no significant effects [Shallcross *et al.* 1991]. Even though no data in the upgrading of the crude oil were disclosed, the examples served as a precedent in the uses of water-soluble metal catalysts that are presented next.

Castanier *et al.* from the same research group described 13 combustion tube runs using four different crudes, i.e., Huntington Beach, Hamaca, and 12°API and 34°API Californian oils and water-soluble metallic compounds [Castanier *et al.* 1992]. The iron-, tin-, and zinc-containing compounds improved the combustion efficiency in all cases. Changes were also observed in the H/C ratio of the fuel, air requirements, and density of the produced crude. For Huntington Beach oil, the amount of fuel increased in the order: Zinc, control, tin, and iron while for the Hamaca crude the sequence was: Control, iron, and tin [Castanier *et al.* 1992].

Several years later, Castanier and coworkers investigated the possible use of *in-situ* combustion to reduce the sulfur content of the produced oil [Castanier and Brigham 2003]. Three combustion tube runs were performed using a matrix containing 95 wt. % Ottawa sand and 5 wt. % fire clay, and a Middle East crude containing 6 wt. % of sulfur. In the control experiment, the average sulfur content of the produced oil was about 3% (50% HDS), and most of the sulfur removed was found as sulfuric acid in the produced water. Tin chloride and iron nitrate were dissolved in water at 5 wt. %, and in both cases, the amount of sulfur in the produced oil was reduced to less than 1 wt. %, i.e., 83% HDS [Castanier and Brigham 2003]. These results showed the catalytic activity of these metal salts in the upgrading of HO/B under *in-situ* combustion conditions.

He and coworkers performed ISU tube runs and ramped temperature oxidation tests to measure the kinetics of combustion [He *et al.* 2005]. They studied heavy and light crude oils from Cymric, CA, and conducted effluent gas analysis and temperature profiles measurements. In all cases, iron and tin additives improved performance including lower activation energy, greater oxygen consumption, lower temperature threshold, and complete oxidation. These results clearly indicate that the water-soluble metal compounds were acting as catalysts for the cracking and oxidation of the crude oils. The authors proposed the cation exchange of metallic salts with clays as a mechanism to create activated sites that enhanced combustion reactions between oxygen and crude oil. Sand and clay surfaces were examined with scanning electron microscopy and gave evidence of cation exchange and alteration of surface properties in the presence of the metallic salts [He *et al.* 2005].

Ramirez-Garnica and coworkers carried out combustion tube lab experiments using a heavy oil from the Gulf of Mexico (12.5°API). Their objectives were to increase the mobility and upgrade the crude inside the reservoir in the presence of organic-soluble metal catalysts [Ramirez-Garnica *et al.* 2007, 2008]. The porous media used in their work consisted of triturated dolomite carbonated rock with 35% porosity and an average particle size of 0.42 mm. The combustion experiments were performed at the approximatively same saturation conditions, i.e., heavy crude oil (~50%) and water (~50%). The catalysts used in the experiments were previously mixed with heavy crude oil and were based on molybdenum, cobalt, nickel, and iron at concentrations of 500 to 750 ppm [Ramirez-Garnica *et al.* 2007, 2008]. The results showed that the Ni-containing solid was the most active of the ones studied [Ramirez-Garnica *et al.* 2008]. The authors found an increase of oil production from 78% (control, no catalyst) to 85% (Ni-compound). They also reported faster combustion front, higher efficiency in the air consumption (from 76 to 197 ft³ air/ft³ hydrocarbons), and higher temperatures throughout the combustion process (~600°C) in the presence of the Ni-compound in comparison with the no-catalyst experiment [Ramirez-Garnica *et al.* 2008].

Using the Ni-compound, the authors reported the upgrading of the crude oil during their combustion lab experiment. Figure 9.4 shows the API gravity, the percentage of desulfurization (% HDS), and the percentage of reduction of asphaltene content (% Red. Asphaltenes) for the heavy oil feed and upgraded products during the *in-situ* combustion run [Ramirez-Garnica *et al.* 2008]. As seen, the API increased from 12.5° to 14.4°API at the first and middle of the run decreasing to 10.9°API at the end of the experiment. Similar behaviors were found for the % HDS and % Red. of asphaltenes (Figure 9.4) and the viscosity and the nitrogen content (results not shown).

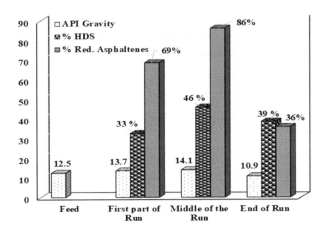

FIGURE 9.4 API gravity, percentage of desulfurization (% HDS), and percentage of reduction of asphaltene content (% Red. Asphaltenes) for the feed and upgraded products during the *in-situ* combustion in the presence of Ni catalyst. Data taken from Ramirez-Garnica *et al.* [2008].

Unfortunately, no comparisons with the properties of the product from the control experiment were carried out.

Jia *et al.* studied the low-temperature catalytic oxidation of heavy crude oil (Xinjiang Oilfield, China) in the presence of three types of catalysts, i.e., copper and manganese naphthenates (oil-soluble), copper dichloride (water-soluble), and copper oxide (dispersed) [Jia *et al.* 2016]. Fine quartz was loaded into the oxidation tube to simulate the porous medium with a porosity of 38% and permeability of 2,100 mD. The crude oil (15.8°API and 3,523 mPa s at 50°C) was mixed with different catalysts (0.5 wt. %) and pumped into the oxidation tube at a constant flow rate throughout the experiment. The dispersed and water-soluble catalysts increase the viscosity of the heavy oil with thickening ratios of 34% and 19%, respectively, while the Cu and Mn oil-soluble catalysts exhibited the excellent capacity to reduce the oil viscosity (65% and 52%, respectively) [Jia *et al.* 2016].

Jia and coworkers also found that that the Cu and Mn catalysts cracked part of heavy components into light components, reduced the asphaltene concentration (74% and 55%, respectively), and decreased the sulfur content (49% and 40%, respectively) [Jia *et al.* 2016]. The authors concluded that the catalysts played an essential role in promoting the oxidation-cracking reactions and could switch the reaction mode from oxygen addition to bond scission [Jia *et al.* 2016].

Based on the results presented in this section, the use of water- and organic-soluble metal compounds as catalysts for *in-situ* combustion process is undoubtedly an alternative to enhance oil production and the properties of the produced oil. The metal centers are acting as oxidation catalysts (i.e., they reduce the energy of activation) and promoting cracking and desulfurization reactions with the concomitant decrease in the viscosity and asphaltene content and, at the same time, increasing the API of the products.

9.3.2 Use of Heterogeneous Catalysts

The utilization of metal supported catalysts to upgrade the HO/B generated by *in-situ* combustion processes has been proposed since 1996. As shown in Table 9.1, Weissman, Moore, and coworkers carried out combustion tube experiments using Middle Eastern heavy crude oil (12.1°API) and the corresponding brine at reservoir conditions [Weissman *et al.* 1996, Moore *et al.* 1999]. As seen, two tests were performed, i.e., a control run without a catalyst and an experiment using a Ni-Mo/Al$_2$O$_3$ catalyst bed located downstream of the combustion zone. The authors used a catalyst suitable for heavy oil residue Hydroprocessing, consisting of 1.6 mm cylindrical alumina extrudates containing about 6.5 wt. % nickel oxide (NiO) and 22.7 wt. % molybdenum oxide (MoO$_3$). The authors

TABLE 9.1

In-Situ Combustion Lab Tests and Upgrading of Heavy Crude Oils and Bitumens in the Presence of Heterogeneous Catalysts

	Feed Characteristics				Combustion Tube			Results				
Catalyst	Crude Oil (Country)	°API (15°C)	Viscosity (mPa s @ °C)[a]		Oil Sat.[b]	Press. (MPa)[c]	Por. (%)[d]	Incr. °API[e]	% Visc. Red.[f]	% DS[g]	% Asp. Conv.[h]	References
Control No Catalyst	Middle Eastern	12.1	490 (@40°C)		67	10.3	40	0.4	45	25	−1	Weissman et al. [1996], Moore et al. [1999]
Ni-Mo/Al$_2$O$_3$								7.9	96	51	72	
Control No Catalyst	Llancanelo (Argentina)	14.8	10,460 (@40°C)		89	10.3	36	2.2	81	−20	6	Cavallaro et al. [2008]
Ni-Mo/Al$_2$O$_3$					69			9.2	99.9	~59	81	
Ni-Mo/Al$_2$O$_3$	Athabasca (Canada)	10.3	7680 (@40°C)		54	10.3	45	12.8	99.9	35	79	Abu et al. [2015a]
Commercial Ni-Mo/Al$_2$O$_3$					48–57	3.45	44	5.0	98	18.1	38	Abu et al. [2015b]
2nd Reg.[i] Ni-Mo/Al$_2$O$_3$								4.4	97	15.2	30	

ND = Not disclosed or not reported.

a. Viscosity (μ) in mPa s measured at the given temperature.

b. Oil saturation in the porous media.

c. Combustion tube pressure.

d. Porosity of the combustion tube.

e. Increase in API gravity of the upgraded crude oil at 60°F with respect to the feed.

f. Percentage of decrease in viscosity of the upgraded crude oil *vs.* the feed calculated as $((\mu_{feed} - \mu_{product}) - \mu_{feed}) \times 100$.

g. Percentage of desulfurization of the upgraded crude oil *vs.* the feed calculated as $((\%S_{feed} - \%S_{product}) - \%S_{feed}) \times 100$. Negative values mean increase in sulfur content.

h. Percentage of reduction of asphaltenes of the upgraded crude oil *vs.* the feed calculated as $((\%Asp_{feed} - \%Asp_{product}) - \%Asp_{feed}) \times 100$. Negative values mean increase in asphaltene content.

i. Catalyst regenerated twice by steam stripping and burning off coke [Abu et al. 2015b].

reported that the total recovery of hydrocarbons, from all parts of the combustion tube system, was close to 100% for both runs and only 5 to 6 wt. % of the oil was consumed as fuel. From the analysis of the generated gases, the most significant differences were the much higher amounts of hydrogen and hydrocarbons produced in the catalyst-containing run than the control counterpart. Hydrogen production was attributed to the water–gas-shift reaction (Equation 7.3) from CO to H_2O [Weissman et al. 1996, Moore et al. 1999].

As seen in Table 9.1, the control experiment gave mild upgrading with an increase of 0.4°API, 45% viscosity reduction, and 25% sulfur removal [Weissman et al. 1996, Moore et al. 1999]. The authors attributed the decrease in sulfur content to the removal of S-compounds by thermal conversion or distillation processes. Alternatively, passing the produced oil over a Hydroprocessing catalyst resulted in a product that was significantly improved as compared to either the original oil or thermally processed oil (control run). The authors reported that the catalytic upgrading led to an increase in ~8°API, 96% viscosity reduction, 51% desulfurization, and 72% of asphaltene conversion. Unfortunately, they observed that the heated catalyst zone accumulated significant amounts of coke which diminished the system performance [Weissman et al. 1996, Moore et al. 1999].

Similarly, Cavallaro and coworkers carried out in-situ combustion tube tests using an Argentinian heavy crude (14.8°API) and a Ni-Mo/Al_2O_3 catalyst [Cavallaro et al. 2008]. The objectives were to confirm the combustion characteristics of the heavy oil at reservoir conditions (see Table 9.1) and to evaluate the potential for in-situ upgrading using the Hydroprocessing catalyst. As in Weissman and Moore [Weissman et al. 1996, Moore et al. 1999], two tests were performed, i.e., a control run and a catalytic test. The overall oil recoveries were 85% and 74%, respectively. The upgrading results are shown in Table 9.1. As seen, an increase in the API of the produced oil was observed from 2.2°API (control) to 9.2°API (catalytic). Similarly, the viscosity reduction increased from 81% to 99.9%, the desulfurization from −20% to 59%, and the asphaltene conversion from 6% to 81%. Furthermore, the authors found a decrease in the nitrogen content from 0.3 wt. % to 0.1 wt. % and an increase in the 400°C distillate fraction from 18 wt. % to 62 wt. % [Cavallaro et al. 2008].

As in Weissman and Moore [Weissman et al. 1996, Moore et al. 1999], Cavallaro et al. proposed that the catalyst efficiently used carbon monoxide generated at the combustion front, via the water–gas-shift reaction (Equation 7.3) to produce hydrogen, which then reacted with the oil leading to HO upgrading. As before, the authors found the presence of a significant amount of coke on the spent catalyst which probably indicated the need for periodic regeneration [Cavallaro et al. 2008].

Abu and coworkers from the University of Calgary carried out dry and normal wet combustion tube tests to study the upgrading potential of Athabasca bitumen using supported Mo and Ni catalyst [Abu et al. 2015a]. The tests were performed at the same pressure (10.3 MPa), preheat temperature (95°C), and ignition temperature (350°C). The catalysts bed for the dry combustion was heated externally to 325°C while that of the normal wet combustion was not so it achieved a temperature of 275°C. Water was co-injected at water-to-air ratio of 1.8 kg water/m^3 of air for the normal wet combustion. Both runs exhibited similar percentages of oil recovery (80–90% vs. OIIP) and combustion behavior as those found for Athabasca bitumen [Xu et al. 2001, Chen et al. 2014].

Abu et al. reported that the best results were obtained for the wet combustion test [Abu et al. 2015a]. As shown in Table 9.1, an increase of ~13°API was found with 99.9% viscosity reduction, 35% desulfurization, and 79% asphaltene conversion. Similarly, the authors reported 67% denitrogenation. As before, the water–gas-shift reaction (Equation 7.3) was proposed to be the source of hydrogen for the upgrading process. The authors recommended the use of wet combustion for field applications because the injected water could transport steam to heat and activate the catalyst bed so that the bitumen upgrading could be achieved [Abu et al. 2015a].

The same research group from the University of Calgary reported the use and regeneration of a commercially supported metal catalyst for the upgrading of Athabasca bitumen under dry-combustion conditions [Abu et al. 2015b]. The tests (Table 9.1) were carried out at the same pressure (3.45 MPa), preheat temperature (95°C), and ignition temperature (350°C). Test 1 used a fresh supported Ni-Mo catalyst. Test 2 used a regenerated catalyst retrieved from Test 1, and Test 3 used a regenerated catalyst recovered from Test 2. The catalysts were regenerated by steam stripping and burning off coke. Table 9.1 shows the results using the fresh catalyst (Test 1) and the second regeneration (Test 3). As seen, similar values of the upgrading parameters were obtained for both experiments, i.e., API increase of 4–5°API, viscosity reduction of 97–99%, desulfurization of 15–18%, and asphaltene conversion of 30–38%. However, the authors reported that the regenerated catalysts (Tests 2 and 3) lost activity toward denitrogenation compared to the fresh solid, i.e., 52% for Test 1, 38.1% for Test 2, and 23.8% Test 3. In general, these results indicated the regenerated catalysts were still active for repeated use for the subsurface upgrading of HO/B [Abu *et al.* 2015b].

Based on the results presented, it can be concluded that the enhanced oil recovery and subsurface upgrading of heavy oils and bitumens using *in-situ* combustion and heterogeneous catalysts have been successfully demonstrated at laboratory scale [Weissman *et al.* 1996, Moore *et al.* 1999, Cavallaro *et al.* 2008, Abu *et al.* 2015a, 2015b]. All the cases used Ni-Mo supported catalysts, and the water–gas-shift reaction (Equation 7.3) was proposed to be the source of H_2 for the upgrading process. Coking deposition seemed to be the primary pathway for catalyst deactivation which should be addressed adequately in future field tests.

9.4 TOE-TO-HEEL AIR INJECTION PROCESS (THAI)

As shown in Figure 9.5, THAI technology is an *in-situ* combustion process that combines a vertical air injection well placed at the toe of a horizontal producer [Greaves 1997, Ayasse and Turta 2001]. This concept was developed in the 1990s by the University of Bath [Greaves 2001] and was the first field tested in early 2006 by Petrobank Energy and Resource Ltd., at Christina Lake, Athabasca tar sands, Alberta, Canada [Hart 2014]. As illustrated in Figure 9.5, a fire is initiated from the vertical injector, and the continuous air injection propagates the combustion front from the toe-position to the heel of the horizontal producer [Xia and Greaves 2002a]. The heavy ends of the crude oil are burned in the reservoir, and lighter ends and thermally upgraded crude oil gravitate toward the horizontal well and are carried to the surface. Gases are also produced, and oil and formation water are vaporized and condensed at the surface facilities. Extensive lab experiments have been performed as discussed in Section 9.4.1.

To further upgrade of the heavy oil, in 1998, the Petroleum Recovery Institute (PRI) of Calgary, Canada, in collaboration with the University of Bath, developed a process called Catalytic Upgrading Process *In-Situ* (CAPRI). As pointed out in Figure 9.5, CAPRI technology involves incorporating

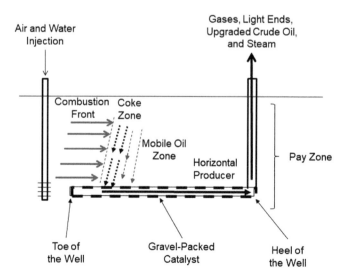

FIGURE 9.5 Schematic representation of the THAI and CAPRI processes.

gravel-packed Co-Mo or Ni-Mo catalyst between the slotted liner and the horizontal wellbore. Several THAI and CAPRI lab and field tests have been performed, and the results are presented in Sections 9.4.2 and 9.4.3, respectively [Greaves and Xia 2001, Ayasse *et al.* 2002, Xia and Greaves 2002a, 2002b]. After those commercial tests were performed, further investigations in the product and catalyst characterization and the mechanism of production have been carried out. This topic is presented in Section 9.4.4.

9.4.1 Lab Results of THAI

Greaves and coworkers performed 3-D lab experiments of the THAI process using Wolf Lake heavy oil sand packs in dry and wet modes [Greaves *et al.* 2000, 2001]. A rectangular stainless-steel combustion cell measuring 0.4 m square by 0.1 m deep was used for these experiments. One hundred and eighty minutes after the fireflood was initiated (dry mode), the authors observed that the combustion front was nearly vertical and located just before the "toe" of the producer well (see Figure 9.5). At 360 minutes, the edge of the combustion front (~500°C) was centered along the horizontal producer well and halfway through the sand pack. The authors reported the presence of a "mobile oil zone," immediately ahead of the combustion front and moving toward the heel of the well [Greaves *et al.* 2000, 2001]. Numerical simulations showed that liquid oil formed at a relatively high mass fraction of the "mobile oil zone." For example, at 20 bars of pressure, more than 81 wt. % and 54 wt. %, of the oil was in the liquid state, at 300°C and 400°C, respectively [Cao *et al.* 2009].

The authors stated that a critical feature of THAI is that the combustion takes place in the top part of the sand pack and does not lead to the uncontrolled gas override condition usually experienced in the conventional ISC process [Greaves *et al.* 2000, 2001]. Oil production peaked at about 300 minutes but declined at a steady rate after that. As shown in Figure 9.6, the overall oil recovery was very high, at 85% vs. the OOIP. From the post-mortem photograph, the authors observed that a clean burned zone was at the top of the matrix.

For the wet combustion run, the authors reported that the cumulative oil recovery was also very high at 86.5% vs. OOIP (Figure 9.6, without internal insulation), and almost 100% of the sand pack was affected by the high-temperature thermal front (> 450°C) [Greaves *et al.* 2000, 2001]. Xia and coworkers carried out the post-mortem study of two THAI experiments, in which the horizontal

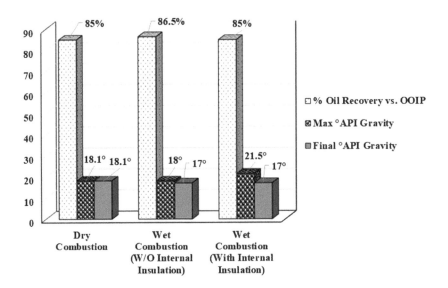

FIGURE 9.6 Percentage of oil recovery vs. OOIP and maximum and final API gravity of the produced oils for the 3-D lab simulations. Data taken from Greaves [2000, 2001].

well was cut open to reveal the extent of heavy oil residue/coke deposition [Xia *et al.* 2005]. The authors attributed the formation of these deposits to heavy residue drainage when the combustion front approaches the horizontal producer. This residue or coke material was formed ahead of the production front but before the "mobile oil zone" (see Figure 9.5) and provided a gas seal, preventing the injected air from channeling through the production well. The authors proposed that the formation of this coke zone inside the horizontal producer was a dynamic process, continuously forming and burning off as the combustion front approached the well [Xia *et al.* 2005].

Figure 9.6 shows the maximum and final API gravity of the produced oils collected during the 3-D lab simulations. As seen, the overall upgrading was similar for the dry and wet combustion experiments (without insulation), with approximately 7° to 8°API above the original crude value (10.1°API). Furthermore, the produced oil viscosity was significantly reduced from 80,000 cSt to 50 cSt [Greaves *et al.* 2005]. The researchers stated that a potential key advantage of THAI was its ability to capture and preserve the full extent of the upgrading produced by thermal cracking and distillation. This phenomenon occurred because mobilized fluids ahead of the combustion front were produced directly into the exposed section of the horizontal producer, rather than being displaced and banked up in the colder downstream regions of the reservoir—which is the situation in a conventional *in-situ* combustion process [Greaves *et al.* 2005].

Greaves and coworkers tried to mimic the adiabatic condition afforded by the under- and overburden rock in an actual reservoir [Greaves *et al.* 2001]. The latter was achieved by fixing a 0.5 cm thick layer of ceramic fiber insulation to the inside surfaces of the combustion cell. The authors reported that this modification proved to be very useful in increasing the reaction temperature by about 50°C and generally promoting a more vigorous and sustained combustion front propagation (wet mode). As seen in Figure 9.6, Greaves *et al.* reported an increase in the API gravity from 10.1° to 21.5°API [Greaves *et al.* 2001]. Furthermore, in-depth analysis of the produced oil showed an increase in the H/C molar ratio from 1.57 to 1.76 and percentages of desulfurization, denitrogenation, and demetallization of 11%, 41%, and 96%, respectively [Greaves *et al.* 2005]. Figure 9.7 shows the distribution of the SARA fractions (in wt. %) from the original Wolf Lake crude oil and the THAI upgraded product [Xia and Greaves 2001, Greaves *et al.* 2005]. As seen, there is a significant reduction in the asphaltenes (80% conversion) with the concomitant increases in the saturates and aromatics. All of these results clearly indicate that there are dramatic changes in the produced oil coming from the THAI process and that those changes are attributed to thermal cracking (see Section 4.1.2) and coking reactions (Section 4.1.3.) at the very high temperature (450–500°C) achieved during combustion.

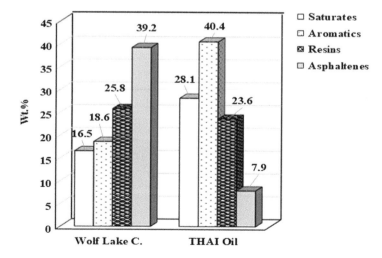

FIGURE 9.7 Weight percentage SARA fractions of the original Wolf Lake crude oil and the THAI upgraded product. Data taken from Greaves [2005].

Xia and coworkers carried out the numerical simulation of the THAI process for heavy oil recovery using the CMG-STARS reservoir simulator [Greaves and Xia 2000, Xia et al. 2005]. In their calculations, the horizontal producer was treated as a discretized wellbore, allowing coke to be formed. A thermal cracking kinetics model was incorporated into the numerical simulations which enabled the prediction of the upgrading effect observed in 3-D combustion cell tests for the Wolf Lake crude oil [Xia et al. 2005].

The authors reported that the produced oil before the start of air injection (t = 300 min) was mainly "heavy oil," i.e., original Wolf Lake crude, but this only represented about 3% of OOIP. After one hour of air injection, the production of the "light oil" pseudo-component became dominant. Overall, the predicted oil recovery was 80.5% OOIP, and the upgrading effect averaged 6.7°API. These calculated values were in good agreement with the experiments performed in the 3-D combustion cell [Xia et al. 2005].

Greaves and coworkers performed a more detail numerical validation study using the results obtained from 3-D combustion cell experiments on Athabasca oil sands under dry-combustion conditions [Greaves et al. 2012]. The dominant chemical reactions were similar to those of conventional in-situ combustion (Equation 9.1 and 9.2). The mathematical model also included the thermal cracking scheme proposed by Phillips et al. in which the Athabasca bitumen was defined by three pseudo-components: Asphaltenes, heavy oil, and distillates [Phillips et al. 1985]. The chemical reactions for the thermal cracking are as follows:

$$\text{Coke} \longleftrightarrow \text{Asphaltenes} \longleftrightarrow \text{Heavy Oils} \longleftrightarrow \text{Distillables} \qquad (9.3)$$

The authors excluded low-temperature oxidation reactions because THAI operates in a high-temperature oxidation mode [Greaves et al. 2012]. They reported excellent agreement between the predicted and experimental oil production rate, the residual coke profile, produced oxygen, and peak combustion temperature. As stated before [Cao et al. 2009], the authors determined that the mobile-oil zone (Figure 9.5) was composed of two distinct oil regions. The first part contained oil produced by thermal cracking of the heavy residue and vaporized lighter oil. The main bulk of the oil produced in THAI came from the second region of the mobile-oil zone. This oil was partially upgraded oil because of the thermally cracked and lighter oil fractions mixing with the original oil when they enter the horizontal producer [Greaves et al. 2012].

9.4.2 Catalytic Upgrading Process In-Situ (CAPRI)

The THAI-CAPRI process involves the addition of a gravel-packed Co-Mo- or Ni-Mo-containing catalyst (as used in a conventional hydrotreatment) between the slotted liner and the horizontal wellbore (see dashed line in Figure 9.5 and discussion Section 8.1.2). In this way, further upgrade of the heavy oil could be obtained [Greaves and Xia 2001, Xia et al. 2002, Xia and Greaves 2002a, 2002b]. Hydroprocessing catalysts such as Mo, Co, Ni, and W supported on alumina, silica, or silica-alumina are commonly used because of their ability to induce hydrocracking, Hydrodesulfurization (HDS), Hydrodenitrogenation (HDN), Hydrodemetallization (HDM), and asphaltenes conversion in the presence of relatively low hydrogen pressure [Weissman 1997]. These catalysts are expected to have mesopores with diameters between 2 and 50 nm to enhance accessibility to large molecules to the active sites and moderate acidity to reduce deactivation by coke [Hart 2014].

The research group at the University of Bath conducted physical simulations of the THAI-CAPRI process in a semi-scaled 3-D combustion cell filled with Wolf Lake crude-containing sand in the presence of fresh commercial catalysts [Greaves and Xia 2001]. The main variables investigated were the type of catalyst (NiMo, CoMo), extrudate or crushed catalyst, and catalyst loading. The authors reported that the lab process was very stable over a ten-hour period, maintaining average peak combustion temperatures of 500 to 600°C [Greaves and Xia 2001, Xia and Greaves 2002a, 2002b].

Overall, the Greaves and coworkers found that dry and wet THAI-CAPRI experiments gave very high oil recovery (85 ± 3 wt. % vs. OOIP) and very substantial upgrading of the produced oil.

Figure 9.8 shows the API gravity and percentages of sulfur of the Wolf Lake crude oil and the THAI-CAPRI upgraded products [Greaves and Xia 2001]. As seen, thermal upgrading alone (no catalyst) achieved a nearly 6.5-point increase in the API gravity of the produced oil with a sulfur reduction from 4.5 wt. % to ~3 wt. %. The use of NiMo or CoMo catalysts (CAPRI) increased the API further to the 22–25°API range. Furthermore, the authors reported a significant reduction of viscosity from 80,000 to 10–60 mPa s (at 20°C) and residue conversion in the 90–100% range [Greaves and Xia 2001, Xia and Greaves 2002a, 2002b].

As reported by other authors [Weissman *et al.* 1996, Moore *et al.* 1999, Cavallaro *et al.* 2008, Abu 2015a, 2015b], the water–gas-shift reaction (Equation 7.3) was proposed to be the source of H_2 for the upgrading process [Xia and Greaves 2002a]. Coking deposition seemed to be the primary pathway for catalyst deactivation which was addressed in future work (Section 9.4.4).

In further investigations, Xia *et al.* performed a series of 3-D combustion cell experiments to physically simulate the THAI-CAPRI process using Lloydminster heavy crude oil (11.9°API) [Xia *et al.* 2002]. As before, stable high-temperature combustion was sustained (500–550°C) with an oil recovery exceeding 79% vs. OOIP. The researchers reported that thermal upgrading (THAI) averaged 18.3°API, i.e., an incremental conversion gain of 6.4 API , with a maximum of 22°API. Higher upgrading was observed using a regenerated CoMo catalyst (CAPRI). The produced oil averaged 23°API with an oil viscosity as low as 20 to 30 mPa s [Xia *et al.* 2002].

Shah and coworkers conducted an experimental study concerning the optimization of catalyst type and operating conditions for use in the THAI-CAPRI process [Shah *et al.* 2011]. Lab tests were carried out using micro-reactors under different temperatures, pressures, and gas environments. Catalysts tested included alumina supported CoMo, NiMo, and ZnO/CuO. The authors found that there was a trade-off in operating temperature between upgrading performance and catalyst lifetime. At 20 bars and 500°C, the API of the produced oil increased from 10°API (Athabasca oil) to 18.9°API, but catalyst lifetime was limited to 1.5 hours. The authors also found that the operation at 420°C was a suitable compromise, with upgrading by an average of 1.6°API and sometimes up to 3°API, with catalyst lifetime extended to 77.5 hours. Asphaltenes, coke, and metal depositions occurred within the first few hours of the reaction, such that the catalyst pore space became blocked. This finding drastically deactivated the catalyst and could limit the potential applications of this technology [Shah *et al.* 2011].

Based on the results presented in the last two sections, it can be concluded that the production and subsurface upgrading of heavy oils and bitumens using THAI and THAI-CAPRI was successfully demonstrated by physical (laboratory scale) and numerical simulations. The next section presents the results of the field tests of these two *in-situ* combustion technologies.

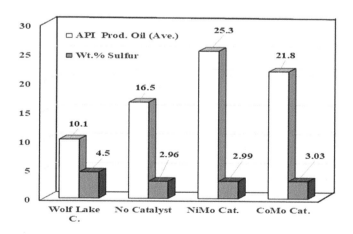

FIGURE 9.8 API gravity and percentages of sulfur of the Wolf Lake crude oil and the THAI-CAPRI upgraded products. Effect of the presence and type of catalyst. Data taken from Greaves [2001b].

9.4.3 FIELD AND SEMI-COMMERCIAL TESTS

THAI technology was pilot tested at Whitesands, Conklin, Alberta, and in semi-commercial scale at Kerrobert, Saskatchewan. Other projects in China and India have been carried out, but no reports have been disclosed yet.

9.4.3.1 Whitesands Experimental Pilot

Petrobank Energy and Resources Ltd., through its wholly owned subsidiary, Whitesands Insitu Ltd., carried out a first field test of the Toe-to-Heel Air Injection process at Conklin, Alberta [Ayasse *et al.* 2005]. As shown in Figure 9.9, the Whitesands Experimental Pilot Project was conducted in the McMurray B formation at a depth of 380 m and used three well pairs of vertical air injectors and horizontal producers with 500 m in length and 100 m spacing horizontally [Petrobank Energy Resources 2010]. Those are labeled A1–P1, A2–P2, and A3–P3. The well P2 was later shut down due to excessive sand production, and a new P3B well was completed at ~15 m from the original P3. A concentric liner system was installed in P3B to locate the heterogeneous catalyst used during the CAPRI pilot test. Also, 12 observation wells were drilled as depicted in Figure 9.9.

Project construction started in May 2006. After a steam preheat phase that lasted three to four months for each well pair, air injection began for the A2, A1, and A3 wells on July 2006, January 20007, and May 2007, respectively [Petrobank Energy Resources 2010]. The project was initially designed for 300 m³/day (1,887 BBL/D), but actual cumulative oil production was significantly lower (~7%) than the one predicted by the reservoir simulations. This finding was attributed to delays caused by sand production issues, the presence of bottom water, and a non-uniform and less thick pay zone. Operations on the P3 well were suspended in April 2008, P1 in April 2009, and P2 in July 2009 [Petrobank Energy Resources 2010].

However, several critical observations and conclusions were made. Firstly, the temperature measurements in function of time and a 4-D seismic survey acquired early in the first quarter of 2008 for the three original well pairs indicated that the combustion front indeed moved from toe to heel. Also, combustion can easily be restarted as shown by A–3 in which air injection resumed after 100 days of suspension, and A–2 resumed after 30 days of shut-in. These results demonstrated the

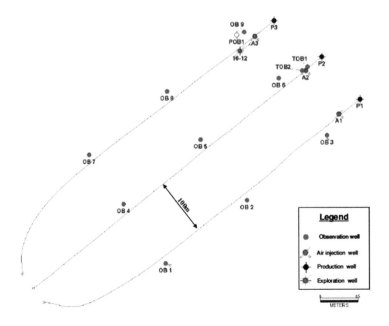

FIGURE 9.9 Project well layout and well spacing for the THAI Whitesands Experimental Pilot Project at Conklin, Alberta.

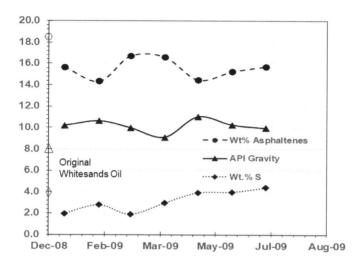

FIGURE 9.10 Average monthly properties (wt. % asphaltenes, API gravity, and wt. % sulfur) of the produced oil (solid symbols) vs. the original Whitesands crude (hollow symbols) of the THAI pilot test at Conklin, Alberta from January to July 2009. Data taken from Petrobank Energy Resources [2010].

feasibility of the toe-to-heel air injection process in real reservoir conditions [Petrobank Energy Resources 2010].

Additionally, the results disclosed showed that field test consistently improved the quality of the produced oil. Figure 9.10 shows the average monthly properties (wt. % asphaltenes, API gravity, and wt. % sulfur) of the upgraded oil (solid symbols) vs. the original Whitesands crude (hollow symbols) from January to July 2009 [Petrobank Energy Resources 2010]. As seen, the percentage of asphaltenes decreased from ~18 wt. % to the 14–16 wt. % range which led to a significant reduction on the viscosity of the produced oil from 1,340 Pa s to 16 Pa s (measured at 20°C). Similarly, the API gravity increased 2 points (from 8°API to 10°API) whereas the percentage of sulfur decreased from 4 wt. % to an average of 3 wt. % (22% HDS). Furthermore, the upgraded oil had a significantly higher concentration of volatiles (9% v/v) and saturated fractions, and the produced gas had no oxygen and ~8% v/v of hydrogen gas. All these results are consistent with a thermal cracking mechanism as predicted by the lab experiments [Greaves *et al.* 2000, 2001, 2008, Xia *et al.* 2005].

As expected, there were many lessons learned from the Whitesands pilot such as higher-than-expected temperatures within the wellbore. Also, a significant sand production was observed so that de-sand vessels were installed to manage the additional solid generation. Finally, a higher-than-expected H_2S production was found that led to the construction of a larger sulfur recovery system compatible with the produced combustion gas [Petrobank Energy Resources 2010].

As mentioned earlier, a concentric liner system was installed in P3B to place the heterogeneous catalyst during the CAPRI pilot test. The concentric liner was disclosed by Ayasse and included horizontal-well liner designs to increase the residence time so that the produced oil could be upgraded by a heterogeneous catalyst during THAI [Ayasse 2011]. The catalyst was designed to operate at 350°C and 450°C, and an additional 3 points in the API gravity were obtained [Jamaluddin *et al.* 2018]. However, catalyst deactivation by coke deposition and bed plugging was reported. Finally, on 19 September 2011, Petrobank announced that all operations of the Whitesands Experimental Pilot Project at Conklin, Alberta were suspended [Mcintosh 2011].

9.4.3.2 Semi-Commercial Project at Kerrobert

The first semi-commercial project of THAI in a conventional heavy oil reservoir was at Kerrobert, Saskatchewan in a 50/50 partnership between Petrobank and Baytex. This project was conducted in the Wasseca Channel (Manville Sand of Lower Cretacic) at a depth of 780 m. Initially, the project consisted of two-well pairs, which were expanded to 12-well pairs in 2011. The drilling and completion of a two-well program concluded in September 2009, and facilities were completed in October 2009. After commencing production in December 2009, the first oil sales occurred in January 2010. For the first well pair, the rates were 250–420 BBL/D total fluid with a 36% to 65% oil cut. By March 2009, the production averaged 123 BBL/D of oil per day with an 81% onstream factor [Petrobank Energy Resources 2010].

At the moment of writing this book, the detailed accounts of this project were not public yet, but generally, it was reported that there were fewer operating problems than those found in the Whitesands Pilot [Turta *et al.* 2018]. The Kerrobert project was conducted for six years. However, the operational performance did not show a proportionality between air injection and oil production. The produced oil from the project consistently averaged 14°API to 17°API during its entire operation [Turta *et al.* 2018].

On March 2012, Petrobank reported that the early-stage production using the THAI process at the Saskatchewan heavy oil property was less than what the original two demonstration wells were producing. This lower estimated production resulted in the negative net present value and therefore caused the elimination of proved reserves of the project [Lewis 2012]. In May 2014, Petrobank Energy and Resources Ltd. and Touchstone Exploration Inc. announced that they were merging. On 22 July 2015, president and chief executive Paul Baay of Touchstone said that THAI was being applied at Kerrobert, Sask., and was producing about 180 BBL/D of heavy oil. However, they were experiencing the same issues as Petrobank. That was that the energy input vs. the oil output made it non-economical [Healing 2015]. In January 2016, Touchstone announced that it had sold its Kerrobert, Saskatchewan facility to Quattro Exploration and Production Ltd. for a total of $4,150,000.

In 2011, Petrobank and Shell Canada (50/50 partnership) announced the second THAI application to conventional heavy oil at Dawson, within the Peace River oil sands region. However, in June 2015, Touchstone made the business decision not to proceed with the THAI pilot at this location due to the low oil prices scenario and the lower-than-expected initial performance results [Shipka 2015].

Turta *et al.* reported that over half a million barrels of upgraded oil were produced from both Canadian THAI projects (Whitesands and Kerrobert) with an increase in the API gravity in the 3–8°API range [Turta *et al.* 2018]. However, the oil rate per well was lower (63–190 BBL/D) than that found in the SAGD process. This issue attempted against the commercialization of this technology. Additionally, THAI-CAPRI suffered from short catalyst lifetime due to rapid deactivation and possibility of catalyst bed plugging. Those issues need to be adequately addressed for the successful implementation of this process.

9.4.3.3 Outside Canada

Two Chinese (Shuguang, Liaohe, and Fengcheng, Xinjiang Oilfield) and two Indian (Balol Field and Lanwa Field) THAI pilots have been carried out by the date of publishing this book [Turta *et al.* 2018]. The authors concluded that, based on the learnings from these six pilots, the technical validity of THAI was confirmed. However, its economic performance needs to be significantly improved for commercial application as a recovery/upgrading technology [Turta *et al.* 2018].

9.4.4 Further Work in CAPRI

During and after the THAI and THAI-CAPRI field tests, a substantial amount of work was carried out on the optimization of the operating conditions, product characterization, and the use of other sources of hydrogen and dispersed catalysts. This section presents the most significant findings of such investigations.

9.4.4.1 Other Sources of Hydrogen

As mentioned, the CAPRI process has been previously investigated using a set of micro-reactors to replicate subsurface upgrading conditions and to optimize the selection of catalyst type, oil, and gas flow rates, reaction temperature, and pressure [Shah *et al.* 2011]. The authors concluded that the potential of the technology could be limited by the deposition of asphaltenes, coke, and metals, which drastically deactivates the catalyst [Shah *et al.* 2011]. Hart and coworkers studied the use of hydrogen, methane, nitrogen, and a blended THAI gas mixture over a Co-Mo/γ-Al$_2$O$_3$ catalyst at a previously determined optimum process reaction temperature of 425°C and pressures of 10 bar and gas-to-oil ratio 50 mL/mL [Hart *et al.* 2014a].

Table 9.2 shows the properties of the THAI feed oil and the effect of the hydrogen source on the CAPRI upgraded products [Hart 2014b]. As seen, the average increases in API gravity of the produced oil were 4° using hydrogen gas, 3° with methane, 2.9° with THAI gas, and 2.7° with nitrogen above the value of 14°API gravity for the feed oil. The viscosity reduction, conversion of 343°C+ residue, and percentages of hydrodesulfurization (% HDS), reduction of asphaltenes, and hydrodemetallization (% HDM) followed the same general trend as the API gravity. Hence, the order of reactivity of these gases during the catalytic upgrading reaction of the THAI feed oil was as follows: H$_2$ > CH$_4$ > THAI gas > N$_2$ [Hart 2013b]. This finding is consistent with the reports of Ovalles *et al.* during the upgrading

TABLE 9.2
Properties of the THAI Feed Oil and the CAPRI Upgraded Products Using a Co-Mo/Al$_2$O$_3$ Catalyst. Effect of the Hydrogen Source Used[a].

Sample	API Gravity (°API)[b]	Viscosity (Pa s)[c]	Wt. Residue 343°C+[d]	Wt. % Sulfur[e]	Wt. % C7-Asphaltenes[f]	V+Ni (ppm)[g]
THAI feed oil	14.1	1,091	43.6	3.52	10.4	149

Exp.	Gas Used	Increase in API Gravity (°API)[h]	% Red. Of Viscosity[i]	% Conv. Of 343°C+[j]	% HDS[k]	% Reduction of C7-Asphaltenes[l]	% HDM (Ni+V)[m]
1	Hydrogen	4.0	92.8	41.0	21	41	16
2	Methane	3.0	90.4	38.1	12	49	27
3	THAI Gas[n]	2.9	89.2	32.6	ND	18	ND
4	Nitrogen	2.7	88.9	29.5	3	11	17

Data taken from Hart [2013b].

ND = Not disclosed or not reported.

[a] Reaction carried out in a continuous reactor at 425°C, 10 bars, gas-to-oil ratio 50 v/v, feed oil rate = 1 mL/min, and weight-hourly-space velocity of 9 h^{-1}.

[b] API gravity of the feed.

[c] Viscosity measured at 20°C.

[d] Weight percentage of the residue 343°C+.

[e] Weight of sulfur as determined by elemental analysis.

[f] Weight percentage of heptane asphaltenes.

[g] Vanadium and nickel content as determined by elemental analysis.

[h] Increase of the API gravity of the CAPRI upgraded product vs. the THAI feed oil.

[i] Percentage of decrease in viscosity of the CAPRI upgraded product vs. the THAI feed oil calculated as $((\mu_{feed} - \mu_{product}) - \mu_{feed}) \times 100$.

[j] Percentage of conversion of the residue 343°C+ of the CAPRI upgraded product vs the feed calculated as $((\%Residue_{feed} - \%Residue_{product}) - \%Residue_{feed}) \times 100$.

[k] Percentage of hydrodesulfurization of the CAPRI upgraded product vs. the THAI feed oil calculated as $((\%S_{feed} - \%S_{product}) - \%S_{feed}) \times 100$.

[l] Percentage of reduction of asphaltenes of the CAPRI upgraded product vs. the THAI feed oil calculated as $((\%Asp_{feed} - \%Asp_{product}) - \%Asp_{feed}) \times 100$.

[m] Percentage of hydrodemetalization (Ni + V) of the CAPRI upgraded product vs. the THAI feed oil calculated as $((\%Metal_{feed} - \%Metal_{product}) - \%Metal_{feed}) \times 100$.

[n] Composition of THAI gas = 13% CO$_2$, 3% CO, 4% CH$_4$, and 80% N$_2$.

of extra-heavy crude oil using Mo- and Fe-based catalysts in the presence of methane as a source of hydrogen [Ovalles *et al.* 1998, 2003]. The mechanism associated with the thermal and catalytic activation of methane was discussed in Sections 4.2.1.5 and 4.2.2.4, respectively.

Hart and coworkers also found that the percentage of loss in specific surface areas due to coke deposition in the different reaction gases were as follows: 57.2% for H_2, 68% for CH_4, and 96% for N_2 relative to the surface area of the fresh catalyst of 214.4 m^2/g [Hart 2013b]. Thus, the upgrading of the THAI feed using hydrogen or methane (H-donor) proceeds via hydrocracking and hydrogenation reactions. Conversely, in a nitrogen atmosphere (inert) most of the upgrading occurred via a carbon rejection route with associated heavy coke deposition on the catalyst surface [Hart 2013b].

In order to prevent coke formation and catalyst deactivation, Hart and coworkers studied the effect of steam addition and steam-to-oil ratio upon upgrading of THAI feed using a Co-Mo/γ-Al_2O_3 catalyst [Hart *et al.* 2014b] at the same reaction conditions as before [Hart *et al.* 2014a], i.e., 425°C and pressures of 10 bar and gas-to-oil ratio 50 mL/mL. The authors observed that the coke content of the spent catalyst was reduced from 17.0 to 11.3 wt. % as the steam-to-oil ratio increased from 0.02 to 0.1 v/v. Similarly, 88–92% viscosity reductions were obtained as the steam-to-oil ratio increased in comparison with the value of 1,091 Pa s for the THAI feed oil. They also found that % HDS and % HDM increased from 3.4% and 16.8% in a steam-free environment to 16–25.6% and 43–70.5%, respectively, over the increasing range of steam flow investigated. The authors attributed the decrease of coke deposition after steam addition to the reaction between H_2O(g) and coke, generating carbon monoxide and hydrogen as shown in Equation 9.4 [Hart *et al.* 2014b].

$$\text{Coke (s)} + H_2O(g) \rightarrow CO(g) + H_2(g) \tag{9.4}$$

Finally, Hart *et al.* studied the use of cyclohexane as hydrogen donor solvent using a finely crushed pelleted Ni-Mo/Al_2O_3 catalyst (2.4 μm) at temperature 425°C, initial pressure 17.5 bar, and a short reaction time of 10 min [Hart *et al.* 2015]. The results using cyclohexane were evaluated against those obtained using hydrogen and nitrogen gas. The authors found that the coke decreased by 6.2–45.4% as the cyclohexane-to-THAI feed oil ratio increased from 0.01 to 0.08 wt./wt. relative to 4.7 wt. % of coke observed without H-donor under nitrogen environment. As the cyclohexane-to-oil ratio increased, the produced oil API gravity and middle distillate fractions (200–343°C) also increased while the viscosity decreased [Hart *et al.* 2015]. Similar results were obtained using a hydrogen atmosphere [Hart and Wood 2018].

9.4.4.2   Effect of Catalysts Type

As mentioned in the previous section, the use of catalysts to enhance downhole upgrading in the THAI process is limited by deactivation due to coking arising from the cracking of heavy oil and metal and asphaltene deposition. To mitigate those issues, Al-Marshed *et al.* studied nanodispersed iron oxide particles ($\leq$ 50 nm) for the catalytic upgrading of THAI feed oil (see properties in Table 9.2) at the following ranges: Metal loading 0.03–0.4 wt. %, temperature 355–425°C, initial H_2 pressure 10–50 bar, and reaction time 20–80 min [Al-Marshed *et al.* 2015]. The researchers found that the optimum conditions were by using iron–metal loading of 0.1 wt. % at a temperature of 425°C, an initial hydrogen pressure of 50 bar, and reaction time 60 min. The properties of upgraded oil at these optimum conditions were 21.1°API gravity, viscosity of 105.8 mPa s, % HDS = 37.5%, and % HDM (Ni + V) = 68.9% [Al-Marshed *et al.* 2015]. However, no information was disclosed on the catalytic activity and deactivation of the Fe oxide nanoparticles as a function of time.

Based on the results presented in the preceding sections, it can be concluded that THAI and THAI-CAPRI are emerging technologies aimed to enhance the recovery and upgrading of heavy oils. They have significant beneficial advantages over the commonly used thermal EOR processes such as high recovery factor, production of valuable products, and lower water usage. However, short catalyst lifetime due to rapid deactivation and possibility of catalyst bed plugging are issues that need to be adequately addressed in the future.

REFERENCES

Abu, I. I., Moore, R. G., Mehta, S. A., Ursenbach, M. G., Mallory, D. G., Pereira Almao, P., Carbognani Ortega, L., 2015a, "Upgrading of Athabasca Bitumen Using Supported Catalyst in Conjunction with In-Situ Combustion", *J. Can. Pet. Technol.*, 54, 220–232.

Abu, I. I., Moore, R. G., Mehta, S. A., Ursenbach, M. G., Mallory, D. G., Pereira-Almao, P., Carbognani Ortega, L., 2015b, "Supported Catalyst Regeneration and Reuse for Upgrading of Athabasca Bitumen in Conjunction with In-Situ Combustion", *J. Can. Pet. Technol.*, 54, 372–386.

Al-Marshed, A., Hart, A., Leeke, G., Greaves, M., Wood, J., 2015, "Optimization of Heavy Oil Upgrading Using Dispersed Nanoparticulate Iron Oxide as a Catalyst", *Energy Fuels*, 29, 6306–6316.

Ayasse, C., 2011 "Well Liner Segments for In Situ Petroleum Upgrading and Recovery, and Method of In Situ Upgrading and Recovery", US Patent No. 7,909,097, 22 March.

Ayasse, C., Bloomer, C., Lyngberg, E., Boddy, W., Donnelly, J., Greaves, M., 2005, "First Field Pilot of the THAI Process", PETSOC-2005-142, presented at Canadian International Petroleum Conference, Calgary, Alberta, 7–9 June.

Ayasse, C., Greaves, M., Turta, A., 2002, "Oilfield *In Situ* Hydrocarbon Upgrading Process", US Patent No. 6,412,557, 2 July.

Ayasse, C., Turta, A., 2001, "Toe-to-Heel Oil Recovery Process", US Patent No. 6,167,966, 2 January.

Belgrave, J. D. M., Moore, R. G., Ursenbach, M. G., Bennion, D. W., 1993, "A Comprehensive Approach to In-Situ Combustion Modeling", *SPE Advanced Technology Series*, 1 (April), 98–107.

Burger, J. G., Sahuquet, B. C., 1972, "Chemical Aspects of In-Situ Combustion", SPE No. 3599, *Soc. Pet. Eng. J.*, 12, 410–422.

Cao, E., Greaves, M., Rigby, S. P., 2009, "Phase Properties of Mobile Oil Zone in the THAI - CAPRI Process", PETSOC-2009-048, presented at Canadian International Petroleum Conference, Calgary, Alberta, 16–18 June.

Castanier, L. M., Baena, C. J., Holt, R. J., Brigham, W. E., Tavares, C., 1992, "In Situ Combustion with Metallic Additives", SPE No. 23708, presented at SPE Latin America Petroleum Engineering Conference, Caracas, Venezuela, 8–11 March.

Castanier, L. M., Brigham, W. E., 2003, "Upgrading of Crude Oil Via In Situ Combustion", *J. Pet. Sci. Eng.*, 39, 125–136, and references therein.

Cavallaro, A. N., Galliano G. R., Moore, R. G., Mehta, S. A., Ursenbach M. G., Zalewski, E., Pereira P., 2008, "In Situ Upgrading of Llancanelo Heavy Oil Using In Situ Combustion and a Downhole Catalyst Bed", *J. Can. Pet. Technol.*, 47, 23–31.

Cavanzo, E. A., Munoz Navarro, S. F., Ordonez, A., Bottia Ramirez, H., 2014, "Kinetics of Wet In-Situ Combustion: A Review of Kinetic Models", SPE No. 171134, presented at SPE Heavy and Extra Heavy Oil Conference: Latin America, Medellín, Colombia, 24–26 September.

Chen, X., Chen, Z., Moore, R. G., Mehta, S. A., Ursenbach, M. G., Harding, T. G., 2014, "Kinetic Modeling of the In-situ Combustion Process for Athabasca Oil Sands", SPE No. 170150, presented at SPE Heavy Oil Conference-Canada, Calgary, Alberta, Canada, 10–12 June.

Freitag, N. P., Verkoczy, B., 2005, "Low-Temperature Oxidation of Oils in Terms of SARA Fractions: Why Simple Reaction Models Do Not Work", *J. Can. Pet. Technol.*, 44, 54–61.

Greaves, M., Dong, L. L., Rigby, S., 2012, "Validation of Toe-to-Heel Air-Injection Bitumen Recovery Using 3D Combustion-Cell Results", *SPE Res. Eval. Eng.*, 15, 72–85.

Greaves, M., Saghr, A. M., Xia, T. X., Turta, A. T., Ayasse, C., 2001, "THAI - New Air Injection Technology for Heavy Oil Recovery and In Situ Upgrading", *J. Can. Pet. Technol.*, 40 (3), 38–47.

Greaves, M., Turta, A. T., 1997, "Oilfield *In Situ* Combustion Process", US Patent 5,626,191, 6 May.

Greaves, M., Xia, T., 2001, "CAPRI-Downhole Catalytic Process for Upgrading Heavy Oil: Produced Oil Properties and Composition", PETSOC-2001-023, presented at Canadian International Petroleum Conf., Calgary, Alberta, 12–14 June.

Greaves, M., Xia, T. X., 2000, "Simulation Studies of THAI Process", PETSOC-2000-084, presented at Canadian International Petroleum Conference, Calgary, Alberta, 4–8 June.

Greaves, M., Xia, T. X., Ayasse, C., 2005, "Underground Upgrading of Heavy Oil Using THAI- 'Toe-to-Heel Air Injection'", SPE No. 97728, presented at SPE Intl. Thermal Operations and Heavy Oil Symposium, Calgary, Alberta, Canada, 1–3 November.

Greaves, M., Xia, T. X., Turta, A. T., Ayasse, C., 2000, "Recent Laboratory Results of THAI and Its Comparison with Other IOR Process", SPE No. 59334 presented at SPE/DOE Improved Oil Recovery Symposium held in Tulsa, OK, 3–5 April.

Guo, K, Li, H., Yu, Z., 2016, "In-Situ Heavy and Extra-Heavy Oil Recovery: A Review", *Fuel*, 185, 886–902.

Hajdo, L. E., Hallam, R. J., Vorndran, L. D. L., 1985, "Hydrogen Generation during In-Situ Combustion", SPE No. 13661, presented at SPE California Regional Meeting, Bakersfield, CA, 27–29 March.

Hart, A., 2014, "The Novel THAI–CAPRI Technology and Its Comparison to Other Thermal Methods for Heavy Oil Recovery and Upgrading", *J. Pet. Exp. Prod. Technol.*, 4, 427–437.

Hart, A., Leeke, G., Greaves, M., Wood, J., 2014a, "Down-Hole Heavy Crude Oil Upgrading by CAPRI: Effect of Hydrogen and Methane Gases upon Upgrading and Coke Formation", *Fuel*, 119, 226–235.

Hart, A., Leeke, G., Greaves, M., Wood, J., 2014b, "Downhole Heavy Crude Oil Upgrading Using CAPRI: Effect of Steam upon Upgrading and Coke Formation", *Energy Fuels*, 28, 1811–1819.

Hart, A., Lewis, C., Thomas White, T., Greaves, M., Wood, J., 2015, "Effect of Cyclohexane as Hydrogen-Donor in Ultradispersed Catalytic Upgrading of Heavy Oil", *Fuel Proc. Technol.*, 138, 724–733.

Hart, A., Wood, J., 2018, "In Situ Catalytic Upgrading of Heavy Crude with CAPRI: Influence of Hydrogen on Catalyst Pore Plugging and Deactivation Due to Coke", *Energies*, 11, 636–654.

Hascakir, B., 2015, "Description of In-Situ Oil Upgrading Mechanism for In-Situ Combustion Based on a Reductionist Chemical Model", SPE No. 175086, presented at SPE Annual Technical Conf. and Exhib., Houston, TX, 28–30 September.

He, B., Chen, Q., Castanier, L. M., Kovscek, A. R., 2005, "Improved In-Situ Combustion Performance with Metallic Salt Additives", SPE No. 93901, presented at SPE Western Regional Meeting, Irvine, CA, 30 March–1 April.

Healing, D., 2015, "Asset Sales Signal Patience Short for Disappointing THAI Heavy Oil Technology", *Calgary Herald*, July 22, downloaded from https://calgaryherald.com/business/energy/asset-sales-signal-end-is-near-for-disappointing-thai-heavy-oil-technology on 8 September 2018.

Huc, A.-Y., Ed., 2011, *Heavy Crude Oils: From Geology to Upgrading: An Overview*, Technip, Paris, France, and references therein.

Jamaluddin, A., Law, D. H.-S., Taylor, S. D., Andersen, S. I., 2018, *Heavy Oil Exploitation*, PennWell, pp 120–121.

Jia, H., Liu, P.-G., Pu, W.-F., Ma, X.-P., Zhang, J., Gan, L., 2016, "In Situ Catalytic Upgrading of Heavy Crude Oil Through Low-Temperature Oxidation", *Pet. Sci.*, 13, 476–488.

Kapadia, P. R., Kallos, M. S., Gates, I. D., 2015, "A Review of Pyrolysis, Aquathermolysis, and Oxidation of Athabasca Bitumen", *Fuel Proc. Technol.*, 131, 270–289, and references therein.

Klock, K., Hascakir, B., 2015, "Simplified Reaction Kinetics Model for In-Situ Combustion", SPE No. 177134, presented at SPE Latin American and Caribbean Petroleum Engineering Conference, Quito, Ecuador, 18–20 November.

Kok, M. V., Karacan, C. O., 2000, "Behavior and Effect of SARA Fractions of Oil during Combustion", *SPE Res. Eval. Eng.*, 3, 380–385.

Lewis, J., 2012, "Petrobank Suffers Setback with THAI", Alberta Oil. Energy Link, March 08.

Manrique, E. J., Thomas, C. P., Ravikiran, R., Izadi Kamouei, M., Lantz, M., Romero, J. L., Alvarado, V., 2010, "EOR: Current Status and Opportunities", SPE No. 130113 presented at SPE Improved Oil Recovery Symposium, Tulsa, OK, 24–28 April.

Marjerrison, D. M., Fassihi, M. R., 1994, "Performance of Morgan Pressure Cycling In-Situ Combustion Project", SPE No. 27793, presented at SPE/DOE Improved Oil Recovery Symposium, Tulsa, OK, 17–20 April.

Mcintosh, J., 2011, "Petrobank Suspends Alberta Recovery Project", *Globe and Mail*, Calgary, downloaded from www.theglobeandmail.com/report-on-business/industry-news/energy-and-resources/petrobank-suspends-alberta-recovery-project/article555584/ on 8 September 2018.

Moore, R. G., Laureshen, C. J., Mehta, S. A., Ursenbach, M. G., Belgrave, J. D. M., Weissman, J. G., Kessler, R. V., 1999, "A Downhole Catalytic Upgrading Process for Heavy Oil Using In Situ Combustion", *J. Can. Pet. Technol.*, 38, 1–8.

Ovalles, C., Filgueiras, E., Morales, A., Rojas, I., de Jesus, J. C., Berrios, I., 1998, "Use of a Dispersed Molybdenum Catalyst and Mechanistic Studies for Upgrading Extra-Heavy Crude Oil Using Methane as Source of Hydrogen", *Energy Fuels*, 12, 379–385.

Ovalles, C., Filgueiras, E., Morales, A., Scott, E. C., Gonzalez, G. F., Pierre, E. B., 2003, "Use of a Dispersed Iron Catalyst for Upgrading Extra-Heavy Crude Oil Using Methane as Source of Hydrogen", *Fuel*, 82, 887–892.

Petrobank Energy Resources, 2010, "Annual Progress Report (2009) for the Whitesand Experimental Project", Revised May 14, IETP 01-019, retrieved from https://content.energy.alberta.ca/xdata/IETP/IETP%20 2009/01-019%20Whitesands%20Experimental%20Project/IETP%20approval%2001_019%20final%20 and%202009%20report.pdf on 7 September 2018.

Phillips, C. R., Haidar, N. I., Poon, Y. C., 1985, "Kinetic Models for the Thermal Cracking of Athabasca Bitumen: The Effect of the Sand Matrix", *Fuel*, 64 (5), 678–691.

Prats, M., 1978, "A Current Appraisal of Thermal Recovery. Society of Petroleum Engineers", *J. Pet. Technol.*, 30, 1129–1136.

Rahnema, H., Barrufet, M., Mamora, D. D., 2012a, "Self-Sustained CAGD Combustion Front Development; Experimental and Numerical Observations", SPE No. 154333, presented at SPE Improved Oil Recovery Symposium, Tulsa, OK, 14–18 April.

Rahnema, H., Barrufet, M., Martinez, J., Mamora, D., 2012b, "Dual Horizontal Well Air Injection Process", SPE No. 153907, presented at SPE Western Regional Meeting, Bakersfield, CA, 21–23 March.

Ramey, H. J., Stamp, V. W., Pebdani, F. N., Mallinson, J. E., 1992, "Case History of South Belridge, California, In-Situ Combustion Oil Recovery", SPE No. 24200, presented at SPE/DOE Enhanced Oil Recovery Symposium, Tulsa, OK, 22–24 April.

Ramirez-Garnica, M. A., Hernandez Perez, J. R., Cabrera-Reyes, M. del C., Schacht-Hernandez, P., Mamora, D. D., 2008, "Increase Oil recovery of Heavy Oil in Combustion Tube Using a New Catalyst Based on Nickel Ionic Solution", SPE No. 117713, presented at International Thermal Operations and Heavy Oil Symposium, Calgary, Alberta, Canada, 20–23 October.

Ramirez-Garnica, M. A., Mamora, D. D., Nares, R., Schacht-Hernandez, P., Mohammad, A. A. A., Cabrera, M., 2007, "Increase Heavy-Oil Production in Combustion Tube Experiments Through the Use of Catalyst", SPE No. 107946, presented at Latin American & Caribbean Petroleum Engineering Conference, Buenos Aires, Argentina, 15–18 April.

Sarathi, P. S., 1999, "In-Situ Combustion Handbook – Principles and Practices", Report No. DOE/PC/91008-0374, Prepared for U.S. Department of Energy by BDM Petroleum Technologies, Bartlesville, OK.

Shah, A. A., Fishwick, R. P., Leeke, G. A., Wood, J., Rigby, S. P., Greaves, M., 2011, "Experimental Optimization of Catalytic Process *In Situ* for Heavy-Oil and Bitumen Upgrading", *J. Can. Pet. Technol.*, 50, 33–47.

Shallcross, D. C., de los Rios, C. F., Castanier, L. M., Brigham, W. E., 1991, "Modifying In-Situ Combustion Performance by the Use of Water-Soluble Additives", SPE No. 19485, *SPE Res. Eng.*, 6, 287–294.

Shipka, J., 2015, *Touchstone Announces Dawson Disposition and Confirms Credit Facility Borrowing Base*, Market Watch, downloaded from www.marketwatch.com/press-release/touchstone-announces-dawson-disposition-and-confirms-credit-facility-borrowing-base-2015-07-22-7160527, on 8 September 2018.

Speight, J. G., 2007, *The Chemistry and Technology of Petroleum*, 4th Ed., CRC Press, Boca Raton, FL, and references therein.

Turta, A. T., Chattopadhyay, S. K., Bhattacharya, R. N., Condrachi, A., Hanson, W., 2007, "Current Status of Commercial *In Situ* Combustion Projects Worldwide", *J. Can. Pet. Technol.*, 46, 8–14.

Turta, A. T., Greaves, M., Starkov, K., 2018, *Toe-To-Heel Air Injection (THAI™) Process*, A T EOR Consulting Inc., downloaded from www.google.com/url?sa=t&rct=j&q=&esrc=s&source=web&cd=1&cad=rja&uact=8&ved=2ahUKEwjvgvqG4qzdAhUl4oMKHTIuD3IQFjAAegQIABAC&url=https%3A%2F%2Fwww.insitucombustion.ca%2Fdocs%2FAdvanced_THAI_Brochure.pdf&usg=AOvVaw0W6QQNZDdFrl7jsLDxBdkQ, on 8 September 2018.

Weissman, J. G., 1997, "Review of Processes for Downhole Catalytic Upgrading of Heavy Crude Oil", *Fuel Proc. Technol.*, 50, 199–213.

Weissman, J. G., Kessler, R. V., Sawicki, R. A., Belgrave, J. D. M., Laureshen, C. J., Metha, S. A., Moore, R. G., Ursenbach, M. G., 1996, "Down-Hole Catalytic Upgrading of Heavy Crude Oil", *Energy Fuels*, 10, 883.

Wichert, G. C., Okazawa, N. E., Moore, R. G., Belgrave, J. D. M., 1995, "In-Situ Upgrading of Heavy Oils by Low-Temperature Oxidation in the Presence of Caustic Additives", SPE No. 30299, presented at SPE International Heavy Oil Symposium, Calgary, Alberta, Canada, 19–21 June.

Xia, T. X., Greaves, M., 2001, "Downhole Upgrading Athabasca Tar Sand Bitumen Using THAI-SARA Analysis", SPE No. 69693 presented at the SPE International Thermal Operations and Heavy Oil Symposium held in Porlamar, Margarita Island, Venezuela, 12 March.

Xia, T. X., Greaves, M., 2002a, "Upgrading Athabasca Tar Sand Using Toe-to-Heel Air Injection", *J. Can. Pet. Technol.*, 41, 51–57.

Xia, T. X., Greaves, M., 2002b, "3-D Physical Model Studies of Downhole Catalytic Upgrading of Wolf Lake Heavy Oil Using THAI", *J. Can. Pet. Technol.*, 41, 58–64.

Xia, T. X., Greaves, M., Turta, A., 2005, "Main Mechanism for Stability of THAI- 'Toe-to-Heel Air Injection'", *J. Can. Pet. Technol.*, 44, 42–48.

Xia, T. X., Greaves, M., Werfilli, W. S., Rathbone, R. R., 2002, "Downhole Conversion of Lloydminster Heavy Oil Using THAI-CAPRI Process", SPE No. 78998, presented at SPE International Thermal Operations and Heavy Oil Symposium and International Horizontal Well Technology Conference, Calgary, Alberta, Canada, 4–7 November.

Xu, H. H., Okazawa, N. E., Moore, R. G., Mehta, S. A., Laureshen, C. J., Ursenbach, M. G., Mallory, D. G., 2001, "In Situ Upgrading of Heavy Oil", *J. Can. Pet. Technol.*, 40, 45–53.

Zhao, R. B., Xia, X. T., Luo, W. W., Shi, Y. L., Diao, C. J., 2015, "Alteration of Heavy Oil Properties under *In-Situ* Combustion: A Field Study", *Energy Fuels*, 29, 6839–6848.

10 New Concepts and Future Trends

As the advantages of subsurface upgrading become clear and the challenges have been addressed, different and new concepts have been disclosed in the patent and open literature. This chapter is dedicated to subsurface upgrading routes that are coming over the horizon and were not included in the traditional pathways listed in Figure 1.8. Even though the production phenomena (Chapter 3) and the chemistry of upgrading (Chapter 4) are basically the same, these new SSU routes bring something novel not found in any of the ones described in the previous chapters. For example, Section 10.1 describes the use of electromagnetic energy to heat the reservoir to the temperatures needed for thermal and catalytic cracking, hydrogen transfer, hydrogenation, or solvent deasphalting. In particular, the latter route has been field tested in Canada in a process called "Effective Solvent Extraction Incorporating Electromagnetic Heating" (ESEIEH™) and is described in Section 10.2. Similarly, another new SSU concept is the use of sonic energy for SSU (Section 10.3). Next, this chapter presents a summary of the key success factors, risks, and challenges of subsurface upgrading in the next decades (Section 10.4). Finally, this book finishes with some concluding remarks (Section 10.4).

10.1 USE OF ELECTROMAGNETIC ENERGY

The use of electromagnetic energy for downhole dielectric heating has found potential applications in enhanced oil recovery (EOR) and subsurface upgrading processes and especially where steam is not the best option [Fanchi 1990, Duncan 1996, Kasevich 1998, Islam 1999, Ovalles *et al.* 2002, Chhetri and Islam 2008, Bientinesi *et al.* 2013, Bera and Babadagli 2015]. Radiations whose frequencies range from 10 to 100 MHz are referred to as radiofrequency (RF) and in the range 300 MHz to 300 GHz as microwaves (MW). In general, the RF/MW heating of formation fluids and porous media can lead to improved mobility of the oleic phase, relative to the aqueous and gas phases, with the concomitant increase in oil production.

Based on the lab experiment and the limited number of pilot tests, several advantages of RF/MW heating technologies can be described. Firstly, electromagnetic energy can be more energy efficient in deep reservoirs because formation heating can be optimized to reduce heavy oil viscosity, whereas saturated steam may far exceed optimum temperatures due to higher reservoir pressures. Additionally, RF/MW is more energy efficient in shallow wells, near thief zones (i.e., gas caps) or thin sands which affect negatively conventional steam flood/fireflood and can be turned "off" during peak power demand to save energy and operating cost. Also, electromagnetic heating could potentially be coupled to direct solar energy or energy from waste which could improve the sustainability of the process. RF at lower frequencies provides more uniform heat distribution than steam because reservoir heterogeneity is not a barrier to RF radiation, whereas low permeability formations are natural obstacles to vapor and *in-situ* combustion processes. Finally, RF/MW heating can reach to very high temperatures needed for the SSU of HO/B, requires no water source, and does not swell clays (Montmorillonite and Illite), like steam condensate.

On the other hand, electromagnetic heating suffers several drawbacks such as high electricity consumption and energy density around the antenna, power losses in the transmission lines, and the relatively high cost of RF/MW power generators. However, as shown in the next section, there is a renewed interest in this form of heating.

10.1.1 Physical and Numerical Simulations

Laboratory measurements have shown that hydrocarbon-containing sands can absorb RF and MW to reach high temperatures (200–400°C) very rapidly in order to upgrade the heavy oils and bitumens [Cambon *et al.* 1978, Milan 1978, Depew *et al.* 1991]. Cambon *et al.* carried out the extraction of crude oil from Canadian tar sand at different times (3–20 min) and powers (from 0 to 2.5 kW at 2.45 GHz) [Cambon *et al.* 1978]. The authors reported up to 86% yield in distillable products with similar quality as those obtained by conventional methods [Cambon *et al.* 1978]. Milan studied the MW absorption of petroleum-containing sands using MW of 2.45 GHz and proposed an empirical model in which the electromagnetic energy has a maximum penetration of 15 m [Milan 1978]. Depew *et al.* studied the thermal decomposition of Canadian tar sands using short-time microwave pulses (3–5 s) and one atmosphere of pressure [Depew *et al.* 1991]. Several light hydrocarbons were detected (C2–C8), and their molar distribution did not change in the presence of water [Depew *et al.* 1991].

C. Jackson carried out heavy oil (11.9°API) upgrading experiments in the presence of ~2wt. % of additives (activated carbon, iron powder, iron oxide, NaOH, and molybdic acid anhydride) at three frequencies (2.45, 6.1, and 6.7 GHz) for 8–10 min [Jackson 2002]. The results showed temperatures up to 330°C and liquid yields from 70 to 90% depending on the frequency, additives, and time of heating. The addition of activated carbon led to 89% viscosity reduction, 7°API uplift, and 23% of HDS [Jackson 2002].

Ovalles *et al.* carried out physical and numerical simulations of electromagnetic heating for a medium crude (24°API), a shallow lake Maracaibo heavy oil (11°API), and a thin pay zone heavy oil (7.7°API) at lab conditions using an MW heated one-dimensional reactor [Ovalles *et al.* 2002]. The authors found temperatures up to 200°C and increases of 86% of cumulative oil production in a shallow heavy oil reservoir using 140 MHz radiofrequency at 50 kW over a ten-year period [Ovalles *et al.* 2002]. The net energy ratio calculated was also 8–20 times (output/input energy).

Salazar applied the RF/MW heating model generated by Ovalles *et al.* [2002] to numerically simulate the subsurface upgrading of heavy oil in the presence of hydrogen donors [Salazar 2003]. As presented in Chapter 7, the latter concept involved the downhole addition of a hydrogen donor (tetralin) under steam injection conditions (280–315°C) and residence times of at least 24 h [Ovalles *et al.* 2003, Ovalles and Rodriguez 2008]. Figure 10.1 shows the API gravity of upgraded crude oil vs. time for the numerical simulation of cyclic steam-H-donor co-injection [Ovalles and Rodriguez 2008] and electromagnetic heating. The RF and MW heating simulations were carried out at 40 MHz/300 kW and 915 MHz/40 kW [Salazar 2003]. As seen, for all cases evaluated, the cumulative API gravity (measured under surface conditions) of the upgraded crude oil reached a maximum value of ~40–65°API at the beginning of production time and decreased with time. This phenomenon was attributed to the generation of a lighter fraction closer to the well by the reactions of the heavy crude oil with the hydrogen donor (tetralin, see Equation 7.14) in the presence of steam or RF/MW heating. These lighter components represent the main fraction of the produced oil at the beginning of the simulation.

For a 120-day production period, the cumulative API of the upgraded oil was found to be approximately 14°API for the cyclic steam-H-donor co-injection [Ovalles and Rodriguez 2008], whereas for the RF and MW cases (Figure 10.1) the API gravities were 19°API and 16°API, respectively (Figure 10.1, dashed arrow). These results were attributed to a more homogeneous temperature distribution in the RF/MW cases in comparison with that obtained for the steam heated case. These numerical simulations indicated that coupling RF or MW with hydrogen donors gave significant advantages compared to those found using conventional steam co-injection technologies.

Bientinesi and coworkers carried out experimental work and numerical simulation of RF/MW heating using quartz sand as low RF absorbance material [Bientinesi *et al.* 2013]. A sample of 2,000 kg of oil-containing sand was heated up to 200°C using a dipolar antenna irradiating at 2.45 GHz and 2 kW. Their lab and modeling results showed that that the presence of the low lossy material (quartz sand) around the antenna was very efficient in lowering the temperature in this critical zone and in better distributing the irradiated energy into the oil-sand [Bientinesi *et al.* 2013].

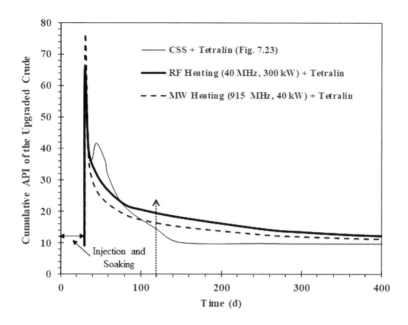

FIGURE 10.1 Cumulative API gravity of upgraded crude oil vs. time for the numerical simulation of cyclic steam-H-donor co-injection (CSS + Tetralin), and RF and MW heating.

Greff and Babadagli investigated the effects of MW radiation (2.45 GHz) on the viscosity reduction of a North Alberta heavy oil at a temperature of 200°C for a period of 5 h [Greff and Babadagli 2011]. Different nanosized metal particles (Fe, Fe(III) oxide, and Cu) were used as catalysts in concentrations ranging 0.1 wt. % to 1 wt. %. The results showed a viscosity reduction in the order of 40% and the generation up to 40 wt. % of gaseous components. The authors found that the amount of vaporized material increased with the concentration of the catalyst [Greff and Babadagli 2011].

Hu and coworkers combined electromagnetic heating (EM) with solvent assisted gravity drainage in laboratory experiments at 2.45 GHz and 700 W [Hu *et al.* 2016]. They studied the effects of initial water saturation, solvent types (n-hexane and n-octane), solvent addition (injection or premixed with oil), a combination of solvent injection and EM heating strategies (simultaneous or alternate), and EM heating power. The results showed that n-octane provides higher vertical displacement efficiency and oil recovery under the same experimental conditions than n-hexane. Alternate MW heating and the solvent injection was more cost-effective than the simultaneous analog due to the lower energy consumption. The authors found temperatures in the 180–200°C range and 19% of asphaltene conversion [Hu *et al.* 2016].

Bogdanov *et al.* numerically simulated RF heating with solvent injection (butane) in a typical Athabasca reservoir using vertical and horizontal antennas [Bogdanov *et al.* 2016]. They observed significant variations of oil recoveries (~10%) depending on the mechanism of solvent mixing with the bitumen, e.g., molecular diffusion or mechanical dispersion [Bogdanov *et al.* 2016].

Based on the previous laboratory-scale tests and numerical simulations, the feasibility of using RF and MW radiations for dielectric heating of oil-containing sands has been successfully demonstrated. This process has potential applications for the subsurface upgrading of HO/B [Ovalles *et al.* 2002, Bera and Babadagli 2015].

10.1.2 Concepts for Downhole Heating

Several concepts have been proposed to take the electromagnetic energy downhole to heat the oil-containing formations using vertical [Haagensen 1965, 1986, Bridges 1979, Sresty *et al.* 1984, 1986, Bridges 1985] and horizontal wells [Warren *et al.* 1996, Kasevich *et al.* 1997, Sahni *et al.* 2000].

For example, Haagensen patented an apparatus to generate the RF and/or MW at the surface and a coaxial or waveguide to take the energy downhole [Haagensen 1965, 1986]. On the other hand, Jeambey uses a downhole MW generator in which the dielectric constants of the mineral formation can be measured and the system optimized to reach very high temperatures (up to 400°C) necessary to achieve subsurface upgrading [Jeambey 1989, 1990].

Kasevich disclosed the use of radiofrequency energy to heat heavy crude oil to enhance the recovery and to promote *in-situ* upgrading [Kasevich 2008]. This invention uses two horizontal or vertical wells to create an "*in-situ* radiofrequency reactor" downhole. The preferred well array is an SAGD configuration that leads to a reduction in the viscosity and mild cracking of the heavy oil to produce the fluids via gravity drainage to the producer. Unfortunately, no examples were given to illustrate the levels of upgrading achieved [Kasevich 2008].

ConocoPhillips and Harris Corp. disclosed several patent applications on the coupling of RF downhole heating with metal-containing activators which also could act as hydrogenation catalysts under SAGD and CCS well configurations [Dreher *et al.* 2013, Madison *et al.* 2014, Wheeler *et al.* 2016]. As shown in Figure 10.2, the activators/catalysts were pumped downhole using the injection well and placed in the area in which the SAGD steam chamber was formed. In this way, the contact between steam, crude oil, and the catalyst was achieved with the simultaneous upgrading of the heavy oil. It was disclosed that RF absorbent material could reach a temperature in the 315–650°C range operating in the 0.1 MHz to 300 MHz frequency range. According to the information reported, the activators/catalysts are based on iron, iron oxide, $AlCl_4^-$, $FeCl_4^-$, $NiCl_3^-$, and $ZnCl_3^-$, nickel, cobalt, steel, magnetite, nickel-zinc ferrite, manganese-zinc ferrite, copper-zinc ferrite, MoS_2, WS_2, CoMoS, and NiMoS, and iron, nickel, and cobalt alloys. Unfortunately, no examples on the upgrading levels achieved or oil recoveries were given [Dreher *et al.* 2013, Madison *et al.* 2014, Wheeler *et al.* 2016].

Blue *et al.* disclosed an apparatus and a process for hydrocarbon recovery and transportation that combines subsurface and wellhead upgrading using RF power with or without SAGD [Blue *et al.* 2015]. Several examples were disclosed using an RF ring antenna to treat a bitumen at 6.78 MHz under different purge gases (N_2, steam, and H_2) for 1.5 h. Other frequencies (7–30 Hz) were also reported. The results showed the formation of hydrocarbon gases and a condensable liquid (~30°API). As shown in Figure 10.3, SARA analysis of the treated residue showed a significant reduction of aromatics from 30% to 1% with the corresponding increase in resins (from 27% to 61%). Asphaltene content remained approximately constant [Blue *et al.* 2015].

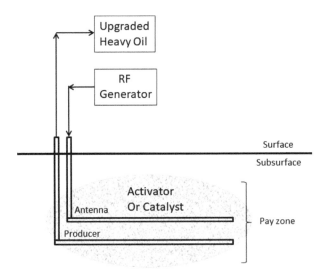

FIGURE 10.2   General diagram of the RF heated upgrading concept using RF activators or catalyst downhole.

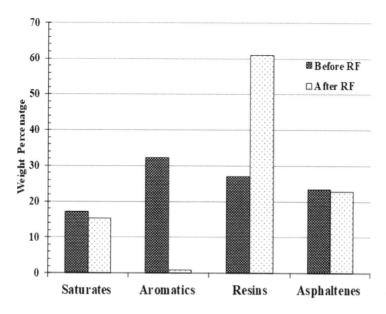

FIGURE 10.3 SARA analysis of the crude oil before and after RF treatment. Data taken from Blue *et al.* [2015].

Dreher and coworkers disclosed a system for enhancing the *in-situ* upgrading of hydrocarbon by implementing an array of radio frequency antennas that can uniformly heat the hydrocarbons *within* a producer well pipe [Dreher *et al.* 2015]. By this way, the optimal temperatures for different Hydroprocessing reactions can be achieved. The invention described a producer well having a catalyst in contact with the hydrocarbons, a plurality of RF antennas inside the casing, and an RF current source connected to the RF antennas by conductive wires. Such a device is used to heat the heavy oil to a temperature sufficient to make upgraded hydrocarbons. As in other patents in the area, no examples were given to illustrate the levels of upgrading achieved [Dreher *et al.* 2015].

10.1.3 FIELD TESTS

Even though there is *no* field test reported in the literature combining electromagnetic heating with subsurface upgrading, there are few but significant examples of several RF/MW field applications [Bridges 1985, Sresty *et al.* 1986, Kasevich *et al.* 1994, Trautman *et al.* 2014, Despande *et al.* 2015, Wise and Patterson 2016]. In a field test in Oklahoma, Bridges and coworkers used three waveguides inserted into the ground to heat a limited amount of an oil-bearing formation [Bridges 1985]. In this field test, a 300 MHz radiation was used to raise the temperature near the wellbore region from 18°C up to 100°C. At 15 ft from the well, a temperature of 33°C was reached. The authors reported an increase in the production of crude oil (6°API), but no additional details were given [Bridges 1985]. Kasevich *et al.* carried out a field test in Bakersfield, California in a 600–700 ft depth reservoir in 1994 [Kasevich *et al.* 1994]. The authors used a 25-kW surface generator (at 13.56 MHz), a coaxial line, and an antenna to bring the energy downhole. Their results demonstrate the technical feasibility of generating, transmitting, and focusing the electromagnetic energy into an oil-bearing formation reaching 120°C at 10 ft from the well in 20 h [Kasevich *et al.* 1994].

Sresty *et al.* conducted laboratory and field scale tests to recover bitumen from tar sand deposits near Vernal, Utah, using RF heating [Sresty *et al.* 1986]. In their trials, the recovery methods included replacing the heated bitumen with sodium silicate solutions, gravity drainage, and autogenously developed steam and hydrocarbon gases. Two small-scale field experiments were carried out in the Asphalt Ridge tar sand deposit near Vernal. They reported that about ~150 bbl of the deposit was heated with RF energy, and about 35% of recovery vs. OOIP was obtained during the three-week test period [Sresty *et al.* 1986].

The field test of the process called "Effective Solvent Extraction Incorporating Electromagnetic Heating ("ESEIEH™")" is described in the next section [Trautman *et al.* 2014, Despande *et al.* 2015, Wise and Patterson 2016]. Other field tests have been reported in the literature and have been reviewed by Bera and Babadagli [2015].

From work described in this section, it can be concluded that significant efforts have been devoted to the development of electromagnetic technologies for thermal enhanced oil recovery. However, no field test combines RF/MW downhole heating with subsurface upgrading concepts. Therefore, it can be safely foreseen that more hybrid processes will be considered in the future with the idea of maximizing the individual advantages of each technology and at the same time, overcoming the disadvantages of each route.

10.2 ESEIEH™ FIELD TEST

Harris, Laricina, Nexen and Suncor formed a Can$33 million consortium called ESEIEH™ for the developing of the Enhanced Solvent Extraction Incorporating Electromagnetic Heating process [Trautman *et al.* 2014, Trautman and MacFarlane 2014, Despande *et al.* 2015, Wise and Patterson 2016]. The consortium is 50% financed by Alberta's Climate Change and Emission Management Corporation (CCEMC). In 2015, Laricina withdrew from the consortium and was replaced by Devon Canada Corp. for the later phases of the project [Bayestehparvin *et al.* 2018].

The concept involves the use of RF heating coupled with hydrocarbon solvent addition (chain lengths of C2 to C5) for enhanced heavy oil recovery. As shown in Figure 10.4, the radiofrequency emitted from the antenna in an SAGD-type of configuration preheats the hydrocarbon-containing formation, allowing the solvent to dissolve the bitumen with the concomitant increase in oil production. The hydrocarbon solvent is added from the top horizontal well (Figure 10.4), and the oil is produced from the bottom. By using this concept, the formation does not have to be brought up to a steam temperature which reduces greenhouse gas emissions. The solvent reduces the viscosity of the bitumen, increasing its mobility, which in turn leads to enhanced heavy oil recovery and possibly subsurface upgrading via solvent deasphalting [Trautman *et al.* 2014, Trautman and MacFarlane 2014, Despande *et al.* 2015, Wise and Patterson 2016].

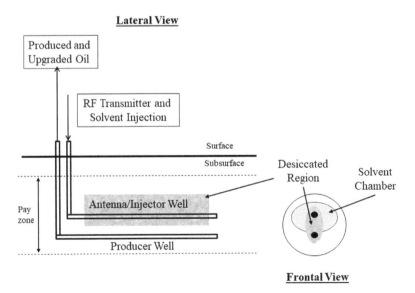

FIGURE 10.4 Lateral and frontal views of the ESEIEH™ process.

Wise and Patterson from Harris Corp. and Devon Canada Corp., respectively, carried out an economic evaluation of this process in three different scenarios [Wise and Patterson 2016]. These scenarios included a green field with dilbit export, a green field with undiluted rail export, and an ESEIEH™ pad added adjacent to an existing SAGD cogeneration facility. The results indicated that the process reached a breakeven point at $50/BBL of WTI, and it is economically viable in a sustained $60/BBL oil price environment [Wise and Patterson 2016].

Up to the moment of writing this book, the development of the ESEIEH™ technology has involved a preliminary work and three phases [Bohn 2018]. In the initial work, small-scale experiments were carried out using an 11-m antenna to heat sand at two different power levels, i.e., 10 kW and 45 kW. In the first test, an increase in the temperature up to ~10°C was observed at 0.5 m from the antenna after one day of operation. At the second power level, up to 35°C was measured at 3 m from the antenna after ~nine days of heating. Modeling work was able to reproduce those results successfully. Next, the other three phases of the technology development path are described.

10.2.1 PHASE 1: PROOF OF CONCEPT

This test was carried out in a built-for-purpose pit at the Suncor North Steepbank Mine during 2011–2013 [Bohn 2018]. The principal objectives were to verify RF energy penetration and absorption rates at reduced scale and validate numerical model predictions, but *no* solvent was used. Also, the goals were to demonstrate the design, deployment, and operation of an RF heating system in a field environment.

The test site had a formation permeability ranging from 60 millidarcy (mD) in a shale layer to as high as 8,800 mD in a clean section of oil sand. The porosity ranged from 29% to 32%, and the oil saturation was as high as 85% in clean oil sand sections. Water content showed a range of 1.3% to 14.7%. All these properties were deemed acceptable for RF heating [Bohn 2018].

The main components of the ESEIEH™ system are the RF antenna ("HeatWave"), solvent injection process ("ESEIEH™"), and the coupled electromagnetic reservoir simulator ("CEMRS") [Despande *et al.* 2015]. The first component was a 12.25 m antenna operating at 6.78 MHz and a maximum power density of 4 kV/m. Harris' design of the coaxial cable had three concentric pipes. The outermost pipe was steel, and the inner pipes that carried the RF energy were copper. Harris filed the inner annulus with nitrogen and circulated mineral oil down the outer annulus and up the inner copper tube as a coolant. All joints were sealed to prevent cross-contamination between the reservoir fluids and those in the coaxial cable. As mentioned, no solvent was added, so a complete evaluation of the ESEIEH™ technology was not carried out in this phase [Jackson *et al.* 2014, Nugent *et al.* 2014, Trautman and MacFarlane 2014, Despande *et al.* 2015, Bohn 2018].

The RF power was initiated on 20 November 2011 and had three stages. Initially, the power was ramped to 28 kW in six days just below 100°C. After ten days of low power, sufficient data were collected to compare with simulations for the pre-desiccated condition. For stage 2, between day 10 and day 14, the power was ramped linearly to the maximum power level of 49 kW for 17 days. Then, the RF power was lowered to 12 kW to maintain formation temperature, and a sampling port was constructed in the annulus at the sand face. After ten days, the power was reset to 49 kW, and the system was run almost continuously for seven days until the middle of day 34 when the test finished (average power 26 kW) [Trautman and MacFarlane 2014, Bohn 2018].

The temperature profiles were followed using two observation wells. The author reported that the temperature increased throughout the test except for a decrease in temperature at the antenna elevation during the low power operation between day 17 and 27. The maximum formation temperature of 127°C occurred approximately at 0.5 m to 1 m below the antenna, not at the antenna centerline [Bohn 2018].

The results of the coupled electromagnetic reservoir simulator were compared with those from the observation wells. The author reported that the correlation was quite good at all times and elevations although the model under-predicted the temperature near the center of the antenna by ~10°C and over-predicted at the tip by ~10°C. In general, good matches with the model predictions were reported [Trautman and MacFarlane 2014].

Despite a short operating period, the consortium considered that all the objectives of the mine face test were met. The ESEIEH™ system was designed, manufactured, and installed in an oil sand reservoir, and RF heating was demonstrated at reasonable power levels. Based on those results, the project was moved to Phase 2 [Jackson *et al.* 2014, Nugent *et al.* 2014, Trautman and MacFarlane 2014, Despande *et al.* 2015, Bohn 2018].

10.2.2 PHASE 2: SMALL-SCALE PILOT AT SUNCOR DOVER

This field test was carried out at the Dover test site between September 2014 and December 2015. The pilot includes equipment and facility integration for a technical demonstration of the full ESEIEH™ process with a 100 m horizontal well pair and three vertical observation wells. The objectives were to measure bitumen production due to RF heating and propane vapor and evaluate the sensitivity to operating conditions such as RF power, solvent (C3) injection rate, pressure, production rate controls, etc. Also, the goals were to validate the numerical modeling, determine functionality, reliability, and efficiency of the pilot hardware, and measure other critical economic indicators including solvent retention, power consumption, and delivery efficiency of EM energy to the reservoir [Bohn 2018].

The ESEIEH™ surface facilities consisted of six major components: 1) 500 kW RF transmitter, 2) dielectric fluid conditioning system, 3) bitumen production handling equipment, 4) 400-bbl product storage tank, 5) propane storage tank, and 6) control room. The producer well was run with a standard slotted liner with a progressive cavity pump for artificial lift. The three observation wells were drilled with RF transparent casing and equipped with pressure and distributed temperature sensors. The construction started in September 2014 and finished in May 2015 [Bohn 2018].

The RF heating commenced on 9 July 2015 and continued without significant issues for several days. Temperatures between 125–150°C were reported [Bohn 2018]. On 13 July 2015, the pilot test suffered a shutdown due to metallic debris deposited during the drilling and completions operations, which impacted surface and subsurface RF heating equipment. After a Root Cause Analysis (RCA), corrective actions were implemented regarding material selection, drilling and construction process, and improving the robustness of the RF equipment [Bohn 2018]. The re-start to the test was expected in the first quarter of 2018.

10.2.3 PHASE 3: CONTINUANCE OF THE SMALL-SCALE PILOT

This phase included modified facilities and well configurations as determined by the RCA study. The modifications provided for the addition of a vertical injection well which permitted operation of the horizontal antenna independently of vaporized propane injection. Up to January 2018, the incurred project costs were approximately CA$75 MM [Bohn 2018]. Based on the previous descriptions, it can be concluded that the technical development of the ESEIEH™ process is still in progress, and the outcome of such investigations will be known in the near future.

10.3 SONICATION

Sonic energy or ultrasound has been a very useful tool in enhancing the reaction rates for a variety of chemical systems [Suslick *et al.* 1999, Thompson and Doraiswamy 1999, Grobas *et al.* 2007, Sawarkar *et al.* 2009, Avvaru *et al.* 2018]. When a liquid is subject to sonication, the cavities within the liquid start oscillating in tune with the sound waves. The formation and subsequent dynamic life of these cavities are called cavitation. When a cavity experiences a compression phase, the local pressure increases while in the rarefaction phase the local pressure decreases. When the local pressure becomes very low as compared to the overall pressure, the cavity implodes. If the pressure during a rarefaction cycle is low enough, localized areas of vaporized liquid can form small bubbles which are known as cavitation bubbles. These cavitation bubbles are at the heart of acoustic cavitation and are responsible for the so-called sonochemical transformations. A series of ultrasound wave cycles causes the bubbles to grow and contract or implode during the subsequent compression phase. Each one of these imploding bubbles is acting like a microreactor, with temperatures reaching over an estimate of about 5,000°C and pressures of over several hundred atmospheres. Therefore, the high local temperatures and pressures, combined with extraordinarily rapid cooling, provide a unique means for driving chemical reactions under extreme operating conditions. While these phenomena are happing at micro-scale, the bulk of the reaction mixture and the vessel are at relatively mild temperatures and pressures [Suslick *et al.* 1999, Thompson and Doraiswamy 1999, Grobas *et al.* 2007, Sawarkar *et al.* 2009, Avvaru *et al.* 2018].

Avvaru and coworkers reviewed the applications of ultrasound energy and cavitation technologies to the petroleum industry [Avvaru *et al.* 2018]. These processes are aimed at improving the well productivity by increasing the permeability of the near wellbore region. Several examples have been reported in the literature such as the breaking of the emulsion generated between invading oil-based mud filtrate and formation water, inorganic scale precipitation, plugging of pore throats by fines migration, and deposition of paraffin and asphaltenes. The authors stated that the main advantages of the ultrasonic technology are low energy consumption, the possibility to treat the near wellbore selectively, no harm to the well and casing, and that these routes are eco-friendly and economically acceptable [Avvaru *et al.* 2018].

Under laboratory conditions, the use of ultrasonic irradiation for the upgrading of heavy crude oils [Sadeghi *et al.* 1994, Gunnerman *et al.* 2003, Varadaraj 2004], asphaltenes [Lin and Yen 1993, Dunn and Yen 2001, Shedid and Attallah 2004], tar sands [Sadeghi *et al.* 1990, 1992], vacuum residues [Sawarkar *et al.* 2009, Kaushik *et al.* 2012], and heavy gas oils [Gopinath *et al.* 2006] is well documented in the literature. Lin and Yen carried out the pioneering work for the ultrasound-assisted asphaltene conversion to gas oil and resin fractions at room temperature and atmospheric pressure [Lin and Yen 1993]. The authors reported that asphaltene upgrading in the presence of ultrasonic energy follows a free radical mechanism, as in the thermal cracking of petroleum residues (Section 4.1.2). Through the combination of sonic energy (cavitation) and surfactant, 35% of the asphaltenes was converted into gas oil and resin fractions at ambient temperature and pressure conditions in 15 min. The authors attributed the relatively high asphaltene conversion to a reduction of asphaltenes aggregation assisted by the ultrasound energy [Lin and Yen 1993].

Acoustic cavitation (sonic energy) can also be successfully employed for the upgrading of the petroleum residues leading to the formation of low boiling, more value-added products. For example, Kumar and coworkers studied the effect of ultrasound (24 kHz) on the upgrading of a vacuum residue in the presence or absence of a surfactant at 30°C and atmospheric pressure for up to 30 min [Kaushik *et al.* 2012]. The authors observed a reduction of the asphaltene content of 48% (from 13.5 wt. % to 7% wt. %) and an increase in the H/C molar ratio from 1.43 to 1.58. Similarly, they reported a viscosity reduction of 26% [Kaushik *et al.* 2012].

Based on the above examples, it is not surprising that the use of ultrasound for the HO/B subsurface upgrading had been proposed in the literature. J. M. Paul and R. M. Davis disclosed a downhole tool for the hydrotreating of heavy crude oils in the presence of sonic energy and a metal hydrogenation catalyst [Paul and Davis 1998]. As shown in Figure 10.5, an acoustic transducer and an acoustic driver are positioned downhole and just below the production tubing. The acoustic driver is coated with a Ni/Zn hydrogenation catalyst. During production operation, water-containing heavy crude oil enters the casing and is subjected to sonic vibrations in the frequency range of 400 to 10 kHz (preferably about 1.25 kHz). At downhole temperatures and pressures, the authors proposed that water generates hydrogen via Equation 10.1, which in turn is used to catalytically hydrotreat the heavy crude oil *in-situ* [Paul and Davis 1998].

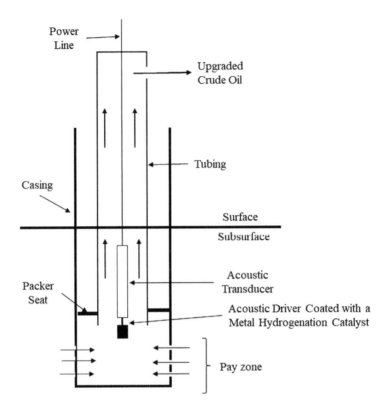

FIGURE 10.5 Schematic diagram of the downhole tool for the hydrotreating of heavy crude oil in the presence of sonic energy and a metal hydrogenation catalyst.

$$2H_2O \xrightarrow[\text{Catalyst}]{\text{Sonic Energy}} 2H_2 + O_2 \qquad (10.1)$$

The authors hydrotreated a Battrum heavy crude oil emulsion containing about 40% v/v of water at the following conditions: 50°C, 100 psig argon pressure, 2.4 g nickel on zinc catalyst/140 ml heavy crude oil emulsion, sonic energy at a frequency of 1.25 kHz, weight hourly space velocity (WHSV) of 233 hour^{-1}, and a reaction time of 15 minutes [Paul and Davis 1998]. The results showed that the asphaltene and resin contents decreased 9% (from 16.19 wt. % to 14.69 wt. %) and 11% (from 41.38 wt. % to 36.71 wt. %), respectively. In the same experiment, the amount of aromatic and saturate fractions increased 19% (from 30.95% to 36.88%) and 2% (from 11.48 wt. % to 11.71 wt. %), respectively. A control experiment carried out in the absence of sonication showed almost no changes in the SARA composition, and oxygen and hydrogen gas were detected in the gas phase [Paul and Davis 1998].

Xu and coworkers studied the sonic-assisted catalytic Aquathermolysis of Gudong heavy crude oil, and 30% produced water in the presence of 0.3wt. % of a copper-sulfonate catalyst at 200°C for 24 h [Xu et al. 2011, 2012]. In batch-wise experiments at 22 kHz, the authors found that the ultrasonic-assisted catalytic Aquathermolysis led to a slight increase in viscosity-reduction values (86.2% vs. 80%) and a decrease in the average molecular weight of heavy oil (418 g/mol vs. 506 g/mol) in comparison with the catalytic-only reactions. Similarly, they observed a reduction of the resin and asphaltene contents (24% and 28% vs. feed, respectively), an improvement on the H/C mol ratio (from 1.733 to 2.271), and percentages of HDS and HDN of 33% and 21%, respectively [Xu and Pu 2011].

Furthermore, the results of displacement experiments showed that ultrasonic-assisted catalytic Aquathermolysis exhibited 54% heavy oil recovery and 80.5% viscosity reduction whereas the catalytic-only experiment led to 46% and 72%, respectively [Xu and Pu 2011]. The author concluded that the combination of ultrasound and catalyst had a synergistic effect, which was attributed to the simultaneous viscosity reduction and upgrading of the heavy oil via Aquathermolysis (see Section 8.2). They also stated that sonication could provide energy for chemical reactions and improve catalyst activity, thus accelerating heavy oil cracking processes [Xu and Pu 2011].

Avvaru and coworkers reported that two pieces of ultrasonic downhole equipment had been developed and field tested [Avvaru *et al.* 2018]. These instruments have power varying from 1 to 20 kW, working frequencies from 1 to 50 kHz, and withstand temperatures and pressures up to +150°C and 50 MPa. Based on the laboratory experiments discussed and the availability of bench and field equipment, more applications of ultrasound-assisted underground upgrading of HO/B can be foreseen in the near future.

10.4   KEY SUCCESS FACTORS, RISKS, AND CHALLENGES

The subsurface upgrading of crude oil and bitumens is a technological area that has many advantages from the technical and economic points of view but also many challenges to overcome. The advantages and disadvantages of SSU processes were presented in Sections 1.5 and 1.6, respectively. The final goal is to move a potential concept from the lab to a deployed reality in petroleum fields around the globe. In this section, several key success factors, risk, and challenges are summarized that are common to all SSU routes covered in Chapters 5 through 9. These issues represent some of the obstacles the project team had to overcome to carry out the development of enhanced oil recovery and underground upgrading technology.

As can be clearly seen from the literature review presented in the previous chapters, one of the most critical challenges for SSU processes is the control and monitoring of the reaction conditions downhole. Several approaches have been proposed to overcome this issue such as the use of wellbore and wellhead sensors (P and T), utilization of observation wells, and more recently wireless technology. Advancements in the geophysical characterization of HO/B reservoirs have a definite contribution to make. In the last decades, 3- and 4-D seismic has been used for detecting the movement of steam/solvent from within a heavy oil reservoir. Seismic attributes have been used for differentiating heated from unheated parts of the porous media and imaging the areas affected by thermal processes. Cross-well electromagnetic induction measurements for fluid saturation have been used successfully to characterize the zones in between wells. It is expected that advancements in all these areas will grow in the future [Chopra *et al.* 2010].

Another SSU operational issue is the poor mixing and/or contact of the reactants at downhole conditions. To diminish the impact, the selection of the additives, catalyst, and solvent should be done wisely. For SAGD and mature CSS wells, water/solvent mixture at the interface is supplied by vapor phase only. Therefore, the preferred way to transport solvent and catalyst to oil interface is in the gas phase. Additionally, the development of highly active catalytic materials (dispersed or nanocatalysts) that requires lower reaction times and mild conditions is critical for the success of the SSU processes.

Environmental impact of additives, catalysts, and solvents used downhole represents another significant challenge to overcome. As mentioned in Section 1.9, there are three significant ecological risks associated with subsurface upgrading processes, i.e., leakage to the surface, migration in the subsurface, and damage to company's reputation, even if there is no impact on the environment. For these reasons, the use of environmentally neutral or benign additives is highly recommended to mitigate the ecological effects of SSU processes.

Another challenge to be faced in the future is that the exploitation of HO/B fields is in remote locations such as Orinoco, Siberia, and China. Thus, the development of local infrastructure jointly with national oil companies is of paramount importance and a vital issue. Furthermore, the lack of expertise of upstream personnel in upgrading processes and the lack of knowledge of downstream engineers in oil production operations represent a significant hurdle for the successful outcome of the SSU project. Despite all the previous risks and challenges, the opportunity for SSU is quite vast, and several HO/B assets around the globe could benefit from subsurface upgrading technologies.

10.4.1 PHYSICAL SEPARATIONS

The most critical challenges for downhole solvent deasphalting (SDA) are to reduce the solvent-to-crude ratio (SvOR) and solvent recycling (Sections 5.2 through 5.7). The large SvOR used in SDA processes could be a significant issue due to the high solvent cost, the large footprint of surface facilities, and relatively high complexity of production operation. Additionally, the separation, purification, and re-injection of the solvent are of critical importance for the technical and economic success of the SSU-SDA technology. The development of additives that can act as asphaltene precipitating agents (Section 5.7) could be an exciting option to reduce the SvOR.

Another essential aspect of deasphalting is the availability of solvent on-site and the cost of storage. Light-hydrocarbon storage tanks can be relatively expensive and generally not available in remote locations. Also, the variability of the solvent composition could be an issue if a maximum efficiency of asphaltene precipitation is desired.

In SSU-SDA processes, wellbore formation damage and reduction in reservoir permeability due to asphaltene precipitation are distinct possibilities. The formation of deposits in surface installations is very plausible. Thus, the use of solvents and asphaltene dispersants to clean and remove deposits is of paramount importance to mitigate and reduce downtime.

10.4.2 THERMAL PROCESSES

For these SSU routes, the most critical issues and uncertainties are the availability of energy at low cost, instability of the upgraded products to asphaltene precipitation and olefin formation, and development of scales through dissolution/transport of formation minerals. To cope with those issues, the use of less expensive or more efficient sources of energy has been suggested such as electrical, electromagnetic (RF/MW), and sonic. The possibility of downhole heating using electricity generated from solar represents an environmentally friendly route. As discussed in Section 10.1, RF/MW radiations are more efficient for deeper and shallower reservoirs than traditional steam injection. Sonication could be an excellent alternative to induce chemical reactions (Section 10.3).

The generation of unstable products could significantly affect the profitability of thermal-only SSU processes. One possibility could be to limit the temperature, residence time, and conversion ($> 25\%$) to avoid asphaltene or olefin precipitation. The use of asphaltene inhibitors and dispersants is also a plausible option.

10.4.3 HYDROGEN ADDITION

The most critical challenges for the hydrogen addition processes are the need to provide the required contact (mixing) between catalyst, hydrogen source, and heated HO/B in the reservoir to make sure the upgrading reactions take place, the availability of inexpensive and abundant energy and hydrogen sources, and cost and activity of catalytic materials. Section 8.1 discussed all these catalyst issues such as the placement for the catalyst downhole (Sections 8.1.1 through 8.1.2), control of operating parameters (Section 8.1.3), and ecological concerns of releasing toxic metals to the environment (Section 8.1.4).

As described in Section 4.2.4 and 8.2.4, there has been considerable interest in the use of highly dispersed metals as nanocatalysts for the SSU of heavy crude oils and bitumens. Nanosized catalysts offer the fundamental advantage of having a significantly superior surface-to-volume ratio. This feature increases the mass transfer of the substrate to the catalytic-active center considerably and at the same time reduces the catalyst-deactivation issues. Thus, nanostructured materials have the potential to generate vastly superior catalytic systems and at the same time, reduce the amount of metal used with the concomitant decrease in operating costs and environmental impacts.

The use of gasification plants to generate H_2 and electricity as well as the use of refinery fractions as hydrogen donors (Section 7.4) are two alternatives that can lead to cost reduction. Also, the need to recover/recycle the hydrogen donor solvent is of paramount importance to improve the economic prospect of the SSU process.

As discussed in Sections 7.2 and 8.2, the thermal and catalytic generation of H_2S takes place during Aquathermolysis. This issue represents another significant challenge for SSU processes because it leads to the use of special metallurgy to handle sulfur-containing gases and thermal stresses simultaneously.

10.4.4 In-Situ Combustion

Significant challenges can be predicted for SSU routes via ISC such as partial oxidation of distillable materials, control of the burning front, possibility of fire and explosion by bringing hot petroleum and air to the surface, corrosivity of exhaust gases, generation of toxic emissions to air (e.g., SO_2, SO_3, and NH_3), and the possibility that polyaromatic hydrocarbons could leach to aquifers. All previous issues are shared with other ISC processes that have been carried out in several parts of the worlds. Among the approaches to overcome those challenges are the use of observation wells, wireless sensors, and 3-D and 4-D seismic to monitor the combustion front and innovative well configurations (injector, producer, and control) to optimize the heavy oil upgrading.

A described in Section 9.4.4, THAI-CAPRI is an emerging technology aimed to enhance the recovery and upgrading of heavy oils by using a catalyst located in the producing well. It has significant beneficial advantages over the commonly used thermal EOR processes such as high recovery factor, production of valuable products, and lower water usage. However, short catalyst lifetime due to rapid deactivation and possibility of catalyst bed plugging are issues that need to be adequately addressed in the future.

10.5 CONCLUDING REMARKS

In the last two decades, the application of new and improved technologies has reduced substantially the operating costs to produce HO/B in different areas of the globe [Chopra *et al.* 2010]. In the near future, the key to sustainable growth and enhanced economic viability lies in the development of even more efficient processes to help sustain a continuous supply of environmentally responsible energy. The abundance of HO/B resources and a stable political climate in most areas around the world will encourage oil companies to decide on the high financial commitments required to become players in the future energy growth.

Several new areas of research can be envisioned that may continue to be developed for the fascinating topic of subsurface upgrading of HO/B. We can propose the following non-exhaustive list:

- Use of SSU in carbonates reservoirs
- SDA in deep mineral formations in which steam is challenged
- New data rich and very powerful analytical characterization for HO/B
- Reservoir numerical simulation including chemical reactions and equation-of-state to model asphaltene precipitation and other physical phenomena
- Combine renewable energies solar/wind with upgrading
- More hybrid and integrated processes (more downhole to downstream facilities)
- Extensive use of nanoparticle as catalysts, asphaltene dispersants, and precipitants
- Fracture-creating methods to inject downhole chemicals, catalysts, hydrogen donors, etc.
- Use of ionic liquids as catalysts for the hydrocarbon cracking at very mild conditions (Section 6.1.3)

- Cheaper and more efficient hydrogen donors to increase the rates and cumulative oil, enhance HO/B upgrading, and reduce the SOR, and SvOR (Sections 7.3 and 7.4)
- Use of RF/MW for SSU of heavy oils, bitumens, and shale oils

Finally, there is little doubt that heavy oils and bitumens will continue to play a role in the energy outlook of the future. The world reserves of heavy crude oils and bitumens represent between 50% and 70% of all hydrocarbons available on the planet [British Petroleum 2016, Faergestad 2016]. Thus, technologies that can transform the HO/B reservoirs in "working refineries" and deliver products with improved quality at a minimal environmental impact and using renewable energy have great potentiality [Huc 2011]. Ultimately, heavy ends and asphaltenes could be discarded *in-situ*, water, solvents, and catalysts could be recycled, and emissions and energy use could be minimized. Some of these futuristic technologies could also be of interest for the recovery of other resources such as conventional oil and potentially shale oils and coal deposits.

REFERENCES

Avvaru, B., Venkateswaran, N., Uppara, P., Iyengar, S. B., Katti, S. S., 2018, "Current Knowledge and Potential Applications of Cavitation Technologies for the Petroleum Industry", *Ultrason. Sonochem.*, 42, 493–507.

Bayestehparvin, B., Farouq Ali, S. M., Abedi, J., 2018, "Solvent-Based and Solvent-Assisted Recovery Processes: State of the Art", SPE No. 179829, *SPE Res. Eval. Eng.*, 1–21 August, and references therein.

Bera, A., Babadagli, T., 2015, "Status of Electromagnetic Heating for Enhanced Heavy Oil/Bitumen Recovery and Future Prospects: A Review", *Appl. Energy*, 151, 206–226, and references therein.

Bientinesi, M., Petarca, L., Cerutti, A., Bandinelli, M., De Simoni, M., Manotti, M., Maddinelli, G., 2013, "A Radiofrequency/Microwave Heating Method for Thermal Heavy Oil Recovery Based on a Novel Tight-Shell Conceptual Design", *J. Pet. Sci. Eng.*, 107, 18–30.

Blue, M. E., Zastrow, L. P., Whitney, R. M., Jackson, R. E., Meyer, J. A., 2015, "Method for Recovering a Hydrocarbon Resource from a Subterranean Formation Including Additional Upgrading at the Wellhead and Related Apparatus", US Patent 9,057,237.

Bogdanov, I., Cambon, S., Mujica, M., Brisset, A., 2016, "Heavy Oil Recovery via Combination of Radio-Frequency Heating with Solvent Injection", SPE No. 180709, presented at SPE Canada Heavy Oil Technical Conference, Calgary, Alberta, Canada, 7–9 June.

Bohn, M., 2018, "ERA Public Facing Report ESEIEH Progress Update. Reporting Period: February 2011 to August 2017", Retrieved from http://eralberta.ca/news/stories/era-publishes-interim-report-eseieh-project/ on 26 October 2018.

Bridges, E. A., 1979, "Method for In-Situ Heat Processing of Hydrocarbonaceus Formations", US Patent 4,140,180.

Bridges, J. E., 1985, "Electromagnetic Stimulation of Heavy Oil Wells", 3rd International Conference on Heavy Crude and Tar Sands, CA, July.

British Petroleum, 2016, *BP Statistical Review of World Energy*, 65th Ed., British Petroleum, London, UK, June.

Cambon, J. L., Kyvana, D., Chavarie, C., Bosisio, R. G., 1978, "Traitement du sable bitumineux par micro-ondes", *Can. J. Chem. Eng.*, 56, 735–742.

Chhetri, A. B., Islam, M. R., 2008, "A Critical Review of Electromagnetic Heating for Enhanced Oil Recovery", *Pet. Sci. Technol.*, 26, 1619–1631, and references therein.

Chopra, S., Lines, L., Schmitt, D. R., Batzle, M., 2010, "Heavy-Oil Reservoirs: Their Characterization and Production", In: *Heavy Oils: Reservoir Characterization and Production Monitoring, Society of Exploration Geophysicists, Geophysical Developments Series*, Vol. 13, S. Chopra, L. Lines, D. R. Schmitt and M. Batzle, Eds., Tulsa, Chapter 1, pp 1–69. https://doi.org/10.1190/1.9781560802235

Depew, M C., Lem, S., Wan, J. K. S., 1991, "Microwave Induced Catalytic Decomposition of Some Alberta Oil Sands and Bitumens", *Res. Chem. Intermed.*, 16, 213–223.

Despande, R., Wright, B. N., Watt, A., 2015. "Techniques for Installing Effective Solvent Extraction Incorporating Electromagnetic Heating ("ESEIEH") Completions". Presented at the World Heavy Oil Conference, Edmonton, Alberta, Canada, Paper No. WHOC15-317.

Dreher, W. R., Allison, J. D., Patton, L. J., Hernandez, V., Parsche, F. E., 2015, "In Situ Radio Frequency Catalytic Upgrading", US Patent 9,004,164.

Dreher, W. R., Wheeler, T. J., Banerjee, D. K., 2013, "In-Situ Upgrading of Heavy Crude Oil in a Production Well Using Radiofrequency or Microwave Radiation and a Catalyst", US Patent 8,365,823.

Duncan, G., 1996, "Enhanced Recovery Engineering", *World Oil*, March, 86–90.

Dunn, K., Yen, T. F., 2001, "A Plausible Reaction Pathway of Asphaltene under Ultrasound", *Fuel Proc. Technol.*, 73, 59–71.

Faergestad, I. M., 2016, "Defining Heavy Oil", Schlumberger, downloaded from www.slb.com/~/media/Files/resources/oilfield_review/defining_series/Defining-Heavy-Oil.ashx on 15 December 2016.

Fanchi, J. R., 1990, "Feasibility of Reservoir Heating by Electromagnetic Irradiation", SPE No. 20483 presented at 65th Annual Technology and Exhibition, New Orleans, LA, 23–26 September.

Gopinath, R., Dalai, A. K., Adjaye, J., 2006, "Effects of Ultrasound Treatment on the Upgradation of Heavy Gas Oil", *Energy Fuels*, 20, 271–277.

Greff, J., Babadagli, T., 2011, "Catalytic Effects of Nano-Size Metal Ions in Breaking Asphaltene Molecules during Thermal Recovery of Heavy-Oil", SPE No. 146604 presented at SPE Annual Technical Conference and Exhibition, Denver, CO, 30 October–2 November.

Grobas, J., Bolivar, C., Scott, C. E., 2007, "Hydrodesulfurization of Benzothiophene and Hydrogenation of Cyclohexene, Biphenyl, and Quinoline, Assisted by Ultrasound, Using Formic Acid as Hydrogen Precursor", *Energy Fuels*, 21, 19–22.

Gunnerman, R. W., Moote, P. S., Cullen, M. T., 2003, "Treatment of Crude Oil Fractions, Fossil Fuels, and Products Thereof with Ultrasound", US Patent Appl. No. 20030051988.

Hu, L., Li, H. A., Babadagli, T., Ahmadloo, M., 2016, "Experimental Investigation of Combined Electromagnetic Heating and Solvent Assisted Gravity Drainage for Heavy Oil Recovery", SPE No. 180747, presented at SPE Canada Heavy Oil Technical Conference, Calgary, Alberta, Canada, 7–9 June.

Haagensen, D. B., 1965, "Oil Well Microwave Tools", US Patent 3,170,519.

Haagensen, D. B., 1986, "Oil Recovery System and Method", US Patent 4,620,593.

Islam, M. R., 1999, "Emerging Technologies in Enhanced Oil Recovery", *Energy Sources*, 21, 97–111, and references therein.

Jackson, C., 2002, "Upgrading a Heavy Oil Using Variable Frequency Microwave Energy", SPE No. 78982 presented at SPE International Thermal Operations and Heavy Oil Symposium and International Horizontal Well Technology Conference, Calgary, Alberta, Canada, 4–7 November.

Jackson, R. E., Daniel, D., Trautman, M., 2014, "An Approach to Frequency Selection for the Operation of In Situ Hydrocarbon RF Heating Systems". World Heavy Oil Conference, New Orleans LA, Paper No. WHOC14-298.

Jeambey, C., 1989, "Apparatus for Recovery of Petroleum from Petroleum Impregnated Media", US Patent 4,187,711.

Jeambey, C. G., 1990, "System for Recovery of Petroleum from Petroleum Impregnated Media", US Patent 4,912,971.

Kasevich, R., 1998, "Understanding the Potential of Radio Frequency Energy", *Chem. Eng. Prog.*, 94, 75–81.

Kasevich, R., 2008, "Method and Apparatus for In-Situ Radiofrequency Assisted Gravity Drainage of Oil (RAGD)", US Patent 7,441,597.

Kasevich, R. S., Price, S. L., Albertson, A., 1997, "Numerical Modeling of Radiofrequency Heating Process for Enhance Oil Production", SPE No. 38311, presented at SPE Western Regional Meeting, Long Beach, CA, 25–27 June.

Kasevich, R. S., Price, S. L., Faust, D. L., Fontaine, M. F., 1994, "Pilot Testing of a Radio Frequency System for Enhanced Oil Recovery from Diatomaceous Earth", SPE No. 28619, presented at 69th Annual Tech. Conf. New Orleans, LA, 25–28 September.

Kaushik, P., Kumar, A., Bhaskar, T., Sharma, Y. K., Tandon, D., Goyal, H. B., 2012, "Ultrasound Cavitation Technique for Up-Gradation of Vacuum Residue", *Fuel Proc. Technol.*, 93, 73–77.

Lin, J.-R., Yen, T. F., 1993, "An Upgrading Process through Cavitation and Surfactant", *Energy Fuels*, 7, 111–118.

Madison, M. J., Banerjee, D. K., Parsche, F. E., Trautman, M. A., 2014, "Simultaneous Conversion and Recovery of Bitumen Using RF", US Patent 8,807,220.

Milan, A. C., 1978, "In Situ Extraction of Bitumen from Alberta's Tar Sand by Microwave Heating", BS Thesis, University of Windsor.

Nugent, K., Blue, M., Whitney, R. Jackson, E., 2014, "Monobore Architecture of a Radio Frequency Heating, Fluid Delivery, and Production System", World Heavy Oil Conference, New Orleans, LA, Paper No. WHOC14-299.

Ovalles, C., Fonseca, A., Lara, A., Alvarado, V., Urrecheaga, K., Ranson, A., Mendoza, H., 2002, "Opportunity of Downhole Dielectric Heating in Venezuela: Three Case Studies Involving, Medium, Heavy, and Extra-Heavy Crude Oil Reservoirs", SPE No. 78980. SPE International Thermal Operations and Heavy Oil Symposium, Alberta, Canada, 4–7 November, and references therein.

Ovalles, C., Rodriguez, H., 2008, "Extra-Heavy Crude Oil Downhole Upgrading Using Hydrogen Donors under Cyclic Steam Injection Conditions. Physical and Numerical Simulation Studies", PETSOC-08-01-43, *J. Can. Pet. Technol.*, 47, 43–51.

Ovalles, C., Vallejos, C., Vasquez, T., Rojas, I., Ehrman, U., Benitez, J. L., Martinez, R., 2003, "Downhole Upgrading of Extra-Heavy Crude Oil Using Hydrogen Donors and Methane under Steam Injection Conditions", *Pet. Sci. Technol.*, 21 (1–2), 255–274.

Paul, J. M., Davis, R. M., 1998, "Method for Hydrotreating and Upgrading Heavy Crude Oil during Production", US Patent No. 5,824,214.

Sadeghi, K. M., Lin, J.-R., Yen, T. F., 1994, "Sonochemical Treatment of Fossil Fuels", *Energy Sources*, 16, 439–449.

Sadeghi, K. M., Sadeghi, M.-A., Kuo, J.-F., Jang, L.-K., Lin, J.-R., Yen, T. F., 1992, "A New Process For Tar Sand Recovery", *Chem. Eng. Commun.*, 117, 191–203.

Sadeghi, K. M., Sadeghi, M.-A., Kuo, J.-F., Jang, L.-K., Yen, T. F., 1990, "A New Tar Sand Recovery Process: Recovery Methods and Characterization of Products", *Energy Sources*, 12, 147–160.

Sahni, A., Kumar, M., Knapp, R. B., 2000, "Electromagnetic Heating for Heavy Oil Reservoirs", SPE No. 62550, presented at SPE/AAPG Western Regional Meeting, Long Beach, CA, 19–23 June.

Salazar, A., 2003, "Simulación Numérica Conceptual del Mejoramiento en Subsuelo de Crudos Pesados y Extrapesados", MSc Thesis, Universidad Central de Venezuela, Caracas, November.

Sawarkar, A. N., Pandit, A. B., Shriniwas, D. S., Joshi, J. B., 2009, "Use of Ultrasound in Petroleum Residue Upgradation", *Can. J. Chem. Eng.*, 87, 329–342.

Shedid, S. A., Attallah, S. R., 2004, "Influences of Ultrasonic Radiation on Asphaltene Behavior with and without Solvent Effects", SPE No. 86473, presented at SPE International Symposium and Exhibition on Formation Damage Control, Lafayette, LA, 18–20 February.

Sresty, G. C., Dev, H., Snow, R. H., Bridges, J. E., 1986, "Recovery of Bitumen and Tar Sand Deposits with Radio Frequency Process", SPE No. 10229, *SPE Res. Eng.*, 1, 85–94.

Sresty, G. C., Snow, R. H., Bridges, J. E., 1984, "Recovery of Liquids Hydrocarbons from Oil Shale by Electromagnetic Heating In Situ", US Patent 4,485,869.

Thompson, L. H., Doraiswamy, L. K., 1999, "Sonochemistry: Science and Engineering", *Ind. Eng. Chem. Res.*, 38, 1215–1249.

Trautman, M., Ehresman, D., Edmunds, N., Taylor, G., Cimolai, M., 2014, "Effective Solvent Extraction System Incorporating Electromagnetic Heating", US Patent No. 8,776,877.

Trautman, M., MacFarlane, B., 2014. "Experimental and Numerical Simulation Results from a Radio Frequency Heating Test in Native Oil Sands at the North Steepbank Mine". Presented at the World Heavy Oil Conference, New Orleans, LA, Paper No. WHOC14-301.

Suslick, K. S., Didenko, Y., Fang, M. M., Hyeon, T., Kolbeck, K. J., McNamara, W. B., Mdleleni, M. M., Wong, M., 1999, "Acoustic Cavitation and Its Chemical Consequences", *Phil. Trans. R. Soc. A*, 357, 335–353.

Varadaraj, R., 2004, "Low Viscosity Hydrocarbon Oils by Sonic Treatment", US Patent Appl. No. 20040232051.

Warren, G. M., Behie, G. A., Tranquilla, J. M., 1996, "Microwave Heating of Horizontal Wells in Heavy Oil with Active Water Drive", SPE No. 37114, presented at Int. Conf. on Horizontal Well Tech., Calgary, Canada, 18–20 November.

Wheeler, T. J., Dreher, W. R., Parsche, F. E., Trautman, M. A., 2016, "Enhanced Recovery and In Situ Upgrading Using RF", US Patent 9,453,400.

Wise, S., Patterson, C., 2016, "Reducing Supply Cost of EESIEH". SPE 180729, presented at the SPE Canada Heavy Oil Technical Conference, Calgary, Alberta, 7–9 June.

Xu, H., Pu, C., Wu, F., 2012, "Low Frequency Vibration Assisted Catalytic Aquathermolysis of Heavy Crude Oil", *Energy Fuels*, 26, 5655–5662.

Xu, H. X., Pu, C. S., 2011, "Experimental Study of Heavy Oil Underground Aquathermolysis Using Catalyst and Ultrasonic", *J. Fuel Chem. Technol.*, 39, 606–610.

Glossary

% Conv. of 1000°F+: Percentage of residue conversion of the fraction with a boiling point greater than 538°C+ (1000°F+)

%DS: Percentages of desulfurization, i.e., reduction of sulfur content of product vs. feed

A: Arrhenius pre-exponential factor

acac: Acetylacetonate

AEBP: Atmospheric Equivalent Boiling Point (°F or °C)

AFM: Atomic Force Microscopy

API: Gravity of the sample as measured by American Petroleum Institute

APPI: Atmospheric pressure photoionization

AR: Atmospheric Residue

Asphaltenes: Fraction of the crude oil that precipitates in paraffins (propane or heptane) and is soluble in aromatics or chlorine-containing solvents (CH_2Cl_2)

Asphaltene-1: Asphaltenes dissolved in the oil phase (Equations 5.16, 5.17, and 5.18)

Asphaltene-2: Asphaltenes deposited (Equations 5.16, 5.17, and 5.18)

BBL: US Barrels (42 Gal or 0.159 m³)

BBL/D: US Barrels per day

BHP: Bottom-hole pressure in psi

BS&W: Basic Sediment and Water

B/d: US Barrels per day

C: Weight fraction of unreacted crude oil

C_{11}: Concentrations of S1

C_{21}: Concentrations of S2

C3: Propane

C3-asphaltenes: Asphaltenes obtained using propane as precipitant solvent

C4: Butane

C5: n-Pentane

C6: n-Hexane

C7: n-Heptane

C7-asphaltenes: Asphaltenes obtained using n-heptane as precipitant solvent

CAGD: Combustion Assisted Gravity Drainage

CAPEX: Capital costs (US$)

CAPRI: Catalytic Upgrading Process *In-Situ* Process

CEMRS: Coupled electromagnetic reservoir simulator

CGO: Coker gas oil

CHOPS: Cold Heavy Oil Production with Sand

CII: Colloidal Instability Index

cP: Centipoise which is equal to one millipascal-second (mPa s)

CSI: Cyclic Steam Injection

CSS: Cyclic Steam Stimulation

CWE: Cold Water Equivalent

DAO: Deasphalted oil

D_{Asp-1}: Density of Asphaltenes-1

DBF: Deep Bed Filtration Model

DBT: Dibenzothiophene

dC/dt: Derivative of the concentration vs. time

D_{DAO}: Density of DAO

D$_{Heavy F.}$: Density of Heavy F.

DBS: Dodecylbenzene sulfonate

DHFI: Dense Hot Fluid Injection process from University of Calgary

DPEP: Deoxophylloerythroetio porphyrins

DSC: Differential Scanning Calorimetry

DMBS: Dimethylbenzenesulfonic ligand

Ea: Arrhenius activation energy in kJ/Mol

ESEIEH: Effective Solvent Extraction Incorporating Electromagnetic Heating

EOR: Enhanced Oil Recovery Process

ELSD: Evaporative Light Scattering Detector

ESP: Electrical submersible pumping

ES-SAGD: Expanded Solvent SAGD

ETIO: Etioporphyrin

FCCS: Fluid catalytic cracking slurry

FT-ICR MS: Fourier transform ion cyclotron resonance mass spectrometry

FEO: Furfural extract oil

GC/MS: Gas Chromatography/Mass Spectroscopy

GHG: Greenhouse gases

GOR: Gas-to-oil ratio

GPC: Gel Permeation Chromatography

GS: Gemini sulfonate surfactant

H/C: Molar hydrogen-to-carbon ratio

HDS: Hydrodesulfurization or removal of sulfur in the presence of hydrogen and catalyst

HDN: Hydrodenitrogenation or removal of nitrogen in the presence of hydrogen and catalyst

HDM: Hydrodemetallization or removal of metals in the presence of hydrogen and catalyst

HDO: Hydrodeoxygenation or removal of oxygen in the presence of hydrogen and catalyst

H-donor: Hydrogen donor

Heavy F. or Heavy Fraction: Pseudo-component used in the asphaltene precipitation model

H-NMR: Proton-Nuclear Magnetic Resonance

HCCO: Distillate fraction used as hydrogen donor

HTO: High Temperature Oxidation

HO: Heavy and Extra-Heavy Crude Oils

HO/B: Heavy crude oils and bitumen

HPLC: High-Performance Liquid Chromatography

Huff and Puff: Cyclic Steam Stimulation or CSS

HVGO: High vacuum gasoil

ICCO: Distillate fraction used as hydrogen donor

IL: Ionic Liquid

IN: Insolubility number which measures the degree of insolubility of the asphaltenes

iOSR: Instantaneous oil/steam ratio (v/v)

ISC: *In-Situ* Combustion

ISUT: *In-Situ* Upgrading Technology from University of Calgary

IUP: Shell *In-Situ* Upgrading Process

k: Arrhenius rate of reaction

k$_1$: Rate constant of reaction of Asphaltene-1 to Asphaltene-2 in the presence of C3

k$_2$: Rate constant of reaction of Asphaltene-2 to Asphaltene-1 in the presence of C3

k$_3$: Rate constant of reaction of Asphaltene-2 to Heavy Fraction in the presence of C3

k$_{12}$: Kinetic constants for the formation of reverse reaction to S1

k$_{21}$: Kinetic constants for the formation of S2

LASER: Liquid Addition to Steam for Enhanced Recovery

LHSV: Linearly hourly space velocity (h^{-1})

LMO: Distillate fraction used as hydrogen donor
LTO: Low Temperature Oxidation
MAC: Multicomponent acrylic copolymer
MCR: Micro-carbon residue
MBD: Thousands of barrels a day
MIS: From its Spanish name "Mejoramiento In Situ", an SSU-SDA process
MMBD: Millions of barrels a day
MMO: Distillate fraction used as hydrogen donor
MS: Mass Spectrometry
MW: Molecular weight in g/mol
NIPER: National Institute of Petroleum and Energy
^{1}H NMR: ^{1}H-Nuclear Magnetic Resonance
NPV: Net Present Value
Nsolv: A solvent deasphalted process, see www.n-solv.com
NMR: Nuclear Magnetic Resonance
o/w: Oil-in-water emulsions
OOIP: Original Oil in Place, generally measured in US barrels (BBL)
OPEX: Operating costs (US$)
P: Overall compatibility of the system also known as P-value
Pa: Peptizability of the asphaltenes
Pc: Critical pressure (psi)
PCP: Progressive Cavity Pumps
Po: Solvent power of the maltenes
P_{oil}: Partial pressure of a volatile oleic phase
P_w: Partial pressure of water
P_T: Total pressure of the system
ppm: Part per million or mg/L
psi: Pounds per square inch
PVT: Pressure, volume, and temperature cell or experiments
r: Reaction rate for the formation of S2
R: Ideal gas constant, 8.3143 J/K Mol
RAISCUP: Residue Assisted *In-Situ* Catalytic Upgrading process from University of Calgary
RCA: Root Cause Analysis
RFM: Resistance Factor Model
ROSE: Residuum Oil Supercritical Extraction Process
% Residue Conv.: Percentage of Residue Conversion (Section 4.4)
S1: Solid asphaltenes 1 as reported by Hammami [2000] and Peramanu [2001]
S2: Solid asphaltenes 2 as reported by Hammami [2000] and Peramanu [2001]
SAGD: Steam Assisted Gravity Drainage
SARA: Analytical methodology that separates *S*aturates, *A*romatics, *R*esins, and *A*sphaltenes
SAXS: Small-angle X-ray scattering
SBN: Solubility blending number
SCF: Standard cubic feet
SCO: Sweet Synthetic Crude Oils
SEM: Scanning Electron Microscopy
SIMS: Secondary Ion Mass Spectrometry
SPO: Distillate fraction used as hydrogen donor
SRP: Sucker rod pumping
STARS: Compositional and thermal reservoir simulator made by CMG
ST-ISUT: Steam co-injected *In-Situ* Upgrading Technology
SSU: Subsurface Upgrading

SSU-SDA: Subsurface Upgrading of Heavy Oils via Solvent Deasphalting
SDA: Solvent Deasphalting Process
SDA Tar: Paraffin-insoluble fraction (asphaltenes) produced by SDA processes
SIMDIS: Simulated distillation
SOR: Steam-to-oil ratio
STB: Stock tank barrels
SvOR: Solvent-to-oil ratio
T: Temperature in K
TAN: Total acid number measured in mg KOH per gram of sample
Tb: Boiling Point
Tc: Critical temperature (°R)
TDS: Total Dissolved Solid in mg/L
THAI: Toe-to-Heel Air Injection Process
TLC-FID: Thin-Layer Chromatography-Flame Ionization Detection
USGS: United States Geological Survey
VAPEX: Vapor Extraction Process for SSU-SDA
VGO: Vacuum Gasoil
VPO: Vapor Phase Osmometry
VR: Vacuum Residue
VSD: Vertical well Steam Drive
W_a: Amount of adsorbed asphaltene
$W_{a,max}$: Maximum amount of asphaltene that can be adsorbed
WGSR: Water–gas-shift reaction
WHSV: Weight Hourly Space Velocity
WOR: Water-to-oil ratio
w/o emulsion: Water-in-Oil emulsions
XRD: X-ray diffraction analysis
x_1: Molar fractions of heavy crude oil or bitumen
y_1: Molar fractions of diluent (naphtha or diesel)

GREEK SYMBOLS

α: Constant
μ: Viscosity in cP or mPa s
μ_{Asp-1}: Viscosity of Asphaltenes-1
μ_{DAO}: Viscosity of DAO
$\mu_{Diluent}$: Viscosity of the diluent (naphtha or diesel) in cP or mPa s
$\mu_{Heavy\ F.}$: Viscosity of Heavy F.
μ_{HO}: Viscosity of the heavy crude oil or bitumen in cP or mPa s
ϕ: Porosity
ϕ_o: Initial porosity

Index

A

Advanced Supercritical Solvent Deasphalting Process, 72
AEBP, *see* Atmospheric equivalent boiling point
Alberta oilsand reservoirs, 22, 24
American Petroleum Institute, 2
    gravity parameter, 2, 3, 8, 20, 21, 119, 120
APPI, *see* Atmospheric pressure photoionization
Aquaconversion process, 87–89
Aquathermolysis
    amphiphilic catalysts, 221–225
    dispersed and nanocatalysts, 225–230
    oil-soluble catalysts, 218–221
    reactions, 184
    on SARA fractions, 179
    sulfur generated by, 177
    typical gas production for, 176
    of Venezuelan oil-sand using watersoluble metal
        catalysts, 213, 214
    water-soluble catalysts, 210–219
Arbitrary demarcation, 3
Archipelago-type molecules, 39
Aromatic condensation, 79
Aromatization, 79–81
Arrhenius equation, 28, 150
Asphaltenes, 27
    carbon atoms, 38
    characterization of, 120
    chemical composition, 28
    heptane-to-bitumen ratio on, 39
    mechanism for, 42
    particle size distributions, 42
    precipitation, 36, 38, 141–142
        deposition models, 130–131
        flocculation models, 130
        models, 129–130
        permeability reductions, 131–132
        viscosity model, 132
    typical properties of, 37
    viscosity, 40
Asphalt Residual Treating Process, 91, 92
ASTM D341, 28
Athabasca oil, 22, 153, 159, 176, 179, 211, 237, 268
Athabasca vacuum residue *vs.* stability value, 106
Atmospheric equivalent boiling point (AEBP), 30, 33, 34
Atmospheric pressure photoionization (APPI), 33
Atomic force microscopy (AFM), 38

B

Biofuels, 1
Bitumens, *see* Heavy crude oils and bitumens
Blending Model, 132
Bond dissociation energy, 75
Bromine number, 104

C

CAGD, *see* Combustion Assisted Gravity Drainage
Canadian Cold Lake Bitumen, 83–84
Canadian crude oil, 116, 121
Canadian oilsands, 23
Canadian petroleum producers, 104
CAPEX, *see* Capital expenses
Capital expenses (CAPEX), 1, 2, 10, 231, 236
Carbonates, 4, 23, 25, 121
Carbon rejection
    coking mechanism, 79–81
    coking processes, 78
    delayed coking, 79, 81–82
    physical separation, 71–72
    solvent deasphalting, 72–74
    thermal conversion, 71–72
    thermal cracking reactions, 74–76
    visbreaking process, 77–78
Catalytic cracking and hydrocracking
    acid catalysis, 90–92
    comparison between, 94
    hydrogenation, 92–93
Catalytic hydrogen addition
    catalyst issues
        catalyst control, 209
        environmental concerns, 209–210
        initial considerations, 207–208
        placement of catalyst downhole, 208–209
        residence times, 209
    catalytic aquathermolysis
        amphiphilic catalysts, 221–225
        dispersed and nanocatalysts, 225–230
        oil-soluble catalysts, 218–221
        water-soluble catalysts, 210–218
    hydrogen donors, catalytic use of
        HDS mechanistic studies, 245–246
        physical simulations, 241–245
    hydrogen gas, 230–231
        early experiments, 231–234
        lab experiments, 240–241
        physical and numerical simulations, 237–240
        porous media, 236–237
        University of Calgary process, 234–236
Catalytic methane activation, 244
CCEMC, *see* Climate Change and Emission Management
        Corporation
Characteristics
    of asphaltenes, 120
    of heavy oil-bearing formations, 19–20
        aqueous phase composition, 23–24
        biogenesis, 20–23
        composition, 20–23
        gas phase composition, 24
        mineral formations, 24–25